Arnold Finck
Fertilizers and Fertilization

Arnold Finck

Fertilizers and Fertilization

Introduction and
Practical Guide to Crop Fertilization

Weinheim · Deerfield Beach, Florida · Basel · 1982

Title of the German edition:
Arnold Finck
Dünger und Düngung: Grundlagen und Anleitung zur Düngung der Kulturpflanzen
Weinheim, New York: Verlag Chemie GmbH, 1979.

Prof. Dr. Arnold Finck
Institute for Plant Nutrition and Soil Science
University of Kiel
D-2300 Kiel

Translated by Magal Translation Agency
Translation Editors: Arnold Finck and Lilian M. Neil

Responsible for production: Hans Jörg Maier

This book contains 33 figures and 83 tables

CIP-Kurztitelaufnahme der Deutschen Bibliothek

Finck, Arnold:
Fertilizers and fertilization: introd. and practical guide to crop fertilization/Arnold Finck.
[Transl. by Magal Transl. Agency]. – Weinheim; Deerfield Beach, Florida; Basel: Verlag
Chemie, 1982.
    Dt. Ausg. u. d. T.: Finck, Arnold: Dünger und Düngung
    ISBN 3-527-25891-4 (Weinheim)
    ISBN 0-89573-052-9 (Deerfield Beach)

Typesetting and printing: Zechnersche Buchdruckerei, D-6720 Speyer
Bookbinding: J. Schäffer OHG, D-6718 Grünstadt
Printed in the Federal Republic of Germany

Preface

From the beginning, this book was compiled not only for local but also for international use. Fertilization is of world-wide importance and many theoretical as well as practical aspects of fertilization are very similar even under different environmental and economic conditions. Furthermore, the aim of fertilization in intensive agriculture is very much the same everywhere, i. e. to achieve an optimum nutrient supply for high-yield crops on soils often of inadequate fertility and under the influence of climatic stress factors.

The concept of this book is based on production conditions in a temperate climate. However, modifications recognizing conditions of arid and humid tropical climates were taken into account, as have most of the world's important crops. The fertilizers mentioned – mainly a selection of types widely applied in Europe – are utilized along similar lines all over the world.

This book, both a basic introduction and a practical guide, is designed to be a source of information on fertilizer use, and is meant to be read in addition to the local particulars available in most countries. Since food for a rapidly growing population must be produced in the future, the information presented here may be a welcome contribution towards this important goal.

Thanks are due to Verlag Chemie who have kindly agreed to produce this English edition of my book which was originally published in German.

Kiel, in Fall 1981

Arnold Finck

Preface to the German Edition

Why a new book on "Fertilizers and Fertilization"? Because until now books on fertilizing problems have offered either practical advice *or* a theoretical introduction. Today, approximately 100 years after the start of any mineral fertilization worthy of the name, it is both possible and necessary to unite these two aims.

For practical purposes in fertilization, it is no longer sufficient to use simple rules drawn from experience or general recipes since, in the course of intensive production, they bring about only a medium yield of crops. In future, precise and comprehensive fertilization will be necessary in order to meet the increased demand with respect to the quantity and quality of the crops. To this end, however, knowledge of new and old fertilizers as well as many basic questions of agricultural chemistry relevant to fertilization is indispensable.

A broader, deeper and perhaps newly oriented *theory* will be necessary in order to achieve increased control of biological production in agriculture. Research considerably increases our knowledge of the complex conditions characteristic of the growth of plants and provides a better basis for the expert use of the necessary chemical products. However, as the theoretical bases are widened, it becomes increasingly necessary to emphasize important problems and to distinguish them from irrelevant ones, as well as to derive the simplest possible guidelines for fertilization, even in the most complicated contexts.

Fertilization in modern agriculture is not an exhausted topic, about which hardly anything new can be said. On the contrary, the application of chemical knowledge and approaches to agriculture in the sphere of fertilization steadily improves production in many ways. Fortunately, requirements as to production, economy, and ecology are largely running on parallel lines. The aim of high yields and healthy food accompanied by preservation or even improvement of soil fertility, without negative effects on the environment, should be worth some effort.

This book aims at giving a survey of the characteristics and uses of fertilizers as well as numerous, easily understandable hints for practical fertilization of commercial crops, within a broad framework including many marginal areas; the emphasis is, at the same time, on the quality of the products grown. Important and problematic questions are dealt with in a scientific manner; where it has been possible and appeared necessary, these have been summarized in a simple manner for practical use.

This book is based on lectures given over a period of twenty years, as well as on innumerable discussions held on the inexhaustible subject of fertilization with expert colleagues in Germany and abroad, particularly with the German Scientific Advisory Board on Fertilization. I would like to thank them all for their useful suggestions.

I owe particular thanks to Verlag Chemie for readily accepting my manuscript, as well as for their preparation of the book in its present attractive form, while considering the author's particular requests, and finally for their many hints with regard to standards. Finally, any book with a two-fold aim must be a compromise. It should be comprehensive and scientifically precise, but, at the same time, it should be easily understandable, attractively laid out, and not too expensive.

Now that this aim has been reached to a certain degree, I hope that further suggestions by readers will contribute to its perfection.

Arnold Finck

Note

Most of the trade-names of fertilizers mentioned in this book are registered trade-marks in many countries; they appear in the text and index without their trade-mark character being indicated. It should not be assumed, however, that anybody may use these marks freely. Moreover, the naming of commercial products in this book neither represents an evaluation nor does it claim to be complete, especially since new products are constantly introduced into, or taken off, the market.

Contents

Units, Abbreviations, Symbols, Factors

Units:

Length and area:	m	meter
	a	are (100 m^2)
	ha	hectare (10,000 m^2)
Mass ("weight"):	g	gram
	kg	kilogram
	dt	deciton (100 kg)
	t	[metric] ton (1 000 kg)
Volume (vol.):	l	liter
	ml	milliliter (0.001 liter)
Time:	h	hour

Contents (concentrations):

%	parts per hundred	(per cent)
%o	parts per thousand	(per mille)
ppm	parts per million	

Heat:	J	joule [1 J = 0.24 cal (calories)]
Pressure:	bar	[1 bar $\approx$ 1 at (atmosphere) = 1 kg/cm^2]

Abbreviations:

FRG	Federal Republic of Germany
EEC	European Economic Community
FAO	Food and Agriculture Organization of the United Nations
conc.	concentrated
LUFA	*see* VDLUFA
M	molar (*see* molar solution), abbreviation for the unit of concentration [mol l^{-1}]

N   normal (*see* normal solution)
org.   organic
R   radical (*see* radicals), general symbol in chemical formulas
rel.   relative
D. M.   dry matter
VDLUFA   Association of German Agricultural Institutes of Investigation and
            Research
dil.   diluted
>   greater than
<   less than

Symbols and relative atomic masses ("atomic weights") of some elements:

Symbol	Element	Relative atomic mass
Al	aluminum	27
B	boron	10.8
C	carbon	12
Ca	calcium	40.1
Cl	chlorine	35.5
Co	cobalt	58.9
Cu	copper	63.5
Fe	iron	55.9
H	hydrogen	1
K	potassium	39.1
Mg	magnesium	24.3
Mn	manganese	54.9
Mo	molybdenum	95.9
N	nitrogen	14
Na	sodium	23
O	oxygen	16
P	phosphorus	31
S	sulfur	32.1
Si	silicon	28.1
Zn	zinc	65.4

Conversion factors:

$$P \underset{\times\, 0.436}{\overset{\times\, 2.29}{\rightleftharpoons}} P_2O_5 \qquad Ca \underset{\times\, 0.715}{\overset{\times\, 1.40}{\rightleftharpoons}} CaO \underset{\times\, 0.56}{\overset{\times\, 1.79}{\rightleftharpoons}} CaCO_3$$

$$K \underset{\times\, 0.83}{\overset{\times\, 1.20}{\rightleftharpoons}} K_2O \qquad Mg \underset{\times\, 0.60}{\overset{\times\, 1.66}{\rightleftharpoons}} MgO \underset{\times\, 0.48}{\overset{\times\, 2.09}{\rightleftharpoons}} MgCO_3$$

1 Introduction

1.1 Introduction to Fertilization

1.1.1 Fundamental Problems of Fertilization

Why Fertilization?

Only very few soils on earth are so well provided with nutrients that high yields of cultivated plants can be obtained over prolonged periods without any fertilization whatsoever. This utopian state, i.e. of abundant harvests without the use of fertilizers, does not seem to have been realized anywhere. In the most favourable cases, nature itself has taken over fertilization, e.g., through annual supplies of mud in the river valleys where ancient human cultures once existed, but ordinarily the situation is quite the reverse, namely there is more or less rapid impoverishment of the soil through the growing of crops. Therefore, even in primitive agricultural systems at least a simple type of fertilization is encountered. Fertilization serves to obtain high crop yields and good quality produce. It is necessary primarily for the following reasons:

- Improvement of the soil as a nutrient substrate.
- Supplementation of the natural, partly deficient, supply of nutrients.
- Replacement of the nutrients removed by harvesting and other losses.

Fertilization is thus always necessary, but it should also be as efficient as possible. The following factors necessitate a purposeful use of fertilizers:

- The need for profitable high-yield production, since only this provides a high net profit and sufficient food in a continuously decreasing area;
- increasing fertilizer prices, due to higher costs of energy and global scarcities of raw materials;
- the requirement to prevent ecological damage through excessive or wrongly applied fertilization.

It follows from these requirements that, on the one hand, fertilization should be restricted to the extent necessary; whereas on the other hand, the aim should be to increase it precisely where deficiencies exist. Deficient fertilization, even today (one hundred years after the beginning of statistically recorded mineral fertilization), is one of the main reasons why the genetic potential of highly bred crops, with possible grain yields of 100 dt/ha, has been only incompletely exploited. Even small errors in fertilization can considerably reduce the yield and, primarily, the net profit when production costs are high. In recognizing the need for better fertilization it is of decisive importance to appreciate the fact that even a high rate of fertilizer application per unit area by itself does not ensure an optimum nutrient supply. Even plants that have a succulent green, i. e., "healthy" appearance, may suffer from some deficiency or imbalance of nutrient supply, thus failing to attain maximum yields or optimum quality.

The subject of fertilization is becoming increasingly complex, so that a knowledge of fertilizers and their uses becomes steadily more important. Fertilization for the sake of high yields can no longer be routinely undertaken according to simple rules, but must be carried out according to extended and improved concepts, as overall production conditions change. The following should be remembered in this connection:

- Changed demands of plants (e. g., new high-yield varieties);
- changed soil conditions (e. g., deeper ploughing, increased drainage, increased liming);
- the need for labor-saving fertilizer application;
- new or economic forms of fertilizers.

High and valuable yields can be obtained only if all growth factors act together optimally. Thus it becomes increasingly important to consider problems of fertilization in conjunction with other problems of crop cultivation, ranging from cultivation of the soil to plant protection. Fertilization is necessary, but can only have an optimum effect as an integral component of all production measures. Fertilization is probably the most important measure for increasing yields, since it has produced yield increases of 50 to 60% during the last century.

The purpose of using fertilizers is to obtain high and valuable yields by improving the supply of nutrients while maintaining or improving the fertility of the soil without harmful effects on the environment.

Fertilize with What?
The fertilizer should provide the plants' needs for optimum growth beyond that which nature supplies. Since growth is also subject to harmful (material and energy) influences, fertilization includes the supply of substances which eliminate harmful effects or reduce their negative influence.

Synopsis 1-1. Possible Fertilizer Uses.

What does the plant need?	Supplementary supply through fertilization
I. energy factors	
light	light fertilization ($=$illumination)
temperature (heat)	heat fertilization ($=$heating)
II. material factors	
1. absorption by leaves	
carbon dioxide (CO_2)	CO_2-gas supply
2. absorption mainly	
by roots	
a) oxygen	fertilization for structural improvement
b) water	irrigation
c) mineral nutrients	
major nutrient elements	nitrogen (N), phosphorus (P), sulfur (S), potassium (K), calcium (Ca), magnesium (Mg)
micronutrient elements	iron (Fe), manganese (Mn), zinc (Zn), copper (Cu), boron (B), chlorine (Cl), molybdenum (Mo)
beneficial elements	e.g., sodium, silicon, cobalt

What harms the plant?	Elimination or reduction of damage through fertilization
1. harmful climatic influences (e.g., cold, heat, storms)	increasing resistance through suitable fertilization
2. toxic substances in the air ("environmental" toxic substances)	
3. toxic soil constituents	
a) natural toxic substances	e.g., fertilization against damage due to soil acidity, salt damage
b) "environmental" toxic substances	e.g., fertilization against excess of heavy metals
4. harmful biotic influences (e.g., attack by diseases)	increasing resistance through fertilization

Besides the supply of mineral and organic substances (fertilization in the narrower sense), it is also possible to include the supply of energy (in a wider sense), e. g., heating as "thermal fertilization".

Nowadays, the various possible uses of fertilization (Synopsis 1-1) are clearly recognized, since the substances needed by plants for their existence are largely known, as are their additional requirements. "Plants", in this sense, should be taken to mean all higher green plants, which, despite their variety, have largely similar demands as regards the type of substances they need. Lower plants (algae, fungi, bacteria) generally do not differ greatly as regards these requirements, but, in some cases, also need other substances.

Fertilization should supply the substances lacking in the soils (substrates) concerned, as the situation requires. Determining these minimum factors may involve cumbersome experimentation with different kinds of fertilization methods, but can be done more quickly and effectively by applying reliable, diagnostic methods. The usual (NPK) fertilization (Chap. 3.2.3) is usually no longer sufficient for obtaining higher yields. The concept of fertilization will, in future, have to include all necessary and beneficial substances, especially the nutrients "beyond NPK".

Once it has been established what it is a plant lacks in a given site, there is usually a choice of fertilizers available as a remedy, whose respective properties, as regards principal and side effects, must be known in order for correct application to be made (choice of correct fertilizer form).

How Much and How to Fertilize?

The correct amount of fertilizer should be such that the supply of nutrients is sufficient for high yields without causing unnecessary enrichment or losses. However, fertilization supplies all the nutrients required by the plant only when the nutrients are fully utilized, as is generally true in hydroculture. In soils, it is usually advisable to aim at a nutrient supply exceeding the immediate demand and to maintain this supply through fertilization.

High-yield plants require abundant nutrients in order to realize their genetic potential during a short vegetation period, i. e., to provide the desired high yields. Thus, in future the amount of fertilizer will by necessity have to match the increasing requirements and removal of nutrients. For this it is not enough to increase the amount of fertilizer proportionally. Thus, if a yield of 30 dt of grain is obtained with a given amount od NPK fertilizer, it will not be possible to obtain 60 dt automatically with double this amount of fertilizer. Requirements will be relatively higher and, most important, will be much more diverse.

The correct amount of fertilizer should be established on the basis of objective criteria. Just as treatment in modern medicine is increasingly based on diagnostic procedures, so should exact diagnosis be used more in agriculture, since it reliably determines what can no longer be estimated accurately by means of data obtained from experience. The quality of fertilization does not depend on the absolute amount of fertilizer used, but rather, on a precise application of fertilizer for specific requirements.

The method of fertilizer application involves many problems, ranging from the correct fertilization time through the labor-saving distribution of the fertilizer on the ground, to its correct introduction into the soil for optimum utilization by the plant. As regards correct fertilizer application, there are certain ideas about an optimum supply to the plants from the agrochemical viewpoint, which may occasionally have to be modified for reasons of labor efficiency or economy.

Correct application of fertilizers should always be meaningfully related to other production measures. Fertilization will be more effective the more the special properties of fertilizers are taken into account in their application.

1.1.2 The Difficult Path to Maximum Yields

Maximum yields can be obtained only if mistakes in fertilization are avoided and the decisive minimum factors of production are recognized and eliminated. Research continually supplies new stimuli to agriculture for both these purposes.

Typical Mistakes of Fertilization in the Past
The following typical mistakes are commonly responsible for reduced yields:

- Neglecting to fertilize with the "basic nutrients" (N, P, K) when the heavy, continuous needs of high-yield plants require easily available mobile nutrients in the soil;
- neglecting to fertilize with other nutrient elements, i. e., magnesium, sulfur, and trace elements (among which manganese, copper, boron, and to some extent zinc primarily play a role);
- one-sided intensive fertilization with N, P, or K, which is not fully effective because of some other deficiency (e. g., intensive fertilization with N is partly ineffective in cases of copper deficiency);
- occasional overfertilization with salt damage, caused by large doses of soluble fertilizers during sowing (salt damge also sometimes caused by irrigation water);
- faulty diagnosis of fertilizer requirements;
- insufficient allowance for the correct fertilizer form with regard to effectiveness and price;
- neglecting the basic conditions of soil fertility (e. g., soil structure and soil reaction);
- insufficient allowance for possible fertilizer losses.

Provided that the soil is worked properly and cultivation methods are suitable, the yield will depend, apart from the weather, on correct fertilization and appropriate plant protection. An important difference between these two methods is the fact that better plant protection usually implies higher expenditure, whereas better fertilization can often be achieved by changing, or sometimes even by reducing the expenditure.

The Search for Minimum Factors

Various factors determine the growth and yield of plants, but, in the final analysis, the minimum factor determines the yield level. Although, strictly speaking, the latter is not decisive alone, any yield can be only as high as permitted by the minimum factor. In any case, it exerts the greatest influence on the yield. The following passage, quoted from an old textbook on fertilization is still valid today: "You should strive to arrange fertilizer application and soil cultivation in such a way that the nutrients (and other) growth factors are not too inhibited by some completely neglected factor."

The yield potential of wheat, maize, and rice is approximately 120 dt/ha for field crops. Yields exceeding 100 dt/ha are already obtained occasionally under ordinary agricultural conditions.

However, the yields of even the best fields are below the yield potential, since either part of the production is destroyed by diseases, etc., or production conditions are not optimum. The cause of yield reduction must be determined and eliminated. The minimum factors in the initial period of mineral fertilization were primarily nitrogen, phosphorus, and potassium. Other nutrients frequently become minimum factors when the supply of these major nutrients has been optimized through fertilization. This could nowadays frequently be the case in the high-yield range. Figuratively, the situation may be represented by the "minimum cask". Just as the water level in a cask cannot be higher than the length of the shortest stave, so the yield cannot be higher than permitted by the minimum factor (Fig. 1-1).

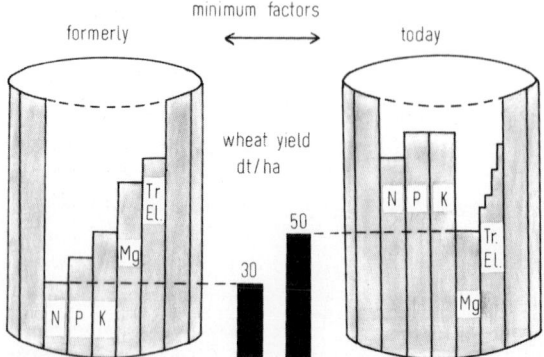

Fig. 1-1. Examples of minimum factors at different production intensities (represented as "minimum cask": shortest stave limits yield; Mg = magnesium, Tr. El. = trace elements [8].

The full yield potential can be realized only if all growth factors (including water supply, etc.) are optimized. This, however, is an ideal situation which, at best, can only be approximated in practice. In any case, it is most profitable to get as close as possible to this condition. Reliable diagnosis is essential for determining the yield-restricting minimum factors.

Practice and Research

The goal of high yields per unit area and high quality can be achieved only through a combination of practice (experimentation) and science (scientific research).

At the *practical* level, the aim is to obtain, as far as possible, prescriptions for the cultivation of certain plants, i.e., precise guidelines for achieving maximum yields and quality. Such prescriptions are obtained on one's own farm, e.g., through field trials in which conclusions are derived from observations and yield measurements. The advantage of this procedure is that it is carried out relatively easily and provides reliable prescriptions. This, however, is countered by the disadvantage of the limited information provided by simple yield trials.

Correlations (apparent connections) are determined first; however, this leaves the question of the basic causes of these connections open. There is a danger of erroneous interpretation, i.e., of insufficient or even incorrect conclusions being drawn from trials that in fact are properly designed (Chap. 6.6.1). Since the reasons for differences in yield cannot always be elucidated, unacceptable generalizations are easily made. Empirical trials are subject to narrow limitations. Progress in agriculture therefore largely depends on advances in research.

Scientific research, in its ideal sense, does not principally enquire about possible uses, but tries to explain natural events. Its method is causal research, i.e., investigating cause and effect. In problems of fertilization, the primary object is to elucidate complex processes in soils and plants that are responsible for growth. The advantage of this procedure is that a precise knowledge of the real relationships provides the possibility of directed intervention. However, causal research is a long-term matter, whereas the farmer with many decisions to make cannot wait for years. On the other hand, basic research is often necessary for further progress; this implies research in many remote fields, involving expensive indirect routes and unavoidable wrong tracks. This inevitably leads to different opinions that may cause scientific arguments because of imprecise results. Investigations must be carried out by specialists and it is often difficult for any one researcher to find the proper place for his special subject in science as a whole. Many theories thus remain incomplete. In this connection we should like to quote *J. v. Schwerz* (c. 1800 at Hohenheim): "Even if agriculture has been transformed from a trade into an art and from an art into a science, experience still remains the touchstone of our knowledge. All speculations and assumptions, all hypotheses and systems are to no avail, if they do not agree with the whole of nature" [26].

Agricultural practice can benefit considerably from scientific research, provided that new theories are first tested and matched to the whole enterprise. This is the main task of consultation, which should also prevent "talking at cross purposes" by practitioner and scientist.

Summary
1. Fertilization is nearly always required for the successful growing of cultivated plants. Its primary purpose is to improve nutrition of the plants.

2. Mistakes in fertilization do not only considerably reduce yield and income but may also have a detrimental effect on the quality of the produce obtained and on the environment.

3. Fertilizer should be carefully applied in intensive cultivation in order to improve the minimum factors. Fertilization should therefore supply the substances which the plants (according to type and quantity) lack for full exploitation of their yield potential and for the desired level of production, or which reduce noxious influences.

4. The basis of well-directed fertilization is a reliable diagnosis of the requirements.

5. The better one knows the fertilizers and the problems involved in their application in theory and practice, the more efficient the fertilization will be.

1.1.3 Nutrients in Soils and Plants

Soil, plants, and fertilizers consist of chemical substances. An understanding of the growth of plants and of their improved nutritional value through fertilization therefore requires the knowledge of some basic agrochemical concepts [7, 9, 15, 17]. Chemical terms are explained in the Appendix.

Nutrients and Nutrient Elements
The plant requires various nutrients for growth. *Nutrients* are substances the plant can absorb and which serve to nourish it. These substances may be molecules, e.g., CO_2 = carbon dioxide, H_2O = water, or ions, electrically charged particles. Ions may be either cations or anions (see Appendix), constituents of nutrient salts, and thus of many fertilizers. The number of possible nutrients is large. They are in part mineral (inorganic), in part organic substances whose relevant constituents are the nutrient elements.

The *nutrient elements* are vital for the nutrition of the plant. They are contained in, or are identical with, the nutrients, apart from the electric charge.

Higher green plants (and thus nearly all cultivated plants) require 16 nutrient elements. These are, carbon (C), oxygen (O), and hydrogen (H), the fundamental constituents of organic matter, and 13 additional mineral nutrient elements, namely:

- the major nutrient elements: N, P, S (absorbed as anions);
  K, Ca, Mg (absorbed as cations);
- the micronutrient elements: Fe, Mn, Zn, Cu (absorbed as cations);
  Cl, B, Mo (absorbed as anions).

(The chemical symbols are explained in synopsis 1–1, p. 3.)

N, P, and K are also termed "basic nutrient elements" (nutrients). Micronutrient elements are frequently called "trace elements". In actual fact, trace elements are all elements present in plants in small amounts (in any case more than 50). Plant growth can, in addition, be promoted by some *beneficial* elements, e. g., Si (silicon) and Na (sodium).

Some other elements are also significant. They are required not by the plant but by man and animals, and should therefore be present in the plant in sufficient amounts.

Nutrients in the Soil

Plants absorb nutrients from nutrient substrates, namely:

- solid nutrient substrates (natural soils, horticultural substrates, etc.), or
- liquid nutrient substrates (nutrient solutions of hydroponics).

The most improtant nutrient substrates are *soils*. Their suitability for plant production is characterized by the soil fertility.

Soil fertility is the capacity of the soil, based on certain properties, to bear fruit, i. e., to provide plant yields in its natural environment: soil productivity or yield potential.

Soil fertility is based primarily on the following important soil properties:

- Depth of soil; deep root penetration should be possible.
- Texture and structure: both factors are decisive for root penetration, water permeability and storage, aeration, and chemical and biological processes.
- Soil reaction is important for the structure and the availability of nutrients and is measured in terms of pH-value: pH = 7 indicates a neutral reaction, low pH-values indicate acid soils, higher values indicate alkaline soils (Chap. 5.1.1).

- Nutrient contents: high contents of nutrient reserves and adequate contents of available nutrients are desirable.
- Humus content and composition: humus, being the organic matter in the soil, is important primarily for a good structure and as a nutrient for soil organisms, mainly microbes; nutrient humus decomposes readily, whereas permanent humus is stable.
- Sorption properties: they characterize the ability of the soil to bind nutrients loosely and thus to store them in available form: storage of nutrients in exchangeable form.
- Content of toxic substances; such inorganic and organic substances should be absent as far as possible.

Nutrients occur in the soil in three bonding forms (Fig. 1–2):

- Water-soluble nutrients in the soil solution; they are not bound and can thus move with the water, being easily available to plants;
- exchangeable nutrients: cations or anions loosely bound by the electrically charged exchange complexes, clay and humus particles, are in general easily available;
- reserve nutrients: they comprise the overwhelming majority of soil nutrients in moderately- to slightly-soluble compounds; a small fraction is readily mobilizable and thus available to a moderate extent, but most of them are in practice unavailable for a long time.

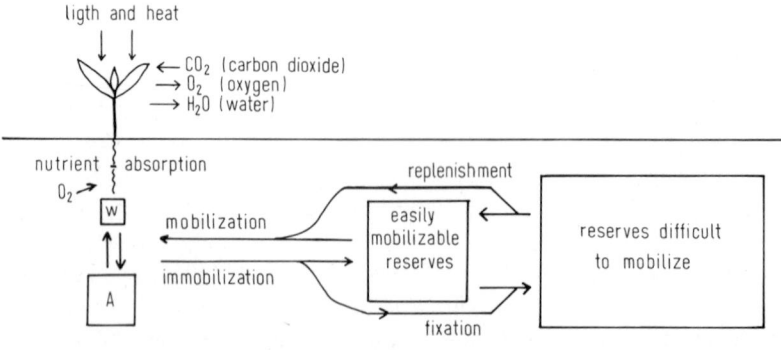

w = water-soluble nutrients
A = exchangeable nutrients

Fig. 1–2. Nutrient forms in the soil and processes of nutrient dynamics.

The *availability* of nutrients in the soil is, in the final analysis, decisive for the plant. Available nutrients are those available to the plant for absorption, i.e., those that can be absorbed. However, the degree of availability may vary within wide limits.

Nutrient dynamics in the soil cause continuous changes in the type of chemical bonds (by large-scale participation of micro-organisms) and thus in nutrient availability (Fig. 1–2).

A soil has *ideal nutrient dynamics* if nutrients can be stored in loosely bound form and are therefore protected against leaching but are still readily available, if excess amounts are buffered, without being fixed, and nutrients can be mobilized and supplied in sufficient amounts in cases of depletion or other losses.

Nutrients in the Plant

Plants absorb nutrients

- through the roots: water, oxygen, mineral and organic substances;
- through the leaves: carbon dioxide, oxygen, water and its solutes, via stomata and micropores in the outer skin.

The absorption of nutrients requires energy, which is made available through "root respiration". This necessitates a proper supply of oxygen to the root (soil aeration). Plants preferentially absorb substances they require. They thus possess a power of selection, and the root serves as an important filter. Toxic substances are largely excluded from absorption; however, this exclusion is not perfect and operates only up to a certain load limit.

Plants can absorb not only the easily available nutrients offered to them, in a manner of speaking, but they can also actively mobilize reserve nutrients, e.g., through the release of acids, which then attack the minerals, or complex-formers and reducing agents, which take up nutrients from compounds containing them. Absorbed nutrients are transported in the plant by water, especially to the principal sites of production, i.e., the leaves.

The *nutrient content of the leaves* is determinant for plant development. The concentrations necessary for optimum production differ for the various plants; however, the far-reaching uniformity of higher plants with respect to types and amounts of nutrients required by them, relative to the active cell substance, is remarkable. An optimum nutrient content of the leaves is a precondition for production: nutrient absorption by necessity thus precedes the production of plant mass. This requires a proper nutrient supply, especially for younger plants. The mineral content then diminishes with increasing age, owing to the *dilution effect*; this should be remembered in plant analysis.

We can distinguish six ranges of nutrient supply to plants, from deficiency to excess, in which important limiting values denote the transitions (Chap. 5.2.2).

Green Plants as Efficient Synthetic Units

Green plants synthesize energy-rich substances from simple basic substances containing little energy, by utilizing solar energy. Photosynthesis involves splitting water molecules with the aid of light radiation. The hydrogen thus released serves to convert carbon dioxide into sugar, while the oxygen is expelled as waste material.

The plant then uses the basic compound, sugar, to synthesize all other substances present in it, mainly

- carbohydrates: numerous sugars, starch, cellulose, etc.;
- fatty substances: true fats and lipoids (substances resembling fats);
- proteins: simple proteins and amino acids as building blocks and composite proteins (proteids);
- enzymes, vitamins, and growth agents activating and regulating the metabolism.

1.2 General Comments on Fertilizers

1.2.1 Definition of Fertilizers

Fertilizers are substances intended for improving the nutrition of plants. Fertilization thus implies the supply of fertilizers to plants or to the nutrient substrate, e.g., to the soil.

Fertilizers are defined in the *fertilizer law* as follows: "Fertilizers are substances intended to be supplied directly or indirectly to crops in order to promote their growth, increase their yield, or improve their quality."

Explanation of the definition: fertilizers should

- promote growth; this usually implies increasing the plant mass, but may sometimes also mean regulation in the sense of restricting the growth of certain plant parts in order to increase the yield;
- increase the yield, with the net yield, obtained as harvested mass, being of primary interest; this may involve vegetative plant organs, e.g., leaves or generative organs, e.g., fruits;
- improve the quality with regard to both commercial quality, market value, and nutritional quality, as well as increase the resistance of the plant to noxious influences of all kinds, (ensuring growth).

Etymological derivation of the term "fertilizer": Formerly, the word in use was "dung" in English. It is derived from the old German word *tung,* meaning a storage pit covered with manure for protection against the cold. From this was derived *tungen* (to cover) and then *dung* (fertilizer) as the cover. This narrow definition of fertilizer as soil cover was then extended to substances introduced into the soil in order to improve it.

The corresponding English term *fertilizer*, which is now accepted internationally, in lieu of the old English word *dung*, was derived from the Latin root *fertil* (fertile). Fertilization thus means to render fertile, and hence largely fits the scientific core of the concept of fertilization.

Designations of Fertilizers

Agriculture has been concerned with fertilizers since antiquity. Many earlier designations of individual fertilizers and their value-determining constituents have remained in use up to the present, although the scientific chemical nomenclature has been adapted to a newer state of knowledge. Fertilizer nomenclature is thus a mixture of old folkloristic designations, historical names from the early stages of chemistry, and modern chemical nomenclature.

The designation of nutrient elements is still only partly internationally uniform, e. g., for phosphorus and some trace elements*. However, there is a trend away from the diversity of national designations to an international nomenclature of maximum uniformity. On the other hand, the Latin-German term *Kalium* should be preferred to the English-French word *potassium/potasse*. Modern chemical designations of the corresponding salts should be preferred to fertilizer terms deriving from the early stages of chemistry, e. g., "ammonium sulfate" should be used instead of "ammoniated sulfuric acid".

1.2.2 Indications of Nutrient Contents

The contents may be indicated in various ways (Synopsis 1–2). However, a good modern nomenclature must satisfy the requirements of exactness, suitability and simplicity. In the case of fertilizers which chiefly contain plant nutrients an obvious, unequivocal reference basis is the indication of the content of nutrient elements. This permits a common denominator for the multiplicity of nutrients formed by a nutrient element to be found, e. g., use of nitrogen content as the basis for the content of nitrogen-containing nutrients (nitrate, ammonium, urea, etc.). Thus, in the case of potassium, the designation K_2O, introduced from inorganic chemistry, by analogy with the designation CaO, is unsuitable because K_2O occurs neither in the soil nor in plants or fertilizers. The new EEC guidelines now legally permit use of the element form; it would thus be possible in the

---

* The German equivalent for nitrogen, i. e. *Stickstoff*, is misleading with respect to plant physiology since nitrogen does not "asphyxiate" (as the term would imply literally) the plants, but on the contrary, is the most important growth stimulant.

FRG to follow the example of many other countries and the FAO, and to introduce this long overdue change. In addition, this would also prevent many misunderstandings and errors.

Applying the principle that, in the final analysis, the active substance should be indicated using a simple reference basis, a different system than indication of the elemental composition is used for fertilizers which mainly serve to improve soils. Soils are improved e.g. by limes whose active substance is calcium hydroxide, which serves to neutralize the acids in the soil. An even simpler reference basis would be its anhydrous form, namely CaO (or MgO for magnesium limes). This reference basis should therefore be preferably used for all lime fertilizers.

Indication of the organic substance as a reference basis is useful in the case of organic fertilizers, since it is precisely the supply of organic compounds that is determinant and not that of pure carbon (C).

Indication of Total Amount or of Active Component
In the final analysis, the user is, interested in the guaranteed, useful, active properties of the fertilizer while the content of ineffective or virtually ineffective substances is unimportant to him. At least the price should depend on the active component. From the user's viewpoint it would therefore be preferable to designate fertilizers only according to the amount of active constituents they contain.

Synopsis 1–2. Description of Fertilizer Content.

Description of the composition of fertilizers containing mainly plant nutrients.

Content indications until now customary in the FRG ("pure nutrient")	Modern content indications now being introduced internationally
N (nitrogen)	N (nitrogen)
P_2O_5 (phosphorus pentoxide)	P (phosphorus)
K_2O (potassium oxide)	K (potassium)
MgO (magnesium oxide)	Mg (magnesium)
Ca (calcium)	Ca (calcium)

Suitable content indications for soil fertilizers

limes	CaO or MgO
organic fertilizers	organic substance

On the other hand, the total content of certain substances in a chemical compound can be defined and determined analytically much more clearly. From the legal aspect it is thus preferable to indicate the total content, the more so since fertilizer law is not an applications law, but a trade law requiring the possibility of unequivocal checking.

Indication of the total content is preferable in all cases where the total content approximately corresponds to the active content. This also applies to fertilizers whose effectiveness is highly dependent on the soil conditions in any given case. However, the users' requirements have been taken into account, especially in the case of many phosphates, by using the content of *active* phosphate as the basis for sale of the fertilizer, instead of the less informative total content.

1.2.3 Classification of Fertilizers

The number of substances suitable as fertilizers is very large; their compositions and origins differ considerably, and there are multiple possibilities for their use. Many fertilizers, in the form of waste matter, have been used since the earliest days of primitive cultivation; others are modern synthetic products of the fertilizer industry. Some fertilizers are produced and used on farms; others are obtained commercially as supplements. Many fertilizers serve primarily to supply the plants with nutrients; others contain hardly any nutrients, but serve to improve the soil and thus indirectly ensure better plant nutrition. From the chemical aspect, fertilizers may be either inorganic compounds (mostly minerals) or organic compounds. They contain either a single or several nutrients [24, 16].

In view of this diversity of fertilizers, it is difficult to establish a classification satisfactory from all points of view. We shall therefore first of all undertake an alternative grouping according to different aspects that partly overlap.

1. Classification of Fertilizers According to Type of Origin

Natural fertilizers are formed in nature and are used in the form in which they occur, without, or with little, processing. Examples are manure (either fresh or decomposed), peat, leaf litter, sludge, ash, lime marl, and crude phosphates.

Artificial fertilizers (synthetic fertilizers) are produced in factories by technical means. This is done either through chemical changes of natural products (e. g., P- und K-fertilizers), or completely synthetically from simple source materials, as is the case with most nitrogen fertilizers.

The term *artificial* fertilizer often creates negative associations of synthetic substitutes. This is completely wrong (Chap. 9.1.2). It would be more correct to speak of man-made products in the positive sense. The term "artificial fertilizer" is now largely avoided because of the possible erroneous interpretation.

2. Classification of Fertilizers According to Source

Farm manures originates on farms. Examples are manure (solid and semi-liquid), compost, straw, and marl from the subsoil of the farmer's own fields.

Commercial fertilizers are obtained through trade channels. Most of them are subject to the regulations of the fertilizer law. This group nowadays includes most fertilizers which are important for additional supplies of nutrients in intensive cultivation.

3. Classification According to Mode of Action

Direct-acting fertilizers (plant fertilizers) contain essential components of available plant nutrients and thus supply them directly to the plants. These are most commercial N-, P-, and K-fertilizers as well as liquid and semi-liquid manure from farms.

Indirect-acting fertilizers (soil fertilizers) primarily improve the nutrient substrate (soil in agriculture, nutrient substrate in horticulture), though they have a certain additional significance as a source of nutrients. Examples are limes, peat, straw.

4. Classification According to Speed of Action

Fast-acting fertilizers are immediately available to the plants, e.g., the water-soluble N- and K-fertilizers, or improve the soil within a short time.

Slow-acting fertilizers are effective only after conversion in the soil. This may be an advantage or a disadvantage, depending on the purpose of the fertilizer.

5. Classification According to Type of Chemical Compound

Organic fertilizers are mostly mixtures of a number of organic compounds, e.g., the natural organic fertilizers such as manure, peat, etc. However, they may also be definite single compounds like some especially slow-acting N-fertilizers or urea.

Mineral fertilizers (inorganic fertilizers) consist of one or more inorganic compounds (salts, oxides, etc.). They mostly contain mineral nutrients or yield them upon conversion. In the list according to type (Chap. 1.2.4), for the sake of a suitable classification, mineral fertilizers also include some compounds that are organic according to their chemistry, but are rapidly converted into mineral substances in the soil, e.g., urea.

6. Classification According to Number of Nutrient Elements

Single-nutrient fertilizers (single fertilizers) are fertilizers containing only one nutrient or essential nutrient (nutrient element), e.g., fertilizers containing nitrogen. The additive term *essential* already indicates that there are various classification possibilities. Thus, the fertilizer type-list groups the potassium-magnesium fertilizers with the potassium single-nutrient fertilizers, although from the agro-chemical point of view it should be included under two-nutrient fertilizers.

Multiple-nutrient fertilizers are fertilizers containing several nutrients. We may distinguish between two-, three-, and up to six-nutrient fertilizers, corresponding to the six major nutrients, or fertilizers with an even larger number of nutrients if trace elements are included. However, the type-list is limited to groupings with, at most, three nutrients. This is advantageous insofar as the three major plant nutrient elements are thus particularly emphasized. Another important concept is that of a *complete fertilizer*. This term has two different meanings. It is usually applied to a fertilizer containing all three major nutrient elements, i.e., complete fertilizer in the narrower sense; it may, however, also be used in a wider sense to indicate a fertilizer containing all the mineral nutrient elements needed by the plant. Multiple-nutrient fertilizers are either mixed fertilizers, obtained by mixing the components, or *complex fertilizers* formed through chemical conversion processes (decomposition processes).

7. Classification According to Amounts Required by the Plant

Major-nutrient fertilizers are fertilizers containing the essential major plant nutrients, which therefore must be supplied to the plant in large amounts. Generally, they are not chemically pure substances and thus, in part, also contain small amounts of micronutrient elements.

Micronutrient fertilizers are fertilizers that contain chiefly micronutrient elements. They are therefore applied in small amounts.

8. Classification According to State of Aggregation

The following are classifications according to physical state:

- solid fertilizers;
- Liquid fertilizers (fertilizer solutions and suspensions);
- gaseous fertilizers (e. g., gaseous ammonia).

9. Classification According to the 1977 Fertilizer Law

The new type-list classifies fertilizers, subject to licensing, into four groups:
  I. Mineral one-nutrient fertilizers;
 II. mineral multiple-nutrient fertilizers;
III. organic and organic-mineral fertilizers;
IV. fertilizers containing micronutrients.

A further series of fertilizers will, in future, only be subject to a designation requirement namely the following *natural and auxiliary substances* (Chap. 1.2.4):

a) Farmyard manures;
b) soil auxiliary substances (e.g., for improving soil structure, soil inoculants, rock powder);

c) growth substrates (e.g., horticultural substrates);
d) plant auxiliary substances (e.g., fertilizers without nutrients, but with a positive action on plants, also compost-processing agents).

10. Attempted Comprehensive Classification of Fertilizers

Comprehensive classification is important and useful, despite the difficulties encountered in definitively grouping the fertilizers. Such a classification should, as far as possible, take the above-mentioned classification principles into account according to their degree of importance. However, it should not get lost in pointless refinements. A pragmatic proposal for comprehensive subdivision according to agrochemical aspects is made in Synopsis 1-3. Classification according to the fertilizer decree, on the other hand, has been slightly simplified for legal and historical reasons.

Synopsis 1-3. Possible Classification of Fertilizers from an Agrochemical Viewpoint.

I. **Mineral fertilizers**
   A. Mineral fertilizers serving primarily to *supply nutrients*
      1. Major-nutrient fertilizers
         a) one-nutrient fertilizers
         b) two-nutrient fertilizers
         c) three-nutrient and other multiple-nutrient fertilizers
      2. Micronutrient fertilizers
         a) Micronutrient fertilizers with one value-determining trace element (one-micronutrient fertilizer)
         b) micronutrient fertilizers with several trace elements
      3. Combined major- and micronutrient fertilizers
      4. Other mineral fertilizers containing substances important for plants, animals or man
   B. Mineral fertilizers serving primarily to *improve soils*
      1. Fertilizers for improving soil reaction
      2. Fertilizers for improving soil structure
      3. Fertilizers acting against toxic excesses
      4. Other fertilizers for soil improvement

II. **Organic fertilizers**
   1. Organic farm manures
   2. Organic commercial fertilizers
   3. Organic-mineral commercial fertilizers
   4. Active-agent fertilizers, etc.

III. **Other fertilizers**
   e.g., carbon dioxide

1.2.4 Fertilizer Legislation

Fertilizers are of great political and economic significance. The trade in commercial fertilizers is therefore regulated by statutory legislation in Germany and many other countries. This legislation only applies explicitly to trade in fertilizers, but not to their use.

Fertilizer legislation in Germany has evolved as follows:

1918: Decree concerning artificial fertilizers
1962/63: Fertilizer Law and Decree for the FRG
1975: EEC Guidelines for co-ordinating the legal regulations for fertilizers in the member states
1977: New fertilizer law with rules for its enforcement in the FRG, to harmonize national and EEC regulations.

In § 1, the new fertilizer law firstly explains the *definitions*. The fertilizer law divides fertilizers in the general sense (for definition, *see* Chap. 1.2.3) into three categories:

a) Fertilizers within the more restricted meaning of the law: this category includes most *mineral and organic commercial fertilizers*, which are divided into four groups (Chap. 1.2.3). The law regulates the trade in fertilizers through licensing, designation regulations, monitoring, etc.

b) Other fertilizers (in a wider sense), which, if they are marketed according to the law, only require to be designated but not licensed. These form a heterogeneous group and are provisionally designated as *"natural and auxiliary substances"* (farmyard manure, soil auxiliary substances, growth substrates, plant auxiliary substances, etc.) (Chap. 1.2.3).

c) *Fertilizers exempt* from regulation by law, such as (town) waste materials, e.g., waste water, sewage sludge, excrement. These fertilizers are subject to the waste-removal law, which also applies to the use of farmyard manure, e.g., liquid, semi-liquid, and solid manure, if the "customary" extent of agricultural fertilization is exceeded.

§ 2 regulates *licensing*. Fertilizers may be marketed only if they conform to a licensed fertilizer type. These types are designated according to their essential constituents, composition, production, etc. Only those fertilizer types are licensed, which, according to the definition of fertilizers, improve the growth, yield, or quality of plants without harming soil fertility, ecology, or animal and human health.

Apart from farmyard manure, fertilizers intended for export and research purposes, as well as fertilizers for lawns and ornamental plants are exempt from licensing. §§ 3 to 6 regulate the designation of fertilizers, tolerances in their contents, trading restrictions, and sampling procedures.

§ 8 states that *trade* in fertilizers is to be monitored by the fertilizer-trade supervision section of the States of the Federal Republic of Germany, in order that the stated valuable properties are maintained to safeguard the user. Contraventions against the legal regulations are considered to be contraventions against public order.

The *fertilizer decree* specifies details of fertilizer licensing [24]. A fertilizer type is licensed only if,

- its effectiveness has been demonstrated in trials, in comparison with the absence of fertilization;
- it is considered to be completely harmless with regard to food products; this applies to both nutrient components and minor components and additives, as well as to undesirable contaminants. Fertilizers are tested thoroughly with regard to their action and harmlessness. Many well-known "natural" fertilizers have not been tested as carefully.

The *type-list* is an essential part of the fertilizer decree. In 1977 the valid type-list included 276 fertilizer types altogether, namely:

  59 mineral one-nutrient fertilizers,
138 mineral multiple-nutrient fertilizers,
  57 organic and organic-mineral fertilizers,
  16 fertilizers with micronutrients,
   6 soil inoculants and soil conditioners.

(*Growth regulators*, formerly included under fertilizers, are now subject to the plant-protection law.)

Certain minimum contents were specified for one-nutrient fertilizers for the sake of type delimitation. On the other hand, contents with tolerances were specified for multiple-nutrient fertilizers.

The number of types was reduced to 113 upon adaptation of the national to the EEC type-list. This substantial decrease is due chiefly to a general stipulation of minimum contents, so that only few types of multiple-nutrient fertilizers remain.

An important innovation is the introduction of *EEC fertilizers*, i. e., fertilizers licensed in all EEC countries. Fertilizers in national type-lists may, under certain conditions, also be licensed as EEC fertilizers. The list of EEC fertilizers up to now only includes types of (mineral) single and multiple-nutrient fertilizers (with major nutrients), without calcium and magnesium fertilizers. It largely conforms to the type-list of the FRG.

Summary of Fertilizers in General

1. Fertilizers (manures) serve directly or indirectly to improve the nutrition of plants and thus to increase growth, yield, and quality.

2. The respective active substances, should be indicated using a simple reference basis with respect to the nutrient content, etc., in fertilizers.

3. All groupings should have a pragmatic orientation, since the multiplicity of fertilizers renders clear and exact classification difficult.

4. Trade in fertilizers is regulated by law, in view of their large national-economic significance. Most fertilizers are subject to licensing after their effectiveness and harmlessness have been tested.

5. Fertilizers licensed nationally may be licensed as EEC fertilizers.

1.3 History of Fertilization

1.3.1 Fertilization in the Pre-scientific Era

The history of fertilization is an essential part of the history of agriculture. The use of fertilizers probably dates back to the beginnings of agriculture more than 5000 years ago. In primitive cultivation (by means of hoeing), man first exploited the natural fertility of the soil during the stone age; however, he obviously very soon recognized the possibilities of improving the growth of crops through the supply of fertilizing substances. Primitive forms of fertilization to improve soil fertility were already common in ancient civilizations (in the countries irrigated by the rivers Nile, Euphrates, Indus, and in China, South America, etc.) [13].

Experience with fertilization in early times was extended and described in classical antiquity (Greece and Rome). *Homer* mentions manure as a fertilizer in the Odyssey, but it was only in ancient Rome that agricultural writers gave comprehensive descriptions of fertilization. Examples are *Cato* (200 BC), *Pliny* and *Columella* (1st century AD). *Cato* states that good husbandry implies good ploughing, good tending, and good fertilization. Certain fertilizers were considered so valuable in antiquity, that their theft was a punishable offense. *Stercutius* in ancient Rome was made immortal by the gods for his invention of fertilization (*Stercutius* or *Sterculus* as a symbolic figure).

In addition to leaving the ground fallow for the regeneration of natural nutrients, the following fertilizers were used in antiquity to supplement soil fertility:

- farmyard manure and compost,
- vegetable and animal waste (straw, blood, etc.)
- human and animal excrement,
- bird excrement deposits (guano),
- silt from rivers and ponds,
- straw and soil material from forests,
- seaweed and other plants, as well as fish wastes from the sea,
- green manure (green plants as fertilizer),
- salt-containing earths,
- ash (from straw, wood, bones, soil material, etc.),
- marl, lime, gypsum.

The principle in fertilizing with these substances is primarily based on closing the nutrient cycle. This permits at least a low yield level to be maintained for long periods. This empirical fertilization in agriculture was used on a world-wide basis approximately up to the 19th century. The theoretical foundation was Aristotle's *humus theory* (350 BC): "The plant obtains nourishment from humus substances, which it absorbs from the soil through the roots; after dying it becomes humus, and humus substances are thus fertilizers."

The definitive formulation is due to *Thaer* (1809), the last important advocate of the humus theory:

"Fertility of the soil in fact depends completely on humus, since apart from water, it alone, nourishes the plants. Just as humus is a product of life, so it is also a condition of life. Without it no individual life can be imagined" [27].

Agricultural Production in Europe up to 1800
In Europe, considerable fertilization experience was adapted from Roman literature. Nevertheless, yields remained low despite all efforts to the contrary. Some better soils did give higher average yields, but significant increases were hardly possible. Yields were at that time mostly measured as yield ratios (also called "grain yields") (Table 1–1).

Table 1–1. Yield ratios for grain in Central Europe from the Middle Ages to the Present Day (ratio of seed quantity to yield) [11].

Period	1 kg seed gave the following yields (kg)	
	average soils	best soils
Middle Ages (12th to 15th century)	3 to 4	—
16th and 17th century	5 to 6	7 to (15)
around 1800	5 to 6	12 to 20
comparison with 1970	30 to 40	

The state of agriculture from the Middle Ages to about 1800 was thus characterized by low and unreliable yields. Bad harvests often caused mass poverty and famine crises. The last famine due to natural causes, in 1846–47, may in part have been responsible for the 1848 revolution in Germany [11].

Soils were largely impoverished, since many nutrient deficiencies could only be eliminated partially despite intensive efforts to close the nutrient cycle. Farmers continually searched for "additional earth" from forest or heath, from the subsoil (marl) or from elsewhere.

The supply of soil material did indeed improve many poor soils, but could not basically eliminate the general nutrient deficiency. Until 1800 grain yields were 5 to 10 dt/ha (with some exceptions on the best soils), the mean being about 8 dt/ha. Soil fertility generally remained at a low level. Part of the growing population (of a given area) was therefore forced to emigrate and look for new living space. This apparently hopeless situation changed considerably after 1800.

1.3.2 The Start and Initial Development of Mineral Fertilization (1840– 1880–1920)

With the beginnings of modern chemistry and a better basic understanding of plant metabolism, fertilization, which until then had been empirical, attained the rank of a natural science. In 1804 *A. v. Humboldt* drew attention to the fertilizing action of guano (and saltpeter) after returning from his expedition to South America. This later led to the first imports of these substances into Europe.

However, use of this commercial fertilizer became significant only after the agrochemical basis had been elucidated, thus providing and presenting a theoretical foundation for the development of modern fertilization (*Liebig*, 1840) [17]. (Synopses 1–4 and 1–5) [21, 23].

The new theory was not readily accepted in practice, especially since the first *patent fertilizer* developed by *Liebig* himself was a complete failure in practical fertilization trials. *Liebig* had feared a loss of nutrients through leaching of the soil and had therefore fused the phosphate and potassium salts used into a water-insoluble substance. Unfortunately, it had thus become unavailable to the plants. It was only when the causes of this initial failure had been clarified, (as was the insignificant effect of bone dust as a fertilizer), that the way was paved for the novel concept of water-soluble fertilizers, the first of these being the superphosphates.

New Fertilizers and Their Applications

Superphosphate, the first artificial fertilizer, was soon produced in many small factories in Europe. This initiated the development and testing of other new fer-

Synopsis 1-4. Great Researchers in German Agrochemistry (Plant Nutrition) in the Classical Era.

Sprengel, Carl	Göttingen	1787–1859 ⎱	founder of mineral
Liebig, Justus v.	Gießen	1803–1873 ⎰	fertilization theory
Knop, Wilhelm	Leipzig	1817–1891 ⎱	exact determination of nutrient requirements of
Sachs, Julius	Bonn	1832–1897 ⎰	plants in nutrient solutions
Wolff, Emil v.	Möckern/Leipzig	1815–1896	plant analysis, importance of nitrogen
Hellriegel, Hermann	Bernburg	1831–1895	exact sand-culture trials, demonstration of nitrogen fixation
Maercker, Max	Halle	1842–1902	introduction of fertilization with potassium
Wagner, Paul	Darmstadt	1843–1930	procedure for fertilizer trials, introduction of *Thomas* phosphate
Mitscherlich, Eilhard	Königsberg	1874–1956	fertilization trials and rules of fertilization

Synopsis 1-5. The Mineral Theory ca. 1900.

The *fundamentals of the mineral theory* were developed between 1830 and 1840 by *Boussingault, Sprengel* and particularly *Liebig*. It implies a far-reaching rejection of the *humus theory*, but in a sense it is also an extension of that theory.

The mineral theory, as it stood at the end of its first phase of development, around 1900, may be summarized as follows:

1. Plants are nourished not by humus, but by minerals.

2. Minerals occur in plants not by coincidence but are necessary constituents for their growth and yield ("Everything constituting the plant promotes its growth").

3. Plants require 10 nutrients (elements) which, apart from carbon, oxygen, and hydrogen, are absorbed from the soil in the form of salts (N, P, K, Ca, Mg, S, Fe).

4. The nutrient requirements of different plants vary and can be determined from the constitution of properly nourished plants.

5. Many soils are only inadequately supplied with nutrients for proper plant growth.

6. The supply of nutrients through fertilization can eliminate nutrient deficiencies.

7. Organic substances (humus) act as fertilizers, since they "improve" the soil and provide mineral nutrients as well as carbon dioxide through their decomposition.

tilizers. In part, these led to the exploitation of new sources of raw material; in part, they represented cheaper forms of fertilizer and thus contributed to a more economical supply of fertilizers to agriculture. The search for raw materials for fertilizers led to the discovery of phosphate deposits in various parts of the world. The hitherto neglected "top-layer salts" of salt mines were recognized as valuable potassium deposits.

However, for a long time the main problem remained the binding of atmospheric nitrogen and thus the mass production of N-fertilizers. The decisive breakthrough to a larger production of N-fertilizers was achieved in 1913 with the synthesis of ammonia by the *Haber-Bosch* process (in the IG-Farben works).

Although the First World War caused some delay, since about 1920 agriculture has had available an abundant choice of N-, P- and K-fertilizers (Synopsis 1-6).

Synopsis 1-6. The Introduction of Modern Fertilizers.

1830 First cargo of saltpeter fertilizer from Chile (to England)
1840 First guano from Peru (to England)
1843 Superphosphate as first "artificial" fertilizer (England, after 1855 also in Germany)
1860 Potassium fertilizer from top-layer salts of salt mines (Germany)
1879 *Thomas* phosphate from iron and steel industry (England)
1890 Ammonium sulfate from coking ammonia as first "artificial" N-fertilizer (Germany)
1905 Calcium cyanamide from atmospheric nitrogen (Germany)
1907 Saltpeter fertilizer from atmospheric nitrogen by arc-gap process (Norway)
1913 Synthetic ammonia from atmospheric nitrogen by *Haber-Bosch* process as a basis for many N-fertilizers (Germany)
1916 Rhenania phosphate through alkaline phosphate decomposition (Germany)
1921 Urea (carbamide) from ammonia (Germany)
1927 Nitrophoska as first NPK fertilizer ("complete fertilizer")
1929 Lime ammonium nitrate as important N-fertilizer

The practical application of fertilizers had meanwhile been repeatedly tested in agricultural experimental stations. Special mention should be made of *Boussingault*'s trials at Pechelbronn (Alsace) after 1836, the first continuous trials by *Lawes* and *Gilbert* at Rothamsted in England, begun about 1843, and establishment of the first German experimental station at Möckern near Leipzig (*Stöckhardt*, 1852).

Some of the new commercial fertilizers were already being increasingly used in some regions between 1840 and 1880, but only after 1880 is it possible to speak of significant fertilization, which can be statistically recorded for the whole of Germany. It therefore appears correct and proper to set the beginning of real fertilization in European agriculture, and thus in the world, at about 1880.

Effects of Mineral Fertilization

Yields almost doubled between 1840 and 1880, from about 8 dt/ha for grain to 14 dt/ha, despite the initially hesitant use of mineral fertilizers until 1880. The main reason for this is the introduction of crop rotation, which was strongly advocated by *Thaer*, in lieu of the old three-crop rotation system, which was almost exclusively based on grain. The increased cultivation of clover, in particular, improved the nitrogen balance of the soil, for the benefit of the plants that followed. The nitrogen-binding action of legumes was, moreover, utilized through green manuring with lupins and serradella, from which poor soils in particular benefitted. A particular advocate of green manuring was *Schultz-Lupitz* (after 1855). However, the general improvement of yields, obtained through crop rotation with clover and root crops became fully effective only after more intensive fertilization was applied.

The change to increasing production, which occurred in agriculture as a result of the new means of production represented by mineral fertilizers, was enormous. The situation of constant scarcity of foodstuffs changed in Europe to a state of abundant supply. It was not only the yields but also the quality of the produce that improved as far as this could be judged by simple means.

The concept of mineral fertilization evolved in Central Europe with Germany as the centre of development. Three basic preconditions coincided there: the challenge of increasingly scarce food for a growing population, a good theoretical foundation provided by the developing natural sciences, and the initiative of leading agrochemists and practical farmers in solving this problem [12, 29, 28, 18, 25, 4, 10].

The great performance of agrochemistry is brought out by the following quotations:

J. v. Liebig (1861)

Imagine that the populations of Europe had grown from 1790 onward at a similar rate to that from 1818 on. In the course of two generations there would have occurred a situation which would have resembled the Middle Ages in its horror. Agriculture then and until a few years ago was completely unable to supply the growing population with the means for its existence in the same proportion ... (Even) in civilized nations famine evokes ruthless, bloodthirsty cruelty, searching for solutions in internal revolutions and external wars. The great wars at the end of the last and at the beginning of this century thus appear to resemble events ordained by natural laws, intended to restore the missing balance between consumption and replacement of foodstuffs [3].

P. Wagner in "40 Jahre Thomasmehl" (40 years of *Thomas* phosphate) (1927):

What was the situation 40 years ago (about 1880)? Plants in the fields and meadows were starved, primarily of phosphoric acid. At the same time, it was impossible to grow clover anywhere in RhineHessen, where today sainfoin, clover, and alfalfa provide the most abundant and reliable yields, since as we have demonstrated, there was a lack of phosphoric acid. The expensive nitrogen ... and potassium were not utilized reliably, since plant nutrition with phosphoric acid was insufficient. Fields and meadows required exceptional amounts of phosphoric acid in order to satisfy the plants. However, exceptionally large doses could not be applied by the farmer, since the material was too expensive. High-intensity fertilization was impossible at those prices; yields had to remain at a low level ... Then *Thomas* phosphate made its appearance. It was offered at very low prices, the price of superphosphate dropped ... Agricultural production increased enormously. Our fields and meadows were rendered capable of processing increasingly large amounts of nitrogen and potassium, thus increasing yields and raising profits.

F. Haber in "Das Zeitalter der Chemie und seine Leistungen" (The Era of Chemistry and Its Performance) (1925)

For more than half a century, potassium from the Stassfurt mines, phosphoric acid from *Thomas* phosphate and from the phosphate deposits of the Pacific Ocean, North Africa, and the southern United States, nitrogen from the deserts of Chile and from coking factories has flowed onto the fields as a stream of abundance and blessing.

D. Prjanishnikov in "Nitrogen in Plant Life" (1945)

At the beginning of this century, our yields (in Russia) were approximately half of those in some countries in Western Europe ... At that time there was a tendency to ascribe this difference to more favourable natural conditions like soil and climate. However, this explanation is completely wrong: on the contrary, soils in the west are naturally inferior to ours; rather, their fertility is a secondary phenomenon, a result of effort and know-how.

1.3.3 Extension of the Fertilization Concept

There were substantial developments in fertilization after the First World War in Europe, and also soon after, in many other parts of the world. The twenties and thirties were, in a sense, the major fertilization "learning period". Many different types of fertilizers were tested on numerous crops in various soils depending on the climate. The concept of simple mineral fertilization thus proved itself. The possibilities for increasing yields appeared to be almost limitless. Mineral fertilization with N, P, and K seemed to guarantee every success.

However, the *successes* were soon found to be accompanied by serious problems caused by the new type of fertilization. Thus, it could not be denied that

- mineral fertilization did not always give the expected results;
- the basically empirical dosage of fertilizers partly caused an imbalance in N, P, and K plant nutrition, with detrimental results;

- mineral fertilization remained rather ineffective if the bases of soil fertility (e.g., fertilization with humus) were neglected;
- sometimes damage was caused by overfertilization;
- at least some fertilizers were not harmless when applied (leaves could be burned, domestic animals and game such as pheasants could be "poisoned" by eating the fertilizer grains);
- intensive fertilization sometimes negatively affected the quality of the produce;
- unknown plant diseases sometimes appeared to be caused by fertilization itself, despite its being "optimum", in which case no growth whatsoever occurred; this was a frequent complaint in the case of oats.

The simple concept of mineral fertilization as practiced until then had obviously to be revised in several respects. Further research, especially by European agrochemists, then clarified various problematic aspects of fertilization.

Allowing for Soil Fertility
An initial, decisive advance was represented by the concept of correct soil reaction. There had been pH-measurements in soils since 1913, but actual research on soil reaction and its significance for fertilization only began after 1920. The application of lime and marl could soon be made on a rational basis; the influence of other fertilizers on soil reaction was recognized [4, 14].

Advances in soil science led to a better understanding of the fertility-determining properties of soils. Investigations of the content and *availability of nutrients* in the soil permitted a more purposeful application of fertilizers. The behavior of fertilizers in the soil, and the conversion and storage of fertilizer nutrients were now considered.

It required decades of research to clarify the manifold significance of *organic fertilizers* for plant nutrition, or at least to keep track of it. It was now found that the value of a good supply of humus to the soil should not be underestimated, although the reasons for this view differed from the former concept of direct plant nutrition with humus. The role of an active "soil life" for fertilizer action, nutrient supply, and yield became increasingly evident. *Cato's* ancient wisdom was rediscovered: good husbandry requires not only fertilization but also good ploughing and tending, i.e., correct working of the soil, correct plant protection, and correct integration (linking) of all these measures ("mineral fertilizer alone is of no use").

Despite this incomplete concept, mineral fertilization was generally quite successful because many farmers continued to adhere to proven rules even after the introduction of the new fertilization methods. Also, many soils themselves supplied several important nutrients due to their good regulatory capacity, and mineral fertilization, even if unintentionally, continued to supply important nutrients to the plants in the form of minor constituents. Practice thus corrected the imperfect theory.

Agrochemists at that time increasingly studied the problems of fertilizer action in detail, mostly directing their investigations toward factual, practical goals. This *practical approach* was their great strength, but also their weakness, since some very important discoveries thus escaped them.

The discovery of Trace Elements

Since the turn of the century, scientists had repeatedly encountered unknown plant diseases. These were obviously linked with nutrition, but their causes could not be elucidated despite the most varied applications of fertilizers. The mineral-nutrition theory in vogue at that time was obviously incomplete (assumed effects of fertilizer "stimulation"). However, the results of fertilization did not explain the causes of these diseases, e.g., the *withertip* or *gray-speck disease* in oats, the *heart rot* in beets, and the peculiar *lick disease* in cows, which seemed to be caused by a deficiency in the grass.

The problem of further nutrient elements could only be elucidated by applying a new theory deriving from plant physiology research. *Hoagland* et al. in the USA investigated in much more detail than hitherto the old problem of *what plants actually require*. The list of essential elements was finally increased by five more: manganese, boron, zinc, copper, and molybdenum (1922-39); chlorine was added later (1954).

Many earlier fertilization failures were explained after the detection of these trace elements. The classical theory's simple form no longer appeared to be the final conclusion, but, on the contrary, proved to be quite incomplete in several respects. It is thus not surprising that, despite the undoubted successes of mineral fertilization during its major development period, criticism continued to grow, pointing to deficiencies and even harmful effects of this type of fertilization and demanding a return to the proven fertilization system. Criticism was undoubtedly justified on some points, and was constructive when aimed at correcting mineral fertilization. On the other hand, the occasional radical rejection of mineral fertilizers as "unnatural" by far overshot the mark and ignored the predominantly positive effects of the simple mineral-nutrition theory itself.

The Route to High-Yield Fertilization

Knowledge of fertilization has increased since 1920 to an extent never before imagined. This phase may be considered to have continued until about 1960. The conditions for plant growth are now largely known; the simple mineral-nutrition theory has evolved into a *comprehensive theory of growth factors*.

The new concept includes complete, or at least far-reaching, knowledge of

- the substances required by, or beneficial to, plants, and other growth factors;
- the requirements of crops with regard to their nutrient substrate (and its diagnostic assessment);
- the fertility properties of soils (and their diagnostic assessment).
- the dangers of incorrect or excessive fertilization;
- reliable methods of assessing fertilizer action (yield, produce quality, etc.).

Fertilization intended to exploit fully the *genetic yield potential* of crops has been possible since 1960 (high-yield fertilization). The minimum factors, which hitherto have limited the yield can now be diagnosed and eliminated by fertilization, so that plant production can be maximized.

However, fertilization is no longer a simple routine measure but requires a good knowledge of fertilizers and fertilization. In the final analysis, a completely controlled nutrient supply will have to be the aim in the production of plants on the nutrient substrate, "soil". This has already been realized in hydroponics (water culture) in scientific experiments, and in intensive branches of horticulture. It involves supplying the continuously required nutrients in the correct dosage and form, thus necessitating appropriate methods of controlling the nutrient supply. The question of fertilization in hydroponics could therefore play a central role at extreme sites, be it on earth, in space vehicles, or on other planets.

In concluding this account of the historical development, the question arises as to whether our present knowledge of plant nutrition and fertilization is "complete". This is hardly likely, as gaps are becoming evident all the time, but a further significant extension of the plant-nutrition concept (within the scope of variable growth factors relevant on earth [6]), is highly improbable. Advances in research may still reveal some surprises as to detail, but the present, obviously largely complete knowledge of fertilization permits large yields of high value to be obtained: yields five to ten times those obtained at the start of mineral fertilization 100 years ago. Converting this effective fertilization concept into practical production methods will be an important task of the years ahead.

1.3.4 Fertilizer Consumption and Yield Increase

The use of modern commercial fertilizers in agricultural production results in increased crop yields. Thus, in Germany, the average yield of winter wheat increased from 14 dt/ha in 1880 to about 50 dt/ha in 1977 (Fig. 1-3).

The increase in peak yields is even more impressive (Table 1-2):

Table 1-2. Average yields and estimated peak yields of winter wheat in Germany and FRG [1].

Year	Grain yield in dt/ha	
	average yield	peak yield
1840	8	(25)
1880	14	(25)
1900	20	(30)
1955	30	60
1967	40	80
1977	50	100

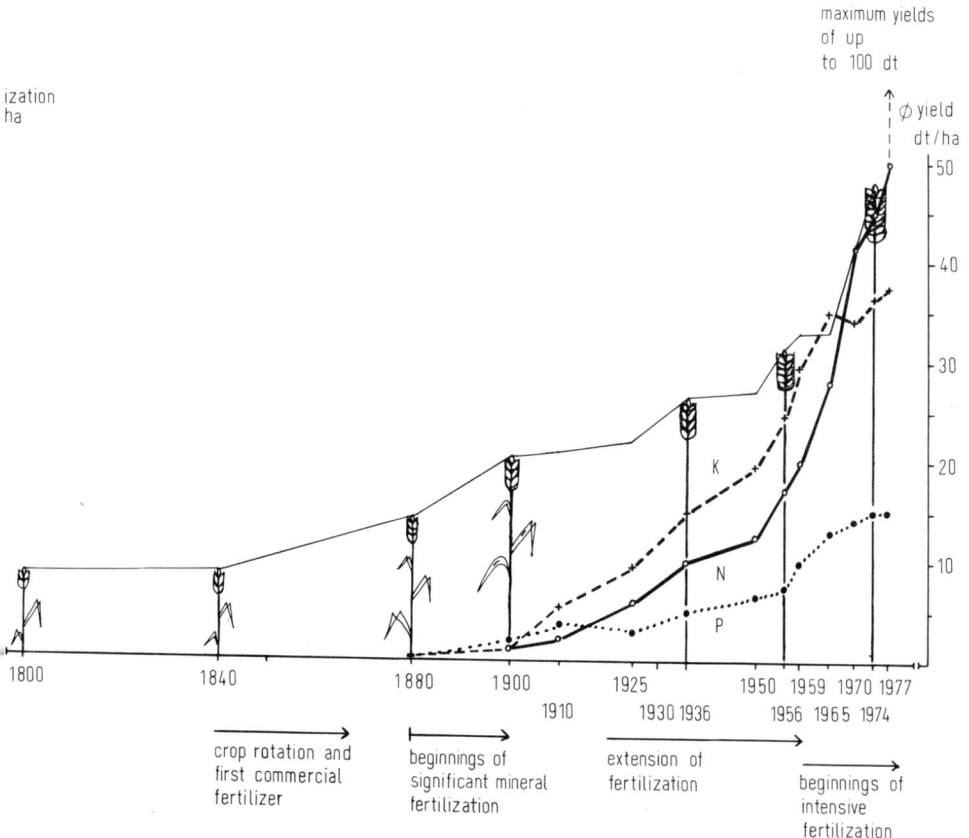

Fig. 1-3. Consumption of commercial fertilizer and yields of winter wheat (Germany and FRG respectively, between 1800 and 1977) according to [1].

Apart from the increased use of N-, P-, and K-fertilizers, it should be stressed that a further important factor in the increased yields is the rise in consumption of lime and other mineral fertilizers.

Fertilizer Production
The increased demand for fertilizers led to rising production figures. Within 100 years, global production of commercial fertilizers has increased from small beginnings to the impressive amount of about 400 million metric tons per year (1976). Details of this development, based on nutrient elements, are given in Table 1-3.

Table 1-3. World production of commercial fertilizer (according to [5], estimated values for 1938).

"Pure nutrient"	million metric tons of "pure nutrient"			
	1938	1955/56	1965/66	1975/76
N	3	7	19	44
P (P_2O_5)	2 (4)	3,7 (8,4)	6,6 (15)	11 (25)
K (K_2O)	2 (2)	5,7 (6,9)	12 (14)	19 (23)

Fertilization and Total Agricultural Production
The effects of better plant nutrition through commercial fertilizers manifest themselves not only in increasing yields of cereals (grains), potatoes, sugar beet, etc., but also in an increase in the total plant mass produced. More leaves, straw, etc. are being produced, so that more feed is available for farm animals. This not only increases the quantities of upgraded animal products, but also of organic fertilizer, which in turn improves the nutrient supply. Increased mineral fertilization thus initiates still more intensified development leading beyond pure yield increases of marketed produce to a relatively larger total population.

Whereas grain yields approximately trebled between 1880 und 1977, *gross soil production* increased five- to sixfold [2].

The Contribution of Fertilization to Yield Increases
The considerable increases in yield and total production were obviously not caused by fertilization alone. Modern plant species have permitted higher nutrient supplies to be utilized; better plant protection has prevented or reduced losses.

The contribution of fertilization to total yield increases between 1850 and 1950 in Germany was approximately 40% for mineral fertilizer. The additional contribution of organic fertilization was about 20% so that a total of 60% can be attributed to improved fertilization [2]. The remaining 40% is due to plant cultivation (15%), plant protection, etc. World-wide, about 50% of the yield increases are attributed to fertilization.

Changes in Nutrient Ratios

The emphasis in fertilization has shifted considerably in the course of time. Thus, fertilization in Germany and the FRG respectively was

- strongly phosphate-oriented around 1900, since a deficiency in phosphates severely limited yields on many soils, while N-fertilizers were still relatively expensive ($N : P_2O : K_2O = 1 : 4.7 : 1.4$);
- strongly potassium-oriented around 1930, since potassium deficiency was widespread and K-fertilizers were cheap ($1 : 1.2 : 1.9$);
- increasingly nitrogen-oriented after 1950/60, since P- und K-supplies to the soils had mean-while attained much higher levels.

The nutrient ratio tends to $1 : 1 : 1$ (*see* Chap. 3.2.3) [20]. There is a clear tendency throughout the world for N-fertilization to predominate in the course of time.

Regional Distribution of Fertilizer Consumption

Fertilization was developed in Central Europe where today consumption per unit area is still highest from the global point of view. However, the situation is considerably different when individual countries are considered (Fig. 1-4).

Even national statistics still do not provide a correct picture of the large differences in fertilizer application in particular localised areas. Wide areas in tropical and subtropical regions receive hardly any mineral fertilizer. Average N-fertilization for fields and gardens is about 30 kg/ha globally, but 540 kg/ha in the Netherlands. More than 1 000 kg/ha of nitrogen is at present being consumed at some intensively worked tropical sites that are cultivated throughout the year.

Fertilization and Food Supply for the Growing Population

Use of modern production methods has now made it possible to obtain very high yields per unit area. However, yields in many regions are still low for a variety of reasons. This causes food shortages in some parts of the world, although sufficient food could be produced on the available area.

Only some 10% of dry land on earth is used for agriculture. This area could be doubled; moreover, grassland (about 20% of the dry land) contributes

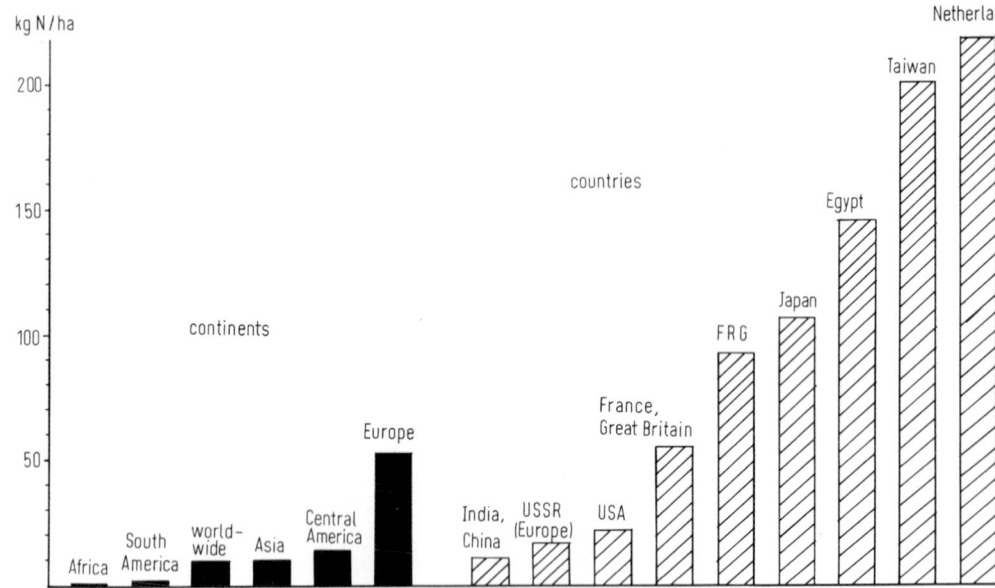

Fig. 1-4. World consumption of nitrogen fertilizers (1975: kg N per ha of agriculturally utilized area, according to [5]).

to food supplies. The food required by about 20 billion people could be produced if the land now being worked, amounting to 1.5 billion hectares, were to be farmed as efficiently as possible with present-day technology. The present population of the earth amounting to 4 billion people, and the 6 billions expected by 2000, could be adequately fed even if only half of this potential area were to be utilized. It is not food production that limits population growth but other factors, such as energy, raw materials, and living space, in the narrower sense.

The supply of *proteins* is, at present, particularly problematic; N-fertilization therefore plays a central role. The use of 1 metric ton of nitrogen as fertilizer can supply the protein required, together with a mixed diet, by 40 persons in one year. Present N-production would thus suffice for almost 2 billion people.

Fertilizer consumption will therefore have to increase in future on many soils, until the maximum yield level, in part already reached, is attained. In recent times yearly rates of increase have been 4 to 8% on a global basis. In any case, fertilization can contribute greatly to the proper nutrition of mankind for a long time to come.

2 Mineral Single-Nutrient Fertilizers

Plants require six major nutrient elements. From the agrochemical point of view these correspond to six single-nutrient fertilizers containing these major nutrients (single-major-nutrient fertilizers). Legal type-lists usually only contain the three single-nutrient fertilizers with the basic nutrient elements (N, P, K), which corresponds to their dominant practical importance in many countries. This chapter deals with these fertilizers and their applications [35, 36, 44, 53, 54, 55, 41, 57].

2.1 Nitrogen Fertilizers (N-Fertilizers)

N-fertilizers are chemical substances that contain the nutrient element nitrogen in absorbable form, chiefly as ammonium or nitrate, or which yield these forms after conversion (see Synopsis 2-1, p. 37).

2.1.1 Origin, Production, and Synopsis

Origin and Reserves

Only a small part of the N-fertilizers is still obtained from deposits in which N-containing salts have accumulated for a long time (e.g., Chile saltpeter). The bulk of commercial N-fertilizers is now produced synthetically from atmospheric nitrogen via ammonia synthesis. Although higher plants are surrounded by large amounts of nitrogen, they cannot split this highly stable molecule and convert it into utilizable forms. Under normal pressure and temperature conditions, this can still only be done by certain micro-organisms in the presence of specific enzymes, e.g., the nodule bacteria in legumes.

Nitrogen reserves are very large. The earth's atmosphere consists of almost 80% nitrogen. Given this and other reserves, the quantities of N-fertilizers produced depend mainly on the energy requirements. Production of 1 kg N in the form of fertilizer requires about 40 000 kJ (kilo-joule) at least, corresponding to the caloric value of 1 kg oil.

Production
N-fertilizers are most commonly produced by binding atmospheric nitrogen (N_2), i.e., converting it into a form that can be utilized by plants. There are various possibilities of doing this, primarily however

- ammonia synthesis (*Haber-Bosch* process);
- calcium cyanamide synthesis;
- nitrate synthesis (arc-gap process).

Production of N-fertilizers via ammonia synthesis is quantitatively the most important method with the ammonia being further processed into various N-fertilizers, for example, lime-ammonium nitrate or urea. Details of their production will be discussed for the individual fertilizer groups separately.

Synopsis of N-Fertilizers
The number of possible N-fertilizers is very large [31, 32, 37]. Many N-containing compounds, for economic and other reasons, are not suitable for use as fertilizers but notwithstanding this, the number which can be used is considerable and constantly increases. Some important N-fertilizers are listed in Table 2–1.

Table 2–1. Important N-mineral fertilizers.

Fertilizer	Name	Chemical formula	N-content %
ammonium fertilizers	a) gaseous ammonia	NH_3	82
	b) ammonia water	NH_3, NH_4OH	.
	c) ammonium sulfate	$(NH_4)_2SO_4$	21
nitrate fertilizers	a) calcium nitrate	$Ca(NO_3)_2$ u. a.	16
	b) sodium nitrate (Chile saltpeter)	$NaNO_3$	16
ammonium nitrate fertilizers	a) ammonium nitrate	NH_4NO_3	35
	b) lime-ammonium nitrate	$NH_4NO_3 + CaCO_3$	26
	c) ammonium sulfate nitrate	$(NH_4)_2SO_4 \cdot NH_4NO_3$	26
amide fertilizers	a) urea	$CO(NH_2)_2$	46
	b) calcium cyanamide	$CaCN_2$	22
ammonium nitrate-amide fertilizers	ammonium nitrate-urea solution	$NH_4NO_3 + CO(NH_2)_2$	28
N-depot fertilizers	several slow-acting fertilizers	(Chap. 2.1.4)	

Note: For the sake of convenience, this list of mineral fertilizers also includes some organic compounds, e.g., urea, which yield mineral forms only after conversion in the soil.

Synopsis 2-1. N-Containing Molecules and Groups.

N	symbol of element nitrogen
N_2	molecular nitrogen (e.g., in the atmosphere)
NH_3	ammonia; simple compound of hydrogen and nitrogen
NH_4^+	ammonium ion; cation, formed from ammonia and water
—NH_2	amino group; typical form of nitrogen in many organic compounds, e.g., amino acids, amides, proteins
NO_3^-	nitrate ion; anion of nitric acid (HNO_3), also called *saltpeter*
NO_2^-	nitrite ion, anion of nitrous acid (HNO_2)

2.1.2 Ammonium and Nitrate Fertilizers

Ammonium Fertilizers

Ammonium fertilizers contain the plant nutrient ammonium or its basic form ammonia as the N-compound (*see* Synopsis 2-2, p. 39).

Gaseous ammonia is a by-product formed from coal in coking plants. Since 1913 it has been synthesized mainly by the *Haber-Bosch* process.

Ammonia is a colorless gas with a pungent odor. When concentrated, it is toxic to plants and man but is a plant nutrient and harmless to man in dilute form (cowshed smell). For transportation it is liquified under pressure and then referred to as liquid ammonia (containing 82% N). However, it always enters the soil as a gas, where it rapidly reacts with water to form ammonium. Ammonia is cheap to produce, and although it is somewhat difficult to apply (Chap. 6.1.3), it accounts for a large part of N-consumption in some countries (e.g., USA, Denmark).

Ammonia water is formed by the introduction of ammonia into water. One liter of water can dissolve about 700 l ammonia at 20 °C. Ammonia-containing fertilizer solutions contain 25 to 40% NH_3 by weight (at least 10% N).

Ammonium sulfate is the oldest synthetic N-Fertilizer. It contains 21% N and is a white, needle-shaped salt. It is readily soluble in water but only slightly hygroscopic (Chap. 6.1.1). The latter property renders it especially suitable for tropical and subtropical regions where it finds its main use. It has a strong acidifying action on the soil (Chap. 2.1.5); its action is sometimes slowed down by the addition of 2% dicyandiamide (Chap. 2.1.5).

The following ammonium fertilizers are, or were, used mainly outside Germany:

- ammonium chloride (NH_4Cl): it forms part of some NPK-fertilizers, is used in Japan as a rice fertilizer, and was formerly used with lime as "lime ammonia";

- ammonium carbonate or bicarbonate, $(NH_4)_2CO_3$ or NH_4HCO_3 respectively: these are also the N-compounds of guano (*see* N-P-fertilizers, Chap. 3.2.1);
- ammonium carbamate (NH_2COONH_4) (carbamate as the amide of carbonic acid).

Conversion of ammonium fertilizers in the soil

The ammonium ion (NH_4^+) can be absorbed directly by plant roots. However, in contrast to nitrate, it is a cation, is thus absorbed by negatively charged colloids in the soil, and its mobility is hence reduced. Only after conversion into nitrate form does it obtain the same mobility and, thus, speed of action as nitrate. *Conversion* of ammonium to nitrate occurs via nitrification by soil bacteria, which secure energy in this process.

Nitrification can be represented in simplified form as follows:

$$NH_4^+ \;+\; O_2 \xrightarrow{\;nitrosomonas\;} NO_2^- \;+\; O_2 \xrightarrow{\;nitrobacter\;} NO_3^-$$

ammonium   oxygen   nitrite   oxygen   nitrate

Conversion occurs within a wide pH-range, the optimum being approximately $pH = 7$. Proper soil aeration is important because of the oxygen requirement. The soil should be moist but not wet (optimum water content $= 40$ to 60%).

The minimum temperature is approximately $6\,^\circ C$. About 50% nitrification occurs in 14 days at $20\,^\circ C$, whereas at $30\,^\circ C$ only half this time is required. Inhibition of nitrification can considerably slow down the speed of action of ammonium fertilizers (Chap. 2.1.5).

Nitrate Fertilizers (Saltpeter Fertilizers)

Nitrate fertilizers contain the plant nutrient nitrate (NO_3^-) as N-compound, i.e., the anion of nitric acid. (Since the German word for HNO_3 is "Salpetersäure" fertilizers containing the NO_3^--constituent are also called "saltpeter".) The oldest nitrate fertilizer is Chile saltpeter which occurs in nature. The largest amount is nowadays accounted for by calcium nitrate. Nitrate fertilizers act particularly quickly (for their production, *see* Synopsis 2–2).

Calcium nitrate in the form of commercial fertilizer contains 16% N and consists of 82% calcium nitrate $(Ca(NO_3)_2)$ and 5% ammonium nitrate in addition to water of crystallization. For use as a fertilizer it is granulated and is a white highly hygroscopic salt, which is easily soluble in water. Calcium nitrate gives an alkaline reaction in the soil (Chap. 2.1.5).

Synthetic *sodium nitrate,* pure sodium nitrate $(NaNO_3)$ obtained by neutralizing nitric acid with caustic soda $(NaOH)$, is a white salt. Its role in fertilization is altogether subordinate to that of calcium nitrate.

Synopsis 2-2. Production of Ammonium and Nitrate Fertilizers.

Ammonia

In the *Haber-Bosch* process the reaction between molecular atmospheric nitrogen and hydrogen, obtained by decomposition of water or from natural gas, etc., takes place at a pressure of 200 atm and a temperature of 550 °C:

$$N_2 \quad + \quad 3\,H_2 \quad \longrightarrow \quad 2\,NH_3$$
atmospheric nitrogen   hydrogen   ammonia

The synthesis of ammonia, being an exothermic reaction, produces heat; however, the production of the two starting materials requires considerably larger quantities of energy, so that production of ammonia is mainly an energy problem.

Ammonium Sulfate

Ammonium sulfate is produced by saturating sulphuric acid with ammonia: $2\,NH_3 + H_2SO_4 \rightarrow (NH_4)_2SO_4$, or by the gypsum process (*Leuna* process):

$$2\,NH_3 \; + \; CaSO_4 \; + \quad CO_2 \quad + H_2O \longrightarrow \quad (NH_4)_2SO_4 \quad + \quad CaCO_3$$
ammonia   gypsum   carbon dioxide   ammonium sulfate   calcium carbonate

Calcium Nitrate

Calcium nitrate is produced in several ways on an industrial scale:

a) Neutralization of nitric acid with calcium carbonate (and partly with NH_3):

$$2\,HNO_3 \; + \; CaCO_3 \; \longrightarrow \; Ca(NO_3)_2 + H_2O + CO_2$$
nitric acid   calcium carbonate (lime)   calcium nitrate

The nitric acid is first produced by the combustion of ammonia: $NH_3 + O_2 \rightarrow HNO_3$.
b) The ODDA process of NP-fertilizer production (Chap. 3.2.1);
c) the arc-gap process with electrical energy by the combustion of N_2 (*Norgesalpeter* in Norway).

Ammonium Nitrate

Ammonium nitrate is produced by neutralizing nitric acid (HNO_3) with ammonia, the acid being obtained from the combustion of ammonia:

$$HNO_3 \; + \; NH_3 \quad \longrightarrow \quad NH_4NO_3$$
nitric acid   ammonia   ammonium nitrate

The granulated fertilizer is then obtained from the highly concentrated solution using the spray method in cooling towers.

Lime-Ammonium Nitrate
Lime-ammonium nitrate is mainly produced by two processes:

a) From *ammonium nitrate* by adding calcium carbonate to the melt; this is followed by granulation in various processes; the addition of a few per cent of powdered substances prevents hardening of the fertilizer by storage pressure.
b) From *calcium nitrate* obtained in large quantities in the ODDA process (Chap. 3.2.1):

$$Ca(NO_3)_2 \cdot 4\,H_2O + 2\,NH_3 + CO_2 \longrightarrow 2\,NH_4NO_3 + CaCO_3 + 3\,H_2O$$

calcium nitrate   ammonia   carbon   ammonium nitrate   calcium
                              dioxide   carbonate

Ammonium Sulfate-Nitrate
Ammonium sulfate-nitrate is produced in two processes:

a) Mixing of liquid salt components and joint crystallization;
b) neutralization of nitric acid and sulfuric acid with ammonia.

Sodium nitrate as a natural product ($=$Chile saltpeter) is obtained from natural saltpeter deposits, mainly in Chile. Naturally formed nitrates accumulated over thousands of years along with other soil salts in the dry desert region between the Andes and the Coastal Cordillera, and hardened to form the *caliche* stratum at a shallow depth. This raw material contains about 20% nitrate. Dissolving processes produce a fertilizer containing about 97% $NaNO_3$ (16% N). The many impurities include 1% NaCl (common salt) and above all the trace elements boron (0.05%) and iodine (0.01%). The boron content rendered Chile saltpeter particularly suitable for fertilizing beets (even before the need for this trace element had been discovered) [45]. (A higher content of potassium perchlorate may be a disadvantage for use as a foliar nutrient. However, the usual 0.3% level is definitely below the limit of toxicity.)

Chile saltpeter is a white, water-soluble, moderately hygroscopic salt. Natural reserves are still considerable, despite continuous mining since 1830. Often more than 2.5 million tons of Chile saltpeter were mined annually during the period of maximum production (1910—28).

Other nitrate fertilizers are two-nutrient fertilizers; e.g., potassium nitrate, magnesium nitrate.

Conversion of nitrate fertilizers in the soil
Nitrate is immediately available to plants without requiring conversion. Near the roots it moves primarily with flowing water to the root surfaces. This is termed *mass flow* (movement through flow).

Ammonium Nitrate Fertilizers

Ammonium nitrate fertilizers contain the two most important N-containing plant nutrients, i.e., nitrate and ammonium. Lime-ammonium nitrate is especially important, and large quantities of it are marketed (for production, *see* Synopsis 2-2, p. 40).

Ammonium nitrate is a useful N-fertilizer. However, it is dangerous (explosion hazard under certain conditions) in pure form, so that trade in it is forbidden or restricted in some countries. Ammonium nitrate is a white, water-soluble, hygroscopic salt containing 35% N. It plays an important role as an essential constituent of many fertilizers.

The explosive ammonium nitrate is rendered completely harmless by adding calcium carbonate to form *lime-ammonium nitrate*. This is a mixture whose chemical formula is $NH_4NO_3 + CaCO_3$ (with varying ratios of the two constituents):

- Lime-ammonium nitrate in the past (21% N) contained 60% NH_4NO_3 and 40% calcium carbonate;
- Lime-ammonium nitrate nowadays (26% N) contains 74% NH_4NO_3 and 26% calcium carbonate.

Lime-ammonium nitrate is a white granulate. Certain manufacturers add dye to give it a specific color, designating it *green- or brown-grain* fertilizer. The N-component is water-soluble, acts quickly or moderately quickly, and acidifies the soil. However, most of the acidic action is compensated for by the added lime (Chap. 2.1.5). NMg-fertilizer is obtained if dolomitic lime is added instead of calcium carbonate (Chap. 3.2.1). Lime-ammonium nitrate is marketed outside Germany under different names, e.g., Cal-Nitro, A-N-L (L indicating *lime*) Nitrolime, calcium-ammonium nitrate, etc.

Gypsum-ammonium nitrate is a variant of lime-ammonium nitrate and contains gypsum ($CaSO_4 \cdot 2H_2O$) instead of calcium carbonate. Since the latter is absent, this fertilizer acidifies the soil more strongly, and hence is only used in a few countries.

Ammonium sulfate-nitrate (ammonium sulfate-saltpeter, abbreviated ASS) was formerly called *Leuna saltpeter*. It was one of the first fertilizers based on synthetic ammonia. Its chemical formula is approximately $(NH_4)_2SO_4 \cdot NH_4NO_3$. It is thus a double salt containing three parts N as ammonium and one part as nitrate. As a fertilizer it is marketed in granulated form containing 26% N. ASS is soluble in water and is slightly hygroscopic. Its role in Germany is subordinate to lime-ammonium nitrate, because of its strong soil-acidifying properties.

Conversion of ammonium nitrate fertilizers in the soil corresponds to that of ammonium or nitrate. Its influence on soil reaction is discussed later (Chap. 2.1.5).

2.1.3 Amide Fertilizers (Urea and Calcium Cyanamide)

Amide fertilizers contain nitrogen in amide form (e. g., urea), or yield *amide-N* after simple conversion (e. g., calcium cyanamide). Amide-N indicates nitrogen in the form of acid amides obtained by coupling acids with amino groups.

Amide-N and intermediate products of decomposition can serve directly as nutrients, but in the soil are mostly converted into the simple nutrients ammonium and nitrate, particularly the latter. The time required for this conversion results in amide fertilizers' slow action. Amide-N, together with ammonium nitrate, is a constituent of fertilizer solutions.

Urea (Carbamide)

The diamide of carbon dioxide, sometimes shortened to carbamide, but most commonly called *urea* (of Latin/English origin), is the simplest solid N-fertilizer to produce. It is also universally applicable. This fertilizer has gained increasing importance in recent years. In some countries it already accounts for the biggest share of all N-fertilizers, and is obviously well on the way to occupying first place.

Urea with a high N-content of 46% is

- a white, organic compound;
- of low specific weight (bulk density only 0.7 kg/l);
- easily soluble in water (1 kg/l at 20 °C);
- as solid fertilizer usually granulated (1 to 2 mm) and stabilized by the addition of diatomaceous earth.

It is slow-acting in the soil because of the required conversion, but it acts directly if applied as a foliar nutrient.

The proportion of the toxic substance, *biuret* (a common by-product)

- is by law limited to 1.2% in the FRG;
- should be less than 0.5% or even only 0.25% if applied as foliar nutrient;
- is sometimes even less than 0.03%.

Urea is, e. g. in the USA, treated or coated with sulfur (*sulfur-coated urea*); this appears to be particularly advantageous in the cultivation of rice; however it is then an NS-fertilizer (Chap. 3.2.1).

Synopsis 2-3. Production of Urea and Calcium Cyanamide.

Carbamide production requires only easily produced and abundantly available starting materials, namely ammonia and carbon dioxide:

$$CO_2 + 2 NH_3 \longrightarrow CO(NH_2)_2 + H_2O$$

carbon dioxide   ammonia   carbamide

Carbamide, being the diamide of carbon dioxide, has the following structural formula:

$$O=C\begin{matrix} OH \\ OH \end{matrix} \xrightarrow[-2H_2O]{+2NH_3} O=C\begin{matrix} NH_2 \\ NH_2 \end{matrix}$$

carbonic acid   carbamide

Carbamide is produced by various processes at a temperature of approximately 150 °C and a pressure of 170 atm. The process is exothermic and yields ammonium carbamate (NH_2COONH_4) as an intermediate product, but considerable energy is required to produce the starting materials. Carbamide production is thus primarily an energy problem. The carbamide solution obtained initially is dried by means of careful evaporation, since otherwise the undesirable *biuret* is easily formed.

$$2 NH_2-CO-NH_2 \xrightarrow[above 100°C]{temperature} NH_2-CO-NH-CO-NH_2 + NH_3$$

carbamide   biuret   ammonia

Since 1905 *calcium cyanamide* has been produced by the *Frank-Caro* process, and later by similar processes at temperatures of about 1 000 °C, as approximated by the following (simplified) equation:

$$CaC_2 + N_2 \longrightarrow CaCN_2 + C$$

calcium carbide   nitrogen   calcium cyanamide   carbon

Cyanamide is the amide of cyanic acid according to the following structural formula:

$$N\equiv C-OH \xrightarrow[-H_2O]{+NH_3} N\equiv C-NH_2$$

cyanic acid   cyanamide

Urea with Additives in Fertilizer Solutions

The tendency to simplify fertilizer application by spraying (Chap. 6.1.3) promotes the development of liquid (nonpressurized) fertilizers.

a) *N-solutions* with ammonium nitrate and urea, e. g., *BASF N-solution (Ensol), Rustica Liquamon 28*. This solution consists of 25% ammonium, 25% nitrate, and 50% urea.

This ensures both immediate action, by the nitrate, and delayed action, by the urea, when applied as soil dressing. The N-solution acts immediately when applied as a foliar nutrient (Chap. 6.1.4).

The solution is obtained by mixing the two components.

The following *properties* are important:

- N-content = 27 to 28% by weight (27 to 28 kg per 100 kg solution);
- N-content = 34 to 36% by volume (34 to 36 kg per 100 liters of solution);
- specific weight = 1.28 at 10°C;
- corrosive action on containers and sprayers, which therefore have to be corrosion-resistant;
- remains liquid down to −17°C;
- can be mixed with micronutrient additives and many plant protective agents.

b) *N-suspension* with calcium nitrate and urea, in which at least four fifths must be nitrate-N.

Conversion of urea in the soil

Decomposition of urea in the soil is effected by microbial enzymes (urease) and is temperature dependent. It is converted to (unstable) ammonium carbonate, then to ammonium, which is usually immediately further converted to nitrate:

$$CO(NH_2)_2 + 2H_2O \xrightarrow{\text{urease}} (NH_4)_2CO_3 \longrightarrow NH_4^+$$

carbamide   water   ammonium carbonate   ammonium

Losses of gaseous ammonia may occur under certain conditions (high soil reaction and temperature) after scattering the fertilizer on the soil surface. Toxic damage, sometimes caused by large carbamide fertilizer doses, may be due to excessive ammonia concentrations at the roots.

Calcium Cyanamide

Calcium cyanamide is an N-fertilizer with herbicidal, fungicidal, etc., side effects. It has lost much of its importance, mainly because of the development of specific, easy to apply herbicides (weed killers), but it is still used as a fertilizer. The composition of calcium cyanamide is given in Table 2-2.

Calcium cyanamide containing approximately 20% nitrogen

- is gray-black (black carbon component);
- has a typical "carbide" smell;
- has a bulk density of 1 to 1.3 kg/l;

- becomes hot on absorption of moisture (contains quicklime);
- burns and causes swelling of the skin, and is particularly toxic when inhaled after the consumption of large amounts of alcohol.

Table 2-2. Composition of calcium cyanamide fertilizers.

Constituent	Content in %
calcium cyanamide	60
CaO (quicklime) or Ca(OH)$_2$*	20
C (carbon)	12
oxides (various)	2–7
carbide	0.1–0.3

* in granulated calcium cyanamide

Calcium cyanamide is marketed in various fertilizer forms:

- calcium cyanamide without oil (finely ground, strongly corrosive);
- calcium cyanamide with oil (less dusty owing to the addition of 1% oil);
- granulated calcium cyanamide;
- granulated calcium cyanamide (This, together with the addition of 1 to 3% calcium nitrate, is the form most often used).

Conversion and corrosive action of calcium cyanamide
The N-action is based on conversion to carbamide and further to ammonium. Calcium cyanamide has a slow fertilizer action because of the conversions required, hence a small amount of nitrate is added to granulated calcium cyanamide.

Calcium cyanamide is *converted* in the soil in three steps:

Step 1, essentially inorganic hydrolysis:

$$N\equiv C-N=Ca + 2 H_2O \longrightarrow N\equiv C-NH_2 + Ca(OH)_2$$

calcium cyanamide   water   cyanamide   calcium hydroxide

Step 2, enzymatic and inorganic conversion, with iron and manganese as catalyzers:

$$N\equiv C-NH_2 + H_2O \longrightarrow CO(NH_2)_2$$

cyanamide   water   carbamide

Step 3: microbial decomposition of carbamide (*see* urea).

This "normal" conversion is to a slight extent accompanied by other processes, e.g., the formation of dicyandiamide (NCNH$_2$)$_2$ from two molecules of cyanamide. A few per cent of dicyandiamide can form during storage under moist conditions. Dicyandiamide can only be slowly converted by plants and inhibits nitrification (Chap. 2.1.5). Calcium cyanamide has a strong lime-effect (theoretical "CaO"-content of about 55%).

The herbicidal action of calcium cyanamide is based on its conversion to toxic cyanamide. The finely powdered, easily dusted form, which should not be inhaled during application, is particularly suitable because of its corrosive action on weed leaves. Addition of oil reduces the formation of dust and also any corrosive action. Thus any corrosive action of the granulated forms that does take place is chiefly restricted to germinating weeds and their roots. Since cyanamide is toxic to all plants, cultivated plants should be sown or planted only after a suitable length of time has elapsed after its use. This duration is considered to be three days for each 100 kg/ha of calcium cyanamide applied.

Other Amide Fertilizers
Oxamide, the diamide of oxalic acid, has recently been licensed as a fertilizer (32% N). It belongs to the N-depot (timed release) fertilizers when suitably coarsely granulated.

2.1.4 N-Depot Fertilizers

Depot fertilizers are particularly slow-acting nitrogen fertilizers. They are necessary since the above-mentioned N-fertilizers, even those considered to be slow-acting, usually act too quickly (Chap. 2.1.5). Depot fertilizers were therefore developed as a "slowly flowing source of nitrogen" in order to better adapt N-supplies to the N-demand of plants. They permit uniform N-supply for longer periods. A single N-application during sowing prevents excessive salt loads at the beginning of growth and deficiencies towards the end [60].

N-fertilizers with especially slow action have been synthesized, particularly in recent decades, by incorporating nitrogen into compunds that are only decomposed with great difficulty. Many such fertilizers are products of the condensation of urea, i.e., they are produced by coupling urea with other molecules. The nitrogen is then located in long chain or ring compounds. A disadvantage of N-depot fertilizers is their high price, and their use is mainly restricted to horticulture. Several N-depot fertilizers are listed in Table 2-3.

N-Depot Fertilizers with Additions of Nitrate or Urea
Combinations of several N-fertilizer forms yield mixtures with action spectra extending over larger periods [34, 39], e. g.,

- combinations with quick and very slow action, e. g., *CD-urea* (Table 2-3) with 4% nitrate N, formerly as Floranid with 28% N;

- combinations with slow and very slow action:
urea-formaldehyde urea (38% N),
urea-isobutylidene diurea (32% N) as *Floranid* with 32% N.

Other combinations with N-depot fertilizers will be discussed under the multiple-nutrient fertilizers with partially long-term action.

Table 2–3. Some N-fertilizers with especially slow action.

Fertilizer type	Trade name	Total N-content in %	Remarks
formalde-hyde urea	Ureaform Nitrozol (URANIA)	38	e.g., 1/3–1/2 depot-N 1/4 acts quickly
crotonyli-dene diurea (CD-urea)	Crotodur (BASF) Cyclo-Di-Urea (Japan)	28	almost solely depot-N
isobutyli-dene diurea (ID-urea)	Isodur (BASF) IBDU (Japan)	28	almost solely depot-N; decomposed slightly faster than CD-urea
oxamide (diamide of oxalic acid)		32	granulation important for depot action, since more easily decomposed than amide

Other N-Fertilizers with Especially Slow Action

The number of N-depot fertilizers being developed, primarily in Japan, the USA, and the FRG, is increasing continually. Some products are not developed beyond the trial stage for economic and other reasons, but some have already found their place in fertilization, and possibilities are still open for others.

Examples are:

- acetaldehyde urea (Urea Z),
- peat containing ammonium,
- ammonium-containing lignin (N-lignin) by processing the lignin waste product obtained in paper production,
- derivatives of prussic acid (HCN) and dicyan $(CN)_2$,
- magnesium ammonium phosphate $(MgNH_4PO_4)$ an inorganic, slightly soluble N-compound (in fact belongs to the three-nutrient fertilizers).

Slow-acting natural organic fertilizers like blood powder, horn powder, etc., will be discussed later under organic fertilizers (Chap. 4.3.3).

The Action of N-Depot Fertilizers

The depot nitrogen of these fertilizers is decomposed gradually by soil microbes, the speed depending on soil reactivity and other conditions. It thus represents a slow-flowing N-source for plants, similar to natural N-containing substances. Decomposition increases, above all, with temperature and thus with rising plant demand. The chief decomposition product, formed via intermediate stages, is ammonium. Since plants can absorb low-molecular weight organic decomposition products, it must be ensured that decomposition will not cause the formation of toxic intermediate products.

The following N-fractions are distinguished in the methodological point of view:

a) soluble in *cold* water: act slowly or rapidly, depending on the N-form;
b) soluble in *hot* water: depot-N with especially slow action;
c) *insoluble* in *hot* water: act extremely slowly and are thus important for fertilization only over long periods, or have practically no significance.

Binding nitrogen more strongly in the fertilizer molecules is only one possibility for slowing down fertilizer action. The possibilities of "braking" the action of water-soluble fertilizers are discussed in Chapter 2.1.5.

Formulas of N-Depot Fertilizers

A. Examples of Urea-Form Components

2-methylene-3-urea: U—CH_2—U'—CH_2—U
3-methylene-4-urea: U—CH_2—U'—CH_2—U'—CH_2—U
U = ureido group (NH_2CONH—)
U' = (—$NHCONH$—)

B. CD-Urea

$$CH_3-HC \overset{\displaystyle -CH_2-}{\underset{\displaystyle HN_{\diagdown CO}\diagup NH}{}} CH-NHCONH_2$$

C. ID-Urea

$$\begin{array}{c} H_3C \\ \diagdown \\ CH-CH \\ \diagup \\ H_3C \end{array} \begin{array}{c} \diagup NHCONH_2 \\ \\ \diagdown NHCONH_2 \end{array}$$

D. Oxamide

$$\begin{array}{c} CO-NH_2 \\ | \\ CO-NH_2 \end{array}$$

2.1.5 Principles of Nitrogen Fertilization

In order to establish the rules to be presented in Chapter 2.1.6 for the practical application of N-fertilizers a knowledge of some basic principles is required (Fig. 2-1).

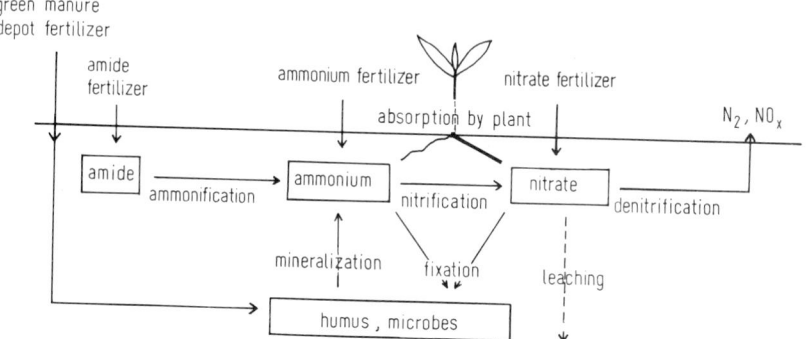

Fig. 2-1. Nitrogen fertilization and fertilizer conversion in soil.

Selection of Correct N-Fertilizer Form

The action of N-fertilizers on the yield is a summation effect of many individual components. The large choice of N-fertilizers and the multiplicity of fertilization purposes make it necessary to determine the N-fertilizer properties affecting growth and yield, to balance them, and to emphasize specific advantages.

Essential factors in the *selection* of N-fertilizers are:

- nutrient form, especially comparing ammonium and nitrate;
- speed of fertilizer action;
- utilization of after-effects, possible losses;
- side effects, minor constituents, changes in soil reaction, other positive or negative effects;
- labour and cost factors.

Effects of Different Nutrient Forms

Most N-fertilizers are approximately equivalent with regard to their N-action in many soils. Plants have scarcely any specific demands for a definite N-form; nitrate and ammonium nitrogen are utilized to approximately the same extent, and all N-fertilizers are in any case largely converted to nitrate in the soil.

Plants should in theory prefer ammonium to nitrate, since ammonium is directly available for protein synthesis after absorption, and nitrate must first be reduced to ammonium, or amino-N, via *nitrate reduction* in root or leaves, with energy expenditure. However, sufficient chemical energy is obviously available from excess photosynthesis production, so that this argument on the basis of energy requirements is invalid. However, precise examination of the alternatives, ammonium and nitrate, is difficult because of unavoidable side effects even in pot-experiments. There are many reasons for assuming that both these major nutrient forms of nitrogen are in principle largely equivalent for the purposes of practical fertilization.

Any remaining *differences* between ammonium and nitrate fertilizers are usually due to other causes. Thus, differences in action occur according to the soil reaction:

- Nitrate fertilizers are superior in highly acid soils (below pH = 5), since they raise the pH-value and thus give additional benefits;
- both forms are roughly equally effective on moderately to slightly acid soils (pH = 5 to 7);
- ammonium is superior in neutral to very slightly alkaline soils (pH = 7 to 7.5), in view of its side effect as a soil acidifier;
- ammonium is generally inferior in more highly alkaline soils (pH above 7.5), since losses of gaseous ammonia occur.

Another reason for the different effects of ammonium and nitrate is claimed to be the *preference* of some plants for one or the other N-form. However, this might also be largely due to side effects. Thus, fertilization with ammonium-N appears to be superior for potatoes. The cause, however, lies less in the better utilization of ammonium than in the soil-acidifying action, which, for example, improves nutrition with manganese.

The apparent inferiority of one fertilizer form may also be due to N-losses, e.g., of nitrate by denitrification in wet soils.

The Speed of Action of N-Fertilizers

A general shortcoming of most N-fertilizers is their very rapid action which is not adapted to the growth rate of plants. This necessitates repeated N-fertilization. Immediate action is possible only by leaf fertilization, since the fertilizer is then absorbed and utilized directly. Almost all N-forms are suitable as foliar nutrients; however, carbamide (urea) has particular advantages because of the low corrosion risk (Chap. 6.1.4).

Differences in the speed of action play an important role in the selection of fertilizers for soil dressing:

- Nitrate fertilizers are *immediately effective* and are therefore especially suitable as "top fertilizers";

- ammonium fertilizers act *moderately quickly* : although ammonium can be absorbed immediately by the roots, it becomes more mobile in the soil only after conversion into nitrate;
- urea and calcium cyanamide *act slowly*;
- N-depot fertilizers containing nitrogen in special organic bonds, or which are "braked" in their speed of action by additives or envelopes have *very slow* and sustained action, apart from farmyard manure.

Measures for Slowing-down the N-Action
One of the principal shortcomings of many N-fertilizers is their very rapid action in comparison to the demand curve of plants. Total N-supplies must therefore either be divided (divided N-doses), or the action of N-fertilizers must be slowed down.

The N-action can be slowed down in various ways:

a) Through *building* the nitrogen *into* molecules that decompose with difficulty, thus forming N-depot fertilizers with various slow speeds of action.

b) Through *braking* the migration of nutrients from the fertilizer grain into the soil. Various possibilities of coating fertilizer granulates with plastic foils, resins, etc., have been tested. There are many possibilities of varying the speed of action, depending on whether the membranes (envelopes) permit diffusion, burst through increased water supply, are perforated, or become permeable only via microbial decomposition. Temperature rises increase both the release of nutrients from coated granulates and requirements of the plants; supply thus occurs by means of a highly significant rhythm. Such fertilizers are therefore also termed "thinking" fertilizers.

c) Through nitrification *inhibition* (Chap. 2.1.2). This permits delaying the conversion of ammonium into quick-acting nitrate. The best known agent for this is *N-Serve* which, in accordance with its name, serves to improve N-utilization.

N-Serve (2-chloro-6-trichloromethylpyridine) is added to ammonium fertilizers in amounts of 0.1 to 0.2%. A concentration of 0.1 to 20 ppm in the soil is required for an effect that lasts four to eight weeks. This substance, in particular, inhibits the bacterium *nitrosomonas* but not microbes that decompose humus or carbamide. Apart from N-Serve many other nitrification inhibitors (*nitrificides*) have meanwhile been developed [56] (especially for rice fertilization in Japan). Dicyanamide (Chap. 2.1.3) also inhibits nitrification. All these substances are effective, but the method is so expensive that in practice the division of N-doses is usually preferred as a simpler procedure.

Utilization and Residual Effects of N-Fertilizers
The purpose of profitable fertilization is to obtain a high-rate utilization of the applied fertilizer even in the first year. Residual effects in the second year play at best a positive role but long-term residual effects should be considered from a different viewpoint, e.g., the slow formation of a soil rich in nutrients. The recovery rate of fertilizers can only be methodically established by approximation (Chap. 5.5.1). The *recovery rate* of N-fertilizers according to many investigations is as follows for soil dressing, [52, 43] (per vegetation period = year):

- 40 to 80% (frequently 50 to 60%) for mineral fertilizers,
- 20 to 30% for farmyard manure.

Values of 80% or even more can be achieved by leaf fertilization. Residual effects may be estimated at about 10% of first-year effects. Total residual effects over many years would thus yield a high total utilization (Chap. 5.5.1). Complete (100%) utilization is impossible, since part of the fertilizer nitrogen is fixed in permanent humus for a long time (this, however, promotes soil fertility), and there are unavoidable losses in liquid and gaseous forms.

Losses through *leaching* occur in humid climates, since the water-soluble nitrate enters the groundwater near the surface together with seepage water and is thus drained off, flows into ditches, and in part also enters deeper groundwater.

Gaseous *ammonia losses* from ammonium and carbamide fertilizers can be caused by the escape of ammonia, e.g., in the case of ammonium fertilizers increasingly at soil pH-values above 7.5. Losses of nitrogen from granulates may attain 10 to 15% during the conversion of urea near the soil surface, especially in hot and windy weather. These N-fertilizers should therefore enter the soil as rapidly as possible; this may occur after a single rainfall. Hardly any ammonia escapes from sulfur-coated urea (Chap. 2.1.3).

Losses by *denitrification* occur under wet soil conditions through microbial decomposition of nitrate (predominantly) to elemental nitrogen (N_2) or nitrogen oxides (N_2O, NO, NO_2) (Chap. 6.2.3). The level of these losses, which occur under reducing conditions, depends greatly on the moisture content of the soil. A few per cent at most are lost in this way in well aerated soils, but losses of 20% or more are quite possible when the soil is highly water-logged.

Overall N-losses are often in the range of 5 to 40 kg/ha, but may be considerably higher. N-losses become smaller, the more precisely the fertilizers are applied. Losses are important not only from the point of view of production technology, since part of the fertilizer is thus lost, but also for environmental protection reasons (Chap. 6.2).

Side Effects of N-Fertilizers

N-fertilizers have various side effects, positive and also negative, apart from the proper N-action, i.e., improved nitrogen supplies to plants. The influence of these side effects should be taken into account in assessing the overall action of fertilizers (Synopsis 2–4).

Synopsis 2–4. Side Effects of N-Fertilizers.

1. Supply of nutrients, besides nitrogen,
   e.g., sulfur, magnesium, calcium, sodium, boron.

2. Changes in soil reaction:
   a) Acidification with positive and negative results; a positive result, for example, might be the mobilization of nutrients from the soil itself.
   b) Increased soil reaction, with positive or negative results; positive results are the reduction of salt damage and the mobilization of molybdenum.

3. Increases in biotic soil activity, with considerable indirect effects on the entire nutrient dynamics.

4. Salt damage accompanying extremely large doses.

5. Damage by minor constituents and products of conversion.

6. Herbicidal and fungicidal side effects of calcium cyanamide.

Some N-fertilizers *supply considerable additional nutrients*. Thus, ammonium sulfate contains more sulfur (24%) than nitrogen (21%). Sulfur deficiencies are now increasingly frequent, since ammonium sulfate has been replaced by urea in many tropical regions. This, however, can be partly compensated through the use of sulfur-coated urea (Chap. 2.1.3). Lime-ammonium nitrate based on dolomitic limestone also supplies magnesium. Chile saltpeter contains sodium and trace elements, especially boron.

Soil acidification through N-fertilizers is especially important as regards changes in soil reaction. This effect is a disadvantage, e.g., if nutrients are fixed in acid soils through further acidification. It may be advantageous in soils with higher pH-values, if nutrients in the soil (e.g., certain trace elements) are better mobilized through acidification, especially in the main period of growth (Chap. 2.1.5). The acid action must usually be compensated for again in order to maintain the optimum range of pH-values; this should be taken into account when prices are compared (price addition for lime compensation).

The intensity of *acid action* of N-fertilizers varies (Table 2–4, p. 54). Most N-fertilizers in aqueous solution or suspension have an acid reaction, but urea is almost neutral. On the other hand, lime-ammonium nitrate, calcium cyanamide,

Table 2–4. Influence of N-fertilizers on soil reaction.

Acidifying fertilizers	1 kg N causes		Remarks
	Acidification eq. H^+	"Lime" loss in kg CaO	
ammonium sulfate	108	−3	in addition 72 eq. H^+ from sulfate component
ammonium sulfate nitrate	72	−2	in addition 72 eq. H^+ from sulfate component
ammonium nitrate ⎫ urea ⎬ ammonia ⎭	36	−1	according to basic formula (Synopsis 2–5)
limme-ammonium nitrate 22% N	0	0	acidification (partly) compensated by lime
lime-ammonium nitrate 26% N	16	−0.4	
pH-increasing fertilizer	reduction of acidity eq. H^+	"Lime" gain in kg CaO	
calcium nitrate	−28	+0.8	differences due to NH_4NO_3-component
calcium nitrate ⎫ sodium nitrate ⎭	−36	+1	according to basic formula
calcium cyanamide	−60	+1.7	additional effect of free CaO

and above all ammonia have an alkaline reaction. This *pH-value of the fertilizer itself*, caused by hydrolysis, determines the effect on the soil or on the liquid nutrient substrate only when the fertilizer is not converted and the nutrients are absorbed by the roots to the extent supplied. However, these two conditions occur only rarely. The pH-value of the fertilizer itself is thus relatively unimportant.

On the other hand, the *physiological reaction* is of interest in practical applications. This is the overall change in pH-value, allowing for conversions. N-fertilizers act on the soil either as

- acidifiers, lowering the pH-value, in the case of ammonium-containing fertilizers, or as
- alkalizers, raising the pH-value, in the case of nitrate fertilizers with alkali or alkaline-earth cations, or fertilizers containing lime.

The causes of changes in pH-value are discussed in Synopsis 2–5.

Salt damage may be due to the N-forms themselves (with extremely large doses), or might be caused by ballast salts not required by the plants, but contained in some fertilizers. Salt damage implies interference with the absorption of water and the supply of minerals (Chap. 6.3.4).

Plants can be harmed by N-fertilizers through *minor constituents* (e. g., higher biuret concentrations in urea, especially when it is applied as foliar nu-

Synopsis 2–5. Changes in Soil Reaction, Caused by N-Fertilizers.

Conversion of N-fertilizers in the soil causes changes in pH-values. *Ammonium fertilizers lower* the *pH* through nitrification.

Acid production is:

1 equivalent ($= 14$ g) N as NH_4^+ yields 1 eq. H^+, i.e., 1 kg N yields 72 eq. ($= 72$ g) H^+, (calculated from equivalent weights, this corresponds to a lime loss of 3.6 kg $CaCO_3$ or 2 kg CaO).

The value of 72 eq. H^+ represents the maximum acidity possible, e. g., in fallow fields without the influence of plants. The acid formed reacts with the soil and thus causes either a loss of lime or lowers the Ca-content of the exchange complex (increase of H^+-ion concentration).

However, maximum acidification does not occur in practice because of the absorption of nutrients by plants: an equivalent amount of alkaline anions (HCO_3^- or OH^-) is released for every nitrate ion absorbed. Maximum acidification is thus reduced to one half under the assumption of 50% utilization as a practical factor for evaluation [30].

Increased pH-values, e. g., through calcium nitrate, are due to the fact that plants absorb nitrate in preference to calcium, so that $Ca(OH)_2$ may be considered to be formed (again through anions released by the roots).

The two effects largely balance each other in the case of lime-ammonium nitrate which contains both ammonium and lime. The former 21% lime-ammonium nitrate at least was therefore an approximately neutral fertilizer.

Ammonia (e. g., as gas) at first reacts as a strong base in the soil, but increasing nitrification very soon causes an acidic action.

The pH-changing action of N-fertilizers can be summarized on the basis of the following fundamental rule, valid for 50% utilization at the usual conversion by nitrification and predominantly downward flow of water in the soil.

Average change in soil reaction through N-fertilizers:

- 1 kg N as ammonium causes acidification of 36 g H^+ or a loss of 1 kg CaO;
- 1 kg N as calcium or sodium nitrate, on the other hand, corresponds to a gain of 1 kg CaO.

trient) or products of decomposition, e.g., ammonia, which has a toxic effect on roots in higher concentrations. This obviously also applies to the fertilizer ammonia.

Herbicidal and fungicidal side effects which should be mentioned are the cyanamide action of calcium cyanamide (Chap. 2.1.3), mainly in combatting weeds, and that of ammonia which is active against harmful fungi in the soil.

Considerations of *labour economy* are often decisive in selecting solid or liquid fertilizers. This problem also exists with other fertilizers and is discussed in detail in Chapter 6.1.

It is natural to choose a fertilizer that is both cheap and appropriate from the professional standpoint. However, it may be advantageous for economic reasons to choose a fertilizer form whose effects are slightly inferior to those of the optimum form, or which is more cumbersome to apply. The price advantage must be weighed against the disadvantages involved.

2.1.6 Practical Use of N-Fertilizers

Nitrogen is the principal stimulant of plant production. The correct use of N-fertilizers is therefore highly important and various aspects should be taken into consideration. The most important agrochemical aspects are presented below to summarize the principles discussed in the preceding chapters.

1. The *basic preconditions* for optimum N-effect are of primary importance. The pH of the soil must be within the correct range (Chap. 5.1.1), so that the necessary conversions in N-dynamics can take place and aeration of the roots is ensured, without which nitrogen cannot be absorbed.

2. The *choice of the N-form* depends on the purpose of application. Here the speed of action and the side effects play an important role.

N-fertilization should supplement the natural N-supply from the soil. It should thus, together with the nitrogen in the soil itself, ensure an N-supply distribution which matches the demands of the plant. Young plants require abundant N-supplies. However, almost all N-fertilizers have a relatively rapid mode of action, compared with plant requirements, and reserves in the soil are usually maximum at the beginning of growth. Thus, when N-fertilization during sowing is intensive, young plants have frequently too much nitrogen available at first and too little later.

The faster acting leaf fertilization is gaining importance as a supplement to soil dressing.

The following possibilities exist for selecting the N-form according to the speed of action, allowing for the various purposes of fertilization:

A. *Rapid elimination* of nitrogen deficiency is possible by

- *leaf fertilization*, which in any case acts immediately (Chap. 6.1.4), or
- *top dressing*, i.e., supplying the plant via the soil with immediately acting nitrate; however, this fertilization method acts rapidly only if there is adequate moisture content in the soil.

B. *Basic fertilization* for sowing or planting may be a first application or complete application:

First application, to supply the plants in the first growth period, may be effected with combined quick- and slow-acting forms, i.e., with immediate action and some slower effects, e.g., lime-ammonium nitrate.

Complete application, to ensure supplies until the harvest, may in principle be effected in the same way as a first application with plants having low N-requirements or short vegetation periods. Additional storage fertilization with very slow-acting depot fertilizers is advisable in the case of plants with long vegetation periods. This may be especially important in horticulture, e.g., for fertilizing potted plants growing in pots or larger containers.

C. *Supplementary* (complementary) *fertilization* of growing plants is carried out after basic fertilization, up to the last late fertilization at the time of ear appearance. Suitable foliar nutrients for this purpose are urea or N-solutions, unless the plants are particularly sensitive to corrosion during certain growth stages. Top dressing with an immediately active fertilizer (nitrate), lime-ammonium nitrate or urea, when application is timely and conversion conditions are good, is also possible.

3. The *optimum time* of N-fertilization is a problem closely linked with that of the optimum N-fertilizer, which in turn is linked with the problem of optimal quantity.

The different speeds of action of N-fertilizers affect the optimum time, although only to a subordinate extent in view of the generally fairly rapid action of most fertilizers applied.
Cultivated plants

- require nitrogen primarily at the time of maximum vegetative growth, i.e., during production of the principal leaf mass;
- can, however, use later N-supplies for increased synthesis of proteins in the reserve organs, e.g., in the grain.

Plants can be supplied with the daily N-requirements via hydroponic culture. Plants grown in the soil must by necessity have a certain reserve from which they can obtain their requirements, and one can be on the safe side by establishing these reserves. The following alternatives are therefore the subject of arguments in practical fertilization, e. g., for winter grains:

- *large N-supplies* to young plants at or before the vegetation period, i. e., fertilization with large safety margins as regards time and quantity. The advantage of this procedure is the prevention of any deficiency phase during the first growth period. Its disadvantage is the (unnecessary) creation of excessive nutrient, which can cause negative effects with respect to the plant's resistance (Chap. 6.5.2) and higher losses through leaching and denitrification.
- *moderate supplies* to young plants at the beginning of the vegetation period, with early supplementary fertilization according to requirements. An advantage is the better utilization of reserves that might be present in the soil; a disadvantage is the danger of a transient nitrogen deficiency.

However, these alternatives become unimportant when the N-supplies in the soil at the beginning of growth are known, i. e., have been determined. Then the first N-application can be suitably large, moderate, or even small, at the time of the start of growth (Chap. 5.4.1).

4. The *optimum N-quantity* is the difference between N-supplies in the soil and N-requirements of the plants.

N-supplies from the soil depend on

- the reserves still remaining in the soil at the end of winter (content of mineral N, to be determined by soil analysis);
- reserves mobilized during the vegetation period, which can be determined only through special incubation methods and largely depend on the weather.

Plants should have abundant nitrogen available particularly at the beginning of growth. However, one-sided intensive N-fertilization is inadvisable for several reasons:

- the utilization rate drops;
- the danger of lodging increases with grain;
- the quality of plant products deteriorates;
- losses by leaching increase.

Intensive N-fertilization thus makes sense only if corresponding supplies of all other nutrients are ensured, obviously with the condition that all other necessary measures are taken. The correct amounts of N-fertilizers are discussed in detail in Chap. 5, or for each plant separately (Chaps. 7 and 8).

5. The *soil-acidifying action* of some N-fertilizers (Chap. 2.1.5) requires special consideration. It is largely considered to be a negative effect because of the cor-

responding lime losses. This implies increased costs owing to the quantities of lime lost. Thus, a fertilizer that has an acidifying action appears to be less economical than a fertilizer of the same price with regard to N-content that does not have an acidifying action. This reasoning (and calculation) is justified if fertilization is considered only within the narrow scope of NPK and lime application.

However, the situation is completely different on sites where the minimum factors are "beyond NPK" (Chap. 1.1.2). Acidifying fertilizers applied precisely during the principal growth period may, by their acidification, activate micronutrients such as iron, manganese, zinc, copper, and boron. Apart from their N-action, they thus cause additional far-reaching improvements in plant nutrition. This possibility is illustrated for manganese in Fig. 2-2.

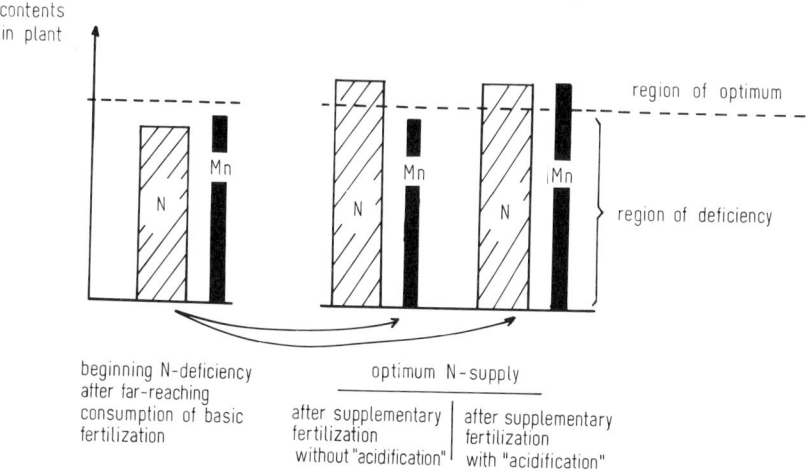

Fig. 2-2. Manganese mobilization by acidifying N-fertilizers (on slightly acid to neutral soils).

Acidification is a great advantage in such cases, whereas the disadvantage of lime losses is almost insignificant.

The following alternatives should always be considered in selecting N-fertilizers:

• Lime-ammonium nitrate or urea should be preferred to calcium nitrate for top dressing when *acidification* is desirable during the principal growth period. This seems to be advisable in cases of acute or even only latent deficiencies of trace elements in soils that have a neutral or only slightly acid reaction.

- Calcium nitrate (or even lime-ammonium nitrate for reasons of price and distribution of effects) should be preferred to more acidifying N-forms (urea, etc.) if acidification only produces *disadvantages.*

6. N-fertilizers should be *applied* correctly so that their action is optimal and losses are prevented. Nitrogen, applied as foliar nutrient (Chap. 6.1.4), reaches the leaves which require it, directly. Most N-forms dissolve in water upon soil dressing, and thus reach the root region together with the water stream. Special working into the soil for the sake of a better mode of action is therefore unnecessary. N-fertilizers are generally very effective via the soil.

However, possible ammonia losses sometimes make it advisable to work the N-fertilizer into the soil, especially in cases of

- fertilizers tending to release ammonia,
- heat and intensive radiation,
- fertilization during dry periods, which causes the fertilizer to stay longer at the surface of the soil,
- soils with neutral or alkaline reaction.

Placement of N-fertilizers is ordinarily not required, but may have advantages in some cases (Chap. 6.1.2). *Salt damage* through excessive concentrations of nitrogen salts near the roots should always be prevented during placement.

7. *N-losses* through leaching are more difficult to prevent than losses of gaseous ammonia. Losses by leaching appear to be unavoidable in humid climates where the water flow is downward. Such losses also occur in nonfertilized soils. N-losses through leaching occur especially in cases of

- large *residues* of water-soluble N-supplies after the harvest,
- extensive precipitation during the winter,
- lack of plant growth (fallow fields) in winter,
- N-application to wet or frozen soils which then remain waterlogged for a considerable time (with additional losses by denitrification).

Erosion of N-fertilizers on slopes seems to be significant only when the soil is frozen. N-losses can be considerably reduced by precise fertilization, which is also desirable on environmental grounds (Chap. 6.2).

Concluding Rules for Practical Application of N-Fertilizers
1. N-fertilizers can act optimally only when the soil has optimum structure and reaction.

2. Proper selection of the N-fertilizer depends on the application purpose. The suitable fertilizer form depends on whether rapid, slow, or combined action is

desired. The correct N-form for basic fertilization depends on whether this is only a first application, or a complete application. For supplementary fertilization of a growing crop it may be advisable not to select a fertilizer with the desired rapid action, but rather one with greater acidification side effects.

3. Young plants especially require abundant N-supplies. This known requirement in practice often necessitates abundant N-fertilization at the beginning of the vegetation period. However, excessive supplies are undesirable for various reasons.

4. The quantity of N-fertilizer should be matched to natural N-reserves in the soil and natural N-supplies from the soil.

5. Soil acidifying N-fertilizers have the dual purpose of N-supplies and increasing natural supplies of practically important trace elements from the soil. This increased mobilization through acidification by N-fertilizers can be highly significant particularly in neutral soils.

6. The application of N-fertilizers causes few problems, since they dissolve easily in water and thus penetrate into the root region. Ammonia losses possible with some fertilizers, e. g., urea, should be prevented.

7. Losses through leaching of nitrogen from the soil are unavoidable, but can be limited considerably by precise N-fertilization.

2.2 Phosphate Fertilizers (P-Fertilizers)

P-fertilizers are chemical substances that contain the nutrient element phosphorus in the nutrient form of absorbable phosphate anions or that yield such phosphate anions after conversion.

With respect to their fertilizing properties, P-fertilizers occupy a broad field ranging from water-soluble salts to compounds insoluble in water but easily mobilized, to slightly soluble substances that can be mobilized only under special site conditions.

The most important components of P-containing substances are listed in Synopsis 2-6.

2.2.1 Origin, Production, and Solubility

Origin and Reserves

The raw materials of P-fertilizers are primarily rock phosphates from phosphate deposits, as well as phosphate-containing ores and other P-compounds (Table 2-5).

Synopsis 2-6. Important P-Containing Substances.

P	chemical symbol of element phosphorus, simple reference basis for P-contents;
H_3PO_4	(ortho-)phosphoric acid whose salts are (ortho-)phosphates;
$H_2PO_4^-$	primary phosphate anion (dihydrogen phosphate, monophosphate);
HPO_4^{2-}	secondary phosphate anion (hydrogen phosphate, diphosphate);
—H_2PO_4	phosphoryl group, phosphoric-acid radical for linking with organic compounds (formation of esters);
P_2O_5	diphosphorus pentoxide, occurs neither in soils nor in plants or fertilizers, but is for historical reasons still used in some countries as a reference basis for P-contents.

Table 2-5. Important inorganic P-compounds in soils and deposits.

Name	Formula	Significance
dicalcium phosphate	$CaHPO_4(2 H_2O)$	still comparatively easily available phosphate for soils
octocalcium phosphate	$Ca_4H(PO_4)_3 \cdot 3 H_2O$	
(tri-)calcium phosphate	$Ca_3(PO_4)_2$	occurs usually as apatite
hydroxide apatite	$3 Ca_3(PO_4)_2 \cdot Ca(OH)_2$	crystalline apatite or crypto-crystalline *phosphorite* of deposits and soils
carbonate apatite	$3 Ca_3(PO_4)_2 \cdot CaCO_3$	
fluor apatite	$3 Ca_3(PO_4)_2 \cdot CaF_2$	
iron phosphate	$FePO_4 \cdot 2 H_2O$	P-compounds in (acid) soils
aluminium phosphate	$AlPO_4 \cdot 2 H_2O$	

Rock phosphates consist of various apatites (Ca-phosphates) of partly magmatic, partly organogenic origin. Weathering and decomposition processes caused the accumulation of apatites from primary minerals or P-containing bones, teeth, etc., of animals (e.g., of saurians). These deposits often occur near the surface where they are worked by open-cast mining. Large deposits exist in

- North Africa (Morocco, Algeria, Tunisia) in the form of organogenic crypto-crystalline *phosphorite,* especially in a soft-earth, finely crystalline form known as *Gafsa* phosphate (Tunisia);
- USA (e.g., *Florida apatite*) in the form of *pebbles*;
- USSR in the form of hard-earth, coarsely crystalline *Kola apatite.*

Phosphate is a scarce raw material. Global reserves are estimated variously, with the data depending on whether visible deposits only are estimated or whether possible reserves are included, and whether currently economically viable deposits (with P-contents above 6%) are considered or if deposits with scarcely half this value are included.

World reserves of rock phosphate are probably in the order of 50 billion tons. This represents a reserve of 5000 million tons P, if calculations are based not on the usual P-content of 12 or 18%, but on a P-content of only 10%, in view of the inclusion of low-content rocks. These reserves will be sufficient for only 500 years at the present annual consumption of about 70 million tons of rock phosphate (10 million tons P). This does not allow for a larger population growth, which would considerably reduce the period mentioned.

However, these pessimistic views are moderated by the fact that phosphate which accounts for about three quarters of total P-consumption is not destroyed by use as fertilizer, but is only diluted and can be recovered from sewage or, at a considerably higher energy expenditure, from seawater. Imports from other planets should also be considered in connection with space travel, in analogy to the carriage of phosphate by sailing vessels half around the globe from the Pacific to Europe.

Production of P-Fertilizers

Mineral P-fertilizers are obtained by chemical treatment or fine grinding of phosphates found in nature. The production of P-fertilizers involves converting rock phosphates, which generally can be utilized to only a small extent in their natural form, into products of considerably higher availability to plants.

Very fine *grinding* of special soft-earth rock phosphates sometimes suffices to provide fertilizers that are mobilizable on many sites. This physical treatment is used in the production of Hyperphos. Grinding "normal soft-earth" rock phosphate yields a fertilizer suitable only for soils having particularly high powers of mobilization, e.g., acid high moors (rock phosphate bog fertilizer). Just grinding hard-earth crude phosphate, consisting of solid rock, provides a fertilizer that is not very effective.

Hard-earth raw materials thus require *chemical decomposition*. There are various possibilities for this:

- Part of the calcium in apatite is replaced by hydrogen: production of super- and triple phosphate;
- part of the calcium in apatite is replaced by sodium: production of Rhenania phosphate;
- the chemical structure of the apatites or other P-compounds is extensively altered (production of Thomas phosphate).

These changes of the apatites create either directly absorbable, water-soluble phosphates or phosphates easily converted into absorbable ions by the mobilization forces of the soil. These are thus all fertilizers with good availability.

Water-soluble *superphosphate* is produced from rock phosphate by treating the latter with concentrated sulphuric acid. Free phosphoric acid is formed first and reacts with the crude phosphate to form Ca-dihydrogenphosphate. A side product is anhydrite (gypsum containing no water). The fluorine contained in the rock phosphate, which would reduce the solubility of the fertilizer, is removed as hydrogen fluoride (HF). Complete decomposition requires several days.

Triple phosphate (= triple superphosphate) is formed when only phosphoric acid is used for treatment.

The simplified equations are as follows:

$$Ca_3(PO_4)_2 \; + \quad H_2SO_4 \quad \rightarrow \quad Ca(H_2PO_4)_2\,(30\%) \quad + CaSO_4\,(50\%)$$

Ca-phosphate   sulphuric acid $\underbrace{\text{Ca-dihydrogen phosphate}}$ gypsum, etc.

$\quad\quad\quad\quad\quad\quad\quad\quad\quad\quad$ superphosphate

$$Ca_3(PO_4)_2 \; + \quad H_3PO_4 \quad \rightarrow \quad Ca(H_2PO_4)_2\,(75\%)$$

Ca-phosphate   phosphoric acid $\underbrace{\text{Ca-dihydrogen phosphate, etc.}}$

$\quad\quad\quad\quad\quad\quad\quad\quad\quad\quad$ triple phosphate

Partly decomposed phosphates (Novaphos, Carolon phosphate) are obtained with smaller amounts of sulphuric acid. The fine fractions of rock phosphates are primarily decomposed in this process, whereas coarser particles are at most slightly changed. This process yields a P-fertilizer containing several components with different P-actions.

Polyphosphates and other water-soluble P-fertilizers are produced by special processes.

High-temperature decomposition to form thermo-phosphates eliminates the need for expensive acid required in acid decomposition. Rhenania phosphate is produced by adding sodium carbonate and sand to rock phosphate and subjecting the mixture to heat treatment in rotary kilns at a temperature of $1200\,°C$, after which the product is ground and granulated. The fluorine remains in the fertilizer, but is converted into insoluble forms (calcium fluoride).

Conversion equation:

$$Ca_5(PO_4)_3F + 2\,Na_2CO_3 + SiO_2 \rightarrow 3\,CaNaPO_4 \cdot Ca_2SiO_4 + NaF + 2\,CO_2$$

fluorapatite   sodium   quartz $\underbrace{\text{Ca-Na-silicophosphate}}$ fluoride, etc.

$\quad\quad\quad\quad\quad$ carbonate   sand $\quad\quad$ Rhenania phosphate

Alpha phosphate (USA), soluble in citric acid, is formed if the fluorine is expelled during high-temperature decomposition, thus saving large amounts of soda.

Melting decomposition is used to separate the P-component from pig iron in the *Thomas-phosphate* process. European iron ores, in contrast to many tropical ores containing only 0.05% P, contain apatite (1 to 2% P) as a minor constituent, which in the blast-furnace process yields pig iron containing 2 to 3% P, in elemental form. These and other impurities are undesirable and have to be removed when pig iron is processed into steel. This is done in Thomas converters by oxidation after the addition of lime and silicate (air at 1600 °C is blown through the charge). Phosphorus is thus oxidized, forming Ca-silicophosphate as Thomas slag. It is decanted and finely ground.

Solubility Data for P-Fertilizers

Evaluation of all N- and K-fertilizers is justifiably based on the total content of nutrients. However, this rule which formally is very clear, would be wrong in the case of P-fertilizers. It is not the total P-content that is decisive in the evaluation of P-fertilizers but rather the components that are active within a given time. The diversity of P-fertilizers and their varying absorbability, depending on site conditions, require differentiation according to the availability to the plant. This is hardly possible in an exact, generally valid form. However, the solubility in certain solvents, determined in the laboratory, gives an indication of the mean effectiveness in the field (as has been demonstrated by field trials). Moreover, designating P-fertilizers with various solvents serves the important purpose of fertilizer-trade checks [54].

P-fertilizers are thus evaluated on the basis of their *active-phosphate* content according to solubility, whereas the difference from the total content is neglected. The only exceptions are rock phosphates and partly decomposed phosphates, whose total contents form the basis for checks, since the effectiveness varies too much with site factors for it to serve as a basis of an average value. However, the effective components of rock phosphates should also be used for evaluating fertilization (Chap. 2.2.3).

Solubility in formic acid has proved to be the best criterion for differentiating rock phosphates according to their effectiveness. However, this method too only provides approximations. Synopsis 2-7 is a list of solvents in the approximate order of their power of attack.

2.2.2 P-Fertilizers and their Properties

Mineral P-fertilizers include a large range of compounds and have different advantages and disadvantages for fertilization.

Synopsis 2-7. Solvents for P-Fertilizers.

Solvent	Field of application
1. water	super- and triple phosphate
2. neutral ammonium citrate (specific weight = 1.09) (according to *Fresenius*)	also for superphosphate; for many fertilizers in USA Rhenania phosphate
3. alkaline ammonium citrate (22%) (according to *Petermann*)	
4. citric acid (2%) (according to *Wagner*)	Thomas phosphate
5. formic acid (2%)	differentiation of rock phosphates
6. concentrated mineral acids (e.g., sulfuric acid, etc.)	determination of total P-content

There are good reasons for arranging P-fertilizers in the order of their *solubility properties* (Chap. 2.2.1). However, this is not necessarily the order of their general effectiveness. A summary of the most important properties of some P-fertilizers is given in Table 2-6.

The following notes supplement the information in Table 2-6: the contents refer to commercial-grade fertilizers. Deviations from these values are quite possible, since the fertilizer decree only stipulates minimum contents for the corresponding fertilizer types. Only those minor constituents that are significant for fertilization or that occur quantitatively most frequently are listed. From the chemical point of view, phosphate fertilizers are highly "impure" products. This is an advantage for fertilization because of the additional supply of micronutrients. However, the problem of "harmful" admixtures such as small amounts of cadmium should not be ignored in future.

Water-Soluble Phosphates

The term *superphosphate* [50, 58] dates from the beginning of mineral fertilization, and is used to indicate the superiority of this fertilizer over rock phosphate. Double-superphosphate and triple (super) phosphate indicate double or triple content.

The evaluated P-components of these fertilizers (Table 2-6) are more than 90% soluble in water, and thus act immediately. Superphosphate also contains 5% diphosphate. Obviously, the gypsum is not easily soluble in water. These fertilizers are coarsely granulated (2 to 5 mm). They can therefore be easily scattered and stored. They act quickly (approximately equal with regard to the P-component) and slightly acidify the soil, although they do so much less than N-fertilizers.

Table 2-6. Selection of phosphate fertilizers.

Trade name	Formula of P-component (%age in fertilizer)	Calculated P-component of fertilizer, soluble in	Content of valued component in % P	Content in % P₂O₅	Minor constituents	Remarks
superphosphate	Ca(H₂PO₄)₂, etc. (30%)	water (at least 93%), remainder in neutral ammonium citrate	8	18	CaSO₄(50%), oxides, H₂O, etc.	grey, coarsely granulated, intermediate form of concentrated superphosphate with at least 11% P
triple phosphate (triple superphosphate)	Ca(H₂PO₄)₂, etc. (85%)		22	50	oxides, H₂O, etc.	
Rhenania phosphate	3 CaNaPO₄·(Ca₂SiO₄) alkaline ammonium (90%)	alkaline ammonium citrate	11	26	Na (12%), oxides, F, etc.	grey, finely granulated
Thomas phosphate	Ca₃(PO₄)₂·x(Ca₂SiO₄) (75%)	citric acid	7	15	CaO (5%), Fe, Mg, Mn, etc.	greyish-black powder, fertilizer with minimum of 4.4% P
Novaphos (partly decomposed rock phosphate)	Ca(H₂PO₄)₂ (40%) + apatite	water (at least 40%), citric acid (about 30%), concentrated acid (about 30%)	7 / 3, 10 (total content)	23 (total content)	CaSO₄, oxides, etc.	grey, granulated
Hyperphos (soft-earth rock phosphate)	apatite	formic acid (up to 80%), (at least up to 55% for fertilizer type)	11 / 2, 13 (total content)	29 (total content)	CaCO₃ (in apatite), etc.	grey powder (90% smaller than 0.063 mm), partly granulated
soft-earth rock phosphate with limited area of application	apatite	formic acid (up to 40–55%) concentrated acid (up to 45–60%)	13 (total content)	30 (total content)		grey powder (90% smaller than 0.16 mm), "bog fertilizer"

Water-soluble phosphates are converted in the soil into typical soil phosphates via dicalcium phosphate and free phosphoric acid. Conversion is primarily to apatites, via defect apatites as precursors, in neutral soils, and to iron and aluminum phosphates in acid soils (Fig. 2–3). The duration of this conversion, which implies reduced mobility, depends not only on the inherent causative factors but also on the contact with fine soil particles. Contact with the soil should be minimum if the water-soluble form is to be maintained to a large degree. This requires *placement* of the fertilizer in "storage nests". Coarse granulation is a step in this direction.

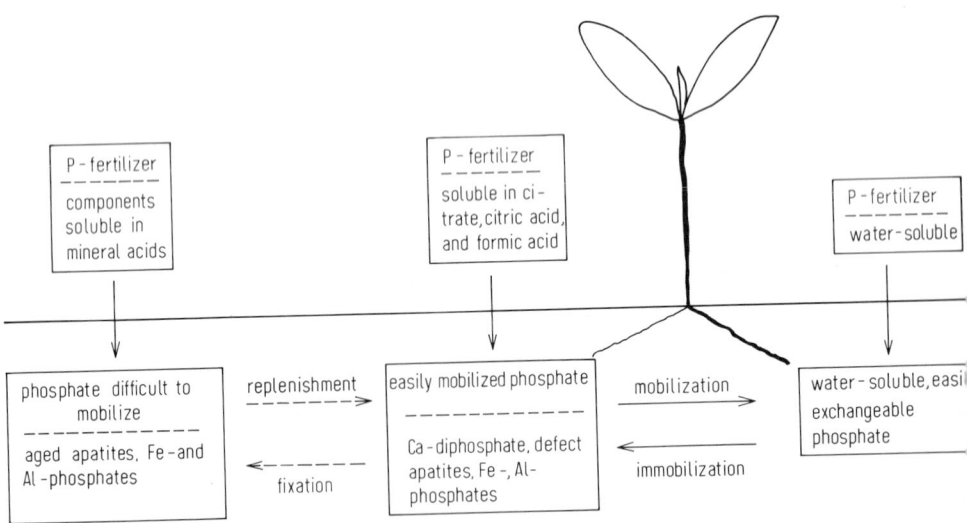

Fig. 2–3. Phosphate fertilization and fertilizer conversions in soil.

The increasing immobilization accompanying conversion reduces availability, but the latter is largely maintained in most soils. Exceptions are soils that fix phosphates to a pronounced degree, e.g., certain tropical soils rich in iron and aluminum, in which phosphate is not only immobilized but even fixed in an unavailable form.

Rhenania Phosphate and Thomas Phosphate
Rhenania phosphate [50, 59] is named after the former Rhenania works near Aachen, West Germany. The valued P-component of this fertilizer is soluble in citrate. Fine grinding of the crude fertilizer (75% of the particles smaller than 0.16 mm) may be followed by loose granulation of the powder for easier application, without reducing its effectiveness. Rhenania phosphate in the form of

calcium phosphate reacts basically in the soil. The CaO "content" computed from the formula is approximately 40% (for the lime action, *see* Chap. 2.2.3).

Thomas phosphate [38] is named after the inventor of the production process, the English metallurgist *Thomas*. Only the P-component soluble in citric acid is calculated. Fine grinding of the fertilizer is important (75% of the particles smaller than 0.16 mm). Subsequent granulation is possible but is omitted since it would reduce the effectiveness. Granulation is therefore carried out in practice only for two-nutrient fertilizers (e.g., Thomas potassium), in which the potassium component causes thorough decomposition of the granulates in the soil.

Thomas phosphate in the form of calcium phosphate reacts basically in the soil. The CaO "content" computed from the formula is approximately 40%, to which 5% free CaO should be added, thus altogether 45% (for the lime action, *see* Chap. 2.2.3).

With reference to the conversion of *Rhenania* and *Thomas phosphates* in the soil it should be noted that in view of their insolubility in water, these substances must first be converted into a water-soluble form (Fig. 2–3). The mobilization power of many soils suffices for this, by acid attack and complex formation by soil organisms and plant roots. These fertilizers thus act more slowly than water-soluble phosphates, but their action, especially that of Rhenania phosphate, must nevertheless be considered to be quite fast. The advantage of a slower conversion of fertilizer phosphates into soil-phosphates consists in the correspondingly slower mobilization, which is an advantage in soils which fix P.

Partly Decomposed Phosphates

The name *Novaphos* [48] is intended to indicate a new trend in phosphate production, i.e., partial decomposition. The products, e.g., *Novaphos* and the similar *Carolon phosphate* may, according to their production process, be considered as fertilizers with several P-forms, namely, superphosphate, a more or less altered, and a barely altered rock phosphate component, on the basis of the respective soft- or hard-earth raw material.

The properties correspond to these components. Good initial action, owing to the water-soluble component, is combined with long-term after-effects. However, the component only soluble in mineral acids might only be important for fertilization under special conditions.

Rock Phosphates

Soft-earth rock phosphates [46, 50], mainly from North Africa, consist of cryptocrystalline apatite, phosphorite, with a crystal size of 0.00001 mm. Because of the "softness" of the rocks and the fineness of the mineral crystals the nutrient can be adequately mobilized through the decomposition power of the soil.

We can distinguish between three groups of rock phosphate fertilizers according to their solubility in formic acid:

- Fertilizers with a large amount of active components (65–80% soluble in formic acid), e.g., *Hyperphos* from Tunisian *Gafsa* phosphate;
- fertilizers with a moderate amount of active components (about 60%);
- fertilizers with a small amount of active components (40 to 55%).

However, the value of rock phosphates depends not only on their solubility properties but also on the mobilization conditions in the respective soils, since the entire work of chemical decomposition has to be effected in the soil. Rock phosphates are mobilized in the soil more intensively, the lower the pH, the better the moistening, and the higher the temperature; this is largely paralleled by biotic activity. Rock phosphates are most suitable below a soil pH of 6 to 6.5. Rock phosphates with low solubility in formic acid are mobilized more than normally in highly acid soils (e.g., high moors). These phosphates are therefore also called *bog fertilizers*.

Rock phosphates have an alkaline reaction in the soil, because of their total-CaO content of 45 to 50%. However, the activity must be taken into account with regard to the effective lime action (Chap. 2.2.3).

Other P-Fertilizers

Dicalcium phosphate ($CaHPO_4$) with at least 17% P is a slow-acting P-fertilizer used as a component of multiple-nutrient fertilizers.

Ammonium phosphates and P-fertilizers with magnesium are two-nutrient fertilizers (Chap. 3.2.1). Other P-containing fertilizers are discussed in the respective sections, e.g., P-containing converter lime in Chapter 4.1.1).

Recent Developments of P-Fertilizers

The scarcity of P-reserves, the decisive and unsolved problem of the low utilization of P-fertilizers, and the desire to obtain P-fertilizers of higher concentrations, suitable for fertilizer solutions, led to various new developments, some of which have already demonstrated their worth.

It is obvious to use the *iron* and *aluminum phosphates* obtained in the third purification stage of town sewage as fertilizers, and thus to close one part of the P-cycle.

The *condensed phosphates*, poly-, meta-, and ultraphosphates [42], are highly concentrated P-fertilizers. They are obtained from (ortho-)phosphoric acid by the removal of water, with diphosphoric acid being formed in the first step:

$$2\,H_3PO_4 \quad -\,H_2O \rightarrow \quad H_4P_2O_7$$

(ortho-)   diphosphoric acid
phosphoric acid   (pyrophosphoric acid)

Continuation of this dehydration finally leads, via tri- and tetraphosphoric acid, to polyphosphoric acid with the general formula $H_{n+2}(P_nO_{3n+1})$ (where n may be a large number). Its alkali or alkaline-earth salts are the *fertilizer polyphosphates*. Complete removal of water leads to the formation of metaphosphoric acid of the general formula $(HPO_3)_n$ yielding salts with ring or chain molecules, e. g., Ca-metaphosphate $(Ca(PO_3)_2)_n$. This is a highly concentrated P-fertilizer containing 28% P. Ultraphosphates are even more highly polymerized, e. g., $(Na_2P_4O_{11})_n$.

Condensed phosphates are in part easily soluble in water and in part insoluble in water. Soluble polyphosphates play a special role in nutrient solutions for liquid fertilization, since no Ca-phosphate is precipitated when they are used in concentrated stock solutions. Condensed phosphates are quickly hydrolized in the soil, i.e., they are again converted into orthophosphate by the addition of water, and can be absorbed by plant roots in this form. Their fertilizer action corresponds largely to that of ordinary P-fertilizers.

Easily soluble *glycidophosphates*, phosphates coupled with sugars, are highly promising for fertilization by irrigation. An extremely high percentage fertilizer would be gaseous *phosphine* (PH_3), which formally corresponds to ammonia (NH_3) but is hardly suitable as a fertilizer because of its toxicity.

A pure P-fertilizer that only contains phosphorus as nutrient element is required for research purposes. Almost none of the ordinary P-fertilizers, not even pure phosphoric acid, because of its very strong acidity, is useful for this purpose. However, an anion exchanger charged with phosphate ions would be suitable, as would, for some purposes, primary Na-phosphate.

2.2.3 Principles of Fertilization with Phosphate

Selection of the Optimal P-Fertilizer

The problem of the optimal P-fertilizer for each case has engaged researchers and experts for hundreds of years. Many thousands of pot and field trials, as well as laboratory tests have been performed for the sake of comparing phosphate fertilizers, in view of the multiplicity of P-fertilizer forms and site conditions. The results form a comprehensive picture of the large number of factors on which P-action depends; a picture, which is rather confusing in its variety. Many individual problems can be separated and solved, but comprehensive assessment of all aspects, which, in the final analysis, is important for practical application, remains difficult. It is impossible to determine the optimal fertilizer form for each crop on every plot by field trials, even when the question of the reliability of data obtained in such field trials of phosphate fertilizers is ignored (*see* Chap. 6.6). Hence, general statements based on critical assessment of previous trials is all the more necessary.

Despite the complexity of the problem, there are now reliable criteria for selecting the correct P-form. However, these too change with general production conditions, e. g., increasing yields, so that a conclusive judgement appears difficult at first.

However, this pessimistic conclusion is moderated or even nullified by a new development. Many formerly disputed problems have largely lost their significance and appear out of date, in view of the trend towards depot fertilization of intensively cultivated plots. First of all, however, we shall discuss the problems involved in assessing the action of P-fertilizers.

The complex effects of P-fertilizers on growth and yield consist of the true *P-action* and the *side effects*. They are tested in comparative experiments whose principal aim is to compare P-forms (tests of true P-action).

The *P-form comparison* is made to clarify the effects of individual P-forms on the yield and to indicate the underlying causes, e. g., quicker action and P-absorption, better utilization, and lastly better effects on the yield. The results of many P-form trials may be summarized as follows with regard to speed of action and the quantities absorbed [51]:

The three tested fertilizers can be arranged in the order expected from their chemical solubility, ranging from quick-acting superphosphate via Rhenania phosphate to the more slow-acting Thomas phosphate, if the fertilizers react only slightly with the substrate, e. g., in pots with sand cultures, and in part in sandy soils.

However, this simple basic pattern is considerably affected by soil factors (Fig. 2-2, p. 68). The absorption of water-soluble fertilizers decreases with the severity of immobilization conditions. Fertilizers not soluble in water, especially Thomas phosphate, are improved by increasing soil mobilization.

Based on the pH-value as index of P-dynamics, water-soluble fertilizers then lead to better P-absorption in the neutral region, whereas fertilizers not soluble in water are often superior in the acid region. From the theoretical point of view, it should be added that small amounts of all P-fertilizers are obviously absorbed *immediately*, i. e., on the first day after application. When quick or slow action is discussed, the reference is obviously to quantitatively significant amounts and relative differences.

Assessment of the P-action finally depends not on the initial or later speed of action and the P-absorption linked with it, but on effects on the yield. This *yield effect* is approximately the same for the three most frequently tested fertilizers, allowing for variations in the results, if all side effects are excluded. This is true for many cultivated plants having medium P-absorption capacities. However, more easily soluble P-fertilizers are superior when rapid initial development of plants with weak root systems require particularly abundant initial P-supplies.

On the other hand, yield effects depend greatly on *soil differences*, not least because of side effects. The following graded sequences can be established based on pH:

- superphosphate acts relatively best on slightly acid and neutral soils;
- Rhenania phosphate occupies an intermediate position;
- Thomas phosphate acts relatively better on light, acid soils.

Problems of Comparing P-Forms

A considerable amount of unclear and partially contradictory data on the relative worth (and action) of P-fertilization is due to inadequate test concepts and insufficiently critical interpretation of results (Chap. 6.6).

Thus, application of the "principle of equal treatment of all test subjects" in many earlier comparative trials is inadequate, although this principle was derived from general testing procedures and was quite correctly used as the basis of many comparative tests. Comparisons were thus carried out of finely ground P-fertilizers, e.g., superphosphate and Thomas phosphate, applied at the same time (frequently at the time of sowing) and by the same method of working into the soil. However, all trials performed according to this principle and their results are largely useless for practical fertilization. The subjects of fertilizer tests should be subject to the same general conditions, but the fertilizers should not be applied in the same, but in the optimum form. Water-soluble fertilizers should be granulated, or placed even more intensively at the time of sowing and applied by minimizing contact with the soil; on the other hand, fertilizers soluble in acids should be finely dispersed and properly mixed with the soil a long time before sowing. Each fertilizer can thus act optimally according to its properties.

It is also a mistake to evaluate P-forms on the basis of P-fertilizer trials in which *side effects* have not been compensated for. Any P-fertilizer has several effects. Strictly speaking, superphosphate is a three-nutrient fertilizer (P, S, Ca) and has a structure-improving effect because of its gypsum component, but it also has a slight soil-acidifying action.

Thomas phosphate contains not only phosphorus but also manganese, magnesium, and silicate, and reacts basically. This possible multiple effect should be taken into account in designing the trials and interpreting their results, even if some possible effects might in practice be unimportant under many conditions. Critically viewed most trials conducted in the past concerning comparison of P-forms yielded no information on the core problem, or the results were wrongly interpreted.

General comparisons of P-fertilizers wiuh respect to their overall effects may, of course, be of value. However, data on possible yield increases, taking only the complex action into account, provide no information at all on whether these are due to different P-forms or are caused by better P-nutrition.

Utilization of P-Fertilizers

The low recovery rate of P-fertilization poses a serious problem. N- and K-fertilizers have a 50% or higher utilization in the first year, (or can at least be utilized); on the other hand, the utilization rate of P-fertilizers is only 15% on average, generally ranging between 10 and 25% [52]. This range applies to mineral and organic P-fertilizers.

Plants differ with regard to utilization according to their root penetration and decomposition capacity. Thus, many vegetable plants and grains have only low utilization rates, whereas clover and root crops attain higher values. Residual effects can be estimated at 1 to 2% per year on average. A utilization rate of 50% may therefore be assumed for longer periods (20 to 30 years) with almost complete utilization over very long periods (Chap. 5.5.1).

Improving Utilization through Placement

The low utilization rate can be improved to some extent with water-insoluble fertilizers by increasing the mobilization capacity; above all with water-soluble P-fertilizers this effect is achieved by reducing immobilization. Water-soluble phosphate must be precisely applied near the roots with minimum soil contact. This provides and preserves "nests" of large phosphate supplies. This placement procedure is more suitable for P-fertilizers than for N- or K-fertilizers, since the risk of damage from high salt concentrations is low.

Placement increases P-absorption, and thus fertilizer utilization, especially under the following conditions [36]:

- Low P-supplies in the soil;
- dry years or arid zones;
- wide spacing of plants;
- high immobilization or even fixation;
- small fertilizer doses, e.g., for starting fertilization;
- plants with short vegetation periods, for which optimum initial P-supplies are a prerequisite for a rapid start to growth;
- plants with deep roots, for better P-supplies to deeper soil layers.

There are many possible techniques which can be used for placement. The use of granulated fertilizers is already an important measure (Fig. 2–4).

The following procedures are possible with specialized machinery:

- Contact fertilization (placement of fertilizer around seed grains);
- row fertilization (placement of fertilizer alongside the seed);
- strip fertilization (placement of fertilizer underneath the seed).

Placement can to a certain extent also be achieved by simple means, e. g., by ploughing under the fertilizer. Placement can improve the utilization of water-soluble fertilizers up to 25% in the first year, if conditions are favourable. However, placement is unimportant under many conditions, e. g., for plants with roots close to one another (grain), or when the soil is moist in humid regions.

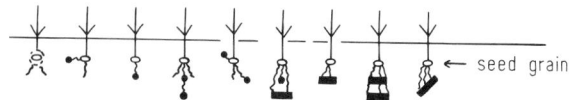

Fig. 2-4. Possibilities of placing water-soluble phosphates through contact, row, and strip fertilization.

Side Effects of P-Fertilizers

Side effects play such an important role in fertilization that P-fertilizers might be termed *multi-purpose fertilizers* (*see* Synopsis 2-8).

The additional *supply of nutrients* by some P-fertilizers is considerable. Thus, besides 8% P, superphosphate contains about 13% sulfur and 20% calcium, and is thus in fact a three-nutrient fertilizer. The Ca-component only plays a minor role in plant nutrition, but in some particular regions superphosphate acts chiefly as a sulfur fertilizer.

Thomas phosphate contains 1 to 2% magnesium and 2 to 4% manganese as an important trace element, which, however, must first be mobilized in the soil. The sodium component of Rhenania phosphate may be important for grassland.

Synopsis 2-8. Side Effects of P-Fertilizers.

1. Nutrient supply (besides phosphate):
   a) nutrient elements: sulfur, calcium, magnesium, manganese, etc.
   b) beneficial substances: sodium, silicate

2. Supply of structure-improving substances:
   lime, calcium as Ca-phosphate and gypsum

3. Changes in soil reaction:
   a) acidification with positive or negative consequences (a positive effect may be the mobilization of nutrients in the soil itself, especially of some trace elements);
   b) increasing soil reaction, with positive or negative consequences (positive effects, e.g., the reduction of acid damage and the mobilization of molybdenum)

4. Immobilization of (necessary and harmful) heavy metals through extremely high phosphate supplies in the soil.

The supply of silicate in suitable P-fertilizers is highly valued in rice cultivation, because it increases resistance to lodging. The soil-acidifying effect of water-soluble P-fertilizers is hardly significant with regard to the lime balance but is important with respect to the supply of trace elements from the soil itself. The reason is that small zones of high acidity (pH-values below 2) are formed near fertilizer granulates, where the mobilization capacity is correspondingly high.

The *lime effect* of many P-fertilizers is a side effect of special, largely positive, significance. The lime effect is usually assessed by evaluating P-fertilizers, like liming materials, on the basis of all basically reacting substances and indicating their possible lime effect as (total) CaO. This procedure is correct, provided that the fertilizers are converted entirely in such a way that the whole Ca-component becomes fully effective as a basically reacting substance. However, the lime effect becomes delayed and incomplete if conversion is slow and Ca-phosphate is formed in the soil itself. There should therefore be a deduction from the total CaO content. Any relevant practical value of the easily converted Rhenania and Thomas phosphates (40 to 45% total "CaO") should lie between one half and the full possible effect. Approximately three quarters of the possible lime effect on average seems to be a useful approximation in practice.

The effective part of rock phosphates, with a theoretical total content of 45 to 50% CaO, would probably be far smaller, and would be fully effective only in extremely acid soils.

2.2.4 Practical Use of P-Fertilizers

This section will provide a concluding summary of the rules governing the use of P-fertilizers, based on the above discussion of the properties of P-fertilizers and the site factors influencing their action.

1. The basic preconditions for optimum P-action are of primary importance. The pH of the soil must be in the correct range (Chap. 5.1.1) and the soil must have a good structure, since only then can the phosphate be optimally effective. Many P-fertilizers also have a lime effect, but they are too expensive to be used as liming materials, however welcome the additional benefits of their lime component may be.

2. The *P-form* should be chosen from the point of view of optimum P-action, allowing for side effects. Selection can be facilitated by grouping in the following four categories, despite the diverse fertilizer properties and conversion conditions:

I.   Water-soluble, chemically converted P-forms, e.g., superphosphate, triple phosphate, ammonium phosphate;

II. easily mobilized, chemically converted P-forms, e.g., Rhenania phosphate, Thomas phosphate;

III. rock phosphate with a large easily mobilizable component, e.g., hyperphosphate;

IV. rock phosphate with a small amount of easily mobilizable component, *bog fertilizer.*

Fertilizers with several active components and multiple-nutrient fertilizers containing phosphate (*see* ammonium phosphate) can also be included in these categories according to their composition. The following suggestions should facilitate the choice of the *correct P-form*:

a) Differences in the action of various P-fertilizers play an important role in *insufficiently supplied* soils with normal decomposition capacity:

• Easily mobilizable forms are generally superior in acid soils;
• water-soluble fertilizers are more suitable for neutral soils (pH approximately 7), with citrate-soluble phosphate being approximately equivalent.

b) The *better a soil is supplied*, the less important is the P-form, since the principal task is replenishment of reserves rather than the fertilization of existing plants. The whole problem of different P-fertilizer forms thus becomes relatively unimportant when the soil is properly supplied. It is thus possible to select the cheapest fertilizer (from categories I to III); value for money must however be determined on the correct basis (*see* under 5). At the same time, an adequate supply to plants with average requirements may imply an inadequate supply to plants with particularly large requirements or low absorption capacity. Such plants, e. g., rapidly growing special crops, should therefore be additionally supplied with quick-acting phosphate at the start of growth if necessary.

c) Soft-earth rock phosphates of category IV give good results on soils with *very high decomposition capacity* (acid high moors, moderately acid but biotically highly active soils, e. g., moist grassland), since even components difficult to mobilize are effective under these conditions. Evaluation according to the total content is therefore most justified for these sites.

3. P-fertilization should be *optimized in its action.*
   The fertilization time is of primary importance:

• Water-soluble P-forms should be, and citrate-soluble P-forms may be, applied at the time of sowing;
• P-forms soluble in acid should be applied a long time before sowing, that is, in autumn for summer crops.

These rules apply primarily to inadequately supplied soils, whereas the application time is of minor importance in the case of well supplied soils. Proper introduction into the soil, especially through placing, can considerably improve the initial action of water-soluble P-forms. An example is placement of ammonium phosphate for the fertilization of maize.

Dividing the P-dose seems to be of little use, although plants have higher P-requirements during two periods, firstly at the start of growth for root development, and later for the development of fruit. However, adequate supplies throughout the period of growth are preferred to late fertilization. This is largely due to the fact that phosphate only penetrates into the soil to a slight extent, even if moistened, and thus hardly reaches the principal root zone during the fertilization of growing crops.

4. *Application of phosphate as a foliar nutrient* is quite possible and is the best method for rapid supplies in case of P-deficiency. However, the quantities to be applied are subject to narrow limits (Chap. 6.1.4).

5. *Price comparison* of P-fertilizers should be based not on the total P-content but on the active components, except for application to soils with very high decomposition capacity. The standard basis is the analytically determined solubility, if the active component undes special production conditions is not known from field trials.

Price comparisons should be based on

• the indicated content in the case of chemically (fully) converted fertilizers;
• the analytically determined active part, i. e., total content less the part insoluble in formic acid or soluble only in mineral acids, in the case of rock phosphates or fertilizers containing forms similar to rock phosphates.

The total. content of rock phosphates should form the basis only if applied to soils with very high decomposition capacity. Price comparison should also allow for important minor constituents according to their effectiveness, e. g., the lime component.

6. The required *amounts* of phosphate fertilizers depend on the state of soil reserves and on the requirements of the plants concerned. The problem of correct P-fertilizer quantities is discussed in detail in Chapter 5.4.2, and for individual plants in Chapters 7 and 8.

7. *Loss* of P-fertilizers through leaching etc. is unimportant from the agricultural point of view. The leached quantities are generally much less than 1 kg/ha of P. Larger losses can occur only under special conditions such as washing-away of the fertilizer from slopes on frozen soil. However, phosphate losses from the soil, and thus introduction into bodies of water, are discussed from the environmental point of view in Chapter 6.2.2.

Summarizing Rules for Practical Application of P-Fertilizers

1. P-fertilizers are fully effective only in soils with good structure and optimum soil reaction.

2. Selection of the correct P-form is important primarily for inadequately supplied soils, and depends largely on the soil reaction. Active components of different analytic solubility are approximately equivalent in well supplied soils, since their primary purpose is to replenish supplies.

3. Correct application should ensure optimum action of P-fertilizers. This requires the proper fertilization time and correct introduction into the soil, mainly because phosphate is relatively immobile in the soil. The method of application is of minor importance for well supplied soils.

4. Application of phosphate as a foliar nutrient only plays a minor role.

5. Price comparison of P-fertilizers for soils with normal decomposition capacity should be based on the analytically determined *effective* content, allowing for important minor constituents.

6. The amounts of P-fertilizers depend on the supply state of the soil and the requirements of the plants considered.

7. Losses of P-fertilizers from the soil are unimportant from the agricultural point of view.

2.3 Potassium Fertilizers (K-Fertilizers)

K-fertilizers are chemical substances containing the nutrient element potassium* in the nutrient form of absorbable potassium cations or yielding K^+-cations after conversion.

Mineral K-fertilizers are predominantly water-soluble salts. The Fertilizer Law states that potassium fertilizers also include potassium-magnesium fertilizers. A survey of the chemical components is given in Synopsis 2-9.

2.3.1 Origin and Production

Crude potassium salts (K-minerals with impurities) were formed in the process of the drying up of seawater in former ocean basins, which occurred largely during the Permian period some 200 million years ago. The seawater salts crystal-

---

* The Latin/German word for the element with symbol K is Kalium. It is derived from the Arab word *el-kali,* meaning ash (*potash* in English, *potasse* in French).

Synopsis 2–9. Important K-Containing Substances.

K   chemical symbol of element potassium, reference basis for potassium contents;

K^+ potassium ion, univalent cation of potassium salts, occurs in soils, plants, and fertilizers;

KCl   potassium chloride, consists of K^+ and the anion Cl^- (chloride);
K_2SO_4 potassium sulfate, consists of K^+ and the anion SO_4^{2-} (sulfate);
K_2CO_3 potassium carbonate, principal salt of plant ash *(potash)*;
K_2O potassium oxide; reference basis for fertilizers to indicate potassium content, but obsolete and unnecessarily involved (should logically be replaced by K).

lized in the order of their solubility, i.e., the common salt (NaCl) stratum is overlain by potassium minerals. In the course of time they were covered by many geological strata and hardened to rock. At many points the potassium rocks later rose nearer to the surface at the top of salt stocks, and are nowadays worked in underground mines. Crude potassium salts are thus natural seawater minerals.

Potassium *reserves* are large. There are large deposits in North America (USA, Canada), etc., apart from the big German deposits that formerly were the most important. The discovery of further deposits is to be expected (in accordance with an old miners' adage, "it is dark in front of the pickax").

Production of K-Fertilizers

K-fertilizers are most commonly produced from crude potassium salts that occur in nature, by grinding or by separating impurities (Table 2–7) [33]. The fertilizer kainite, which is a natural mixture of different minerals, is obtained by grinding K-containing rocks. The K-content is correspondingly low.

The production of high-concentration *fertilizers based on KCl* from K-minerals, necessitates far-reaching separation of the salt impurities, which are

Table 2–7. Minerals of potassium rocks.

Mineral	Formula	Rock
sylvine	KCl	
kainite	$KCl{\cdot}MgSO_4{\cdot}3\,H_2O$	+ NaCl = kainitite
carnallite	$KCl{\cdot}MgCl_2{\cdot}6\,H_2O$	
kieserite	$MgSO_4{\cdot}H_2O$	carnallitite
common salt	NaCl	

generally detrimental to fertilization in these quantities. Solvation processes exploit the different solubilities of the salts in water for separation:

- *Magnesium chloride* ($MgCl_2$) can be separated by solution in cold water;
- *common salt* (NaCl) is approximately equally soluble in cold and hot water;
- *potassium chloride* (KCl) is more soluble in hot water; its concentration (in comparison with NaCl) can therefore be increased by heating the solution, with subsequent crystallization upon cooling.

The newer *flotation process* consists in separating KCl crystals from supersaturated solutions by the addition of organic flotation agents, such as fatty amines, with which they rise to the surface upon introduction of air.

$$K_2SO_4 \cdot MgSO_4 + 2KCl \rightarrow K_2SO_4 + MgCl_2$$

potash magnesia   K-sulfate   Mg-chloride

Potash-containing industrial wastes, e.g., *potash filter dust* from cement production, obtained from converted potassium compounds of limestone, etc., are now also available.

2.3.2 K-Fertilizers and their Properties

The fertilizer kainite, not to be confused with the mineral kainite, consists of ground crude potassium salts of indeterminate composition and contains at least 8% K. A special type of marketed fertilizer is *crude potassium salt enriched* with KCl, which contains at least 15% K. Because of its high content of ballast salts, kainite is now only of minor or limited importance. An especially finely ground form *(Hederich kainite)* was formerly used against weeds, since this salt attacked the leaves.

The K-fertilizers based on KCl (Table 2–8), mostly used now, are [44]:

- *Grade-40 potassium*, i.e., fertilizer with a "K_2O-content" of 40%;
- *grade-50 potassium*, i.e., fertilizer with a "K_2O-content" of 50%.

Grade-60 potassium also plays a role, especially as a highly concentrated mixture component for the production of other fertilizers. The maximum content attainable with pure KCl is about 52% K (63% K_2O). K-fertilizers are largely granulated to improve application; coarse granulation is quite practical in view of the good solubility in water. Accompanying salts may have positive or negative effects, depending on type and intended use.

Potassium sulfate is available for plants sensitive to chloride, such as potatoes and many horticultural plants. The Cl-content must not exceed 3%.

Table 2-8. Selection of potassium fertilizers.

Trade name (fertilizer type)	Formula of K-component (content in fertilizer)	Content* in % K	Content* in % K_2O	Minor constituents	Properties
crude potassium salt (Kainite)	KCl (20%)	11	13	NaCl, $MgCl_2$, etc.	white or slightly colored, water-soluble salt
potassium chloride (grade-40 potassium salt)	KCl (63%)	33	40	NaCl, etc. (20 to 30%)	
potassium chloride (grade-50 potassium salt)	KCl (80%)	42	50	NaCl, etc. (10 to 15%)	white or slightly colored, water-soluble (often granulated) salts
potassium chloride (grade-60 potassium salt)	KCl (96%)	50	60	NaCl, etc. (1 to 3%)	
potassium sulfate	K_2SO_4 (93%)	42	50	sulfur (18%)	white, water-soluble salt
residue potash	K_2SO_4, K_2CO_3, etc.	>17	>20		e.g., potash filter dust, water-soluble potassium component

* all K-contents refer to water-soluble potassium.

The Fertilizer Law states that K-fertilizers also include the KMg-fertilizers discussed in Chapter 3.2.1, e.g., magnesia kainite, granulated potash with magnesium (grade-40 potassium with Mg), and potassium magnesia (KMg-sulfate). *Residue potash*, e.g., potash filter dust, consists of several components, chiefly K-sulfate and -carbonate. It must be free from harmful impurities. *Potassium nitrate* is important in special cases, but belongs to the NK-fertilizers (Chap. 3.2.1). *Other potassium fertilizers* only play subordinate roles.

The above-mentioned K-fertilizers are water-soluble and therefore act immediately. This advantage from the nutrient action aspect is balanced by the risk of "salt damage" when fertilizer doses are large; this applies in particular to sensitive young plants.

The development of slow-acting K-fertilizers is therefore of some importance, although much less so than for N-fertilizers.

The following *slow-acting* K-fertilizers may be considered:

- Less soluble double salts;
- fritted glass, which is very finely ground glass containing potassium;
- K-salts coated with foils.

Rock powder might be thought to be a very slow-acting K-fertilizer, if its K-action were not almost negligible even with the fine grinding. Potassium feldspar does contain about 8% K, but mineral stability must also be taken into account. Ewen less potassium can be set free through weathering from potassium feldspar than from lime feldspar containing only 1%. Hence, rock powder does not have significant K-fertilizer action for practical fertilization, considering the quantities applied.

"Pure" K-fertilizers containing only potassium ions are of interest for research purposes. Hardly any ordinary K-fertilizers, including potassium hydroxide which has a strongly alkaline reaction, can be used for this purpose. However, exchange complexes charged with K-ions would be suitable, as would potassium nitrate, for special purposes, i.e., if fertilization with nitrates is carried out in any case.

2.3.3 Application of K-Fertilizers

Application of K-fertilizers, in contrast to that of N- and P-fertilizers, poses relatively fewer problems and can therefore be considered in brief.

K-Fertilizer Effect

All potassium fertilizers may be considered to be approximately equivalent with respect to their potassium-fertilizer effect. The K-component is water-soluble, so that K-fertilizers are available immediately and thus act rapidly.

In the soil, potassium fertilizer first enters the soil solution, but most of it is then directly sorbed on the exchange complexes. In this form it is stored as loosely bound potassium, being protected against losses and easily available to plants, corresponding to the potassium mobilized from soil minerals. Potassium is partly immobilized in intermediate layers of clay minerals. This reduces availability but also creates a buffer-stock against losses. Complete fixation of potassium occurs only in rare soil forms.

Potassium fertilizers can be applied as a reserve or early on before sowing, if the soil has at least a moderate storage capacity.

Side Effects of K-Fertilizers

Proper application of potassium fertilizers primarily requires that allowance be made for side effects due to the accompanying anions, cations, possible salt load, and effects on the pH.

As regards accompanying anions, the choice with potassium fertilizers is mainly between chloride and sulfate, the chloride form being cheaper. However, exclusive application of KCl is inadvisable because of the sensitivity of some plants to high chloride absorption. These plants, which are sensitive to Cl and therefore prefer *sulfate* (Chap. 3.4.1), should therefore be fertilized with potassium in the form of sulfate (potassium sulfate or potassium-magnesium sulfate).

On the other hand, potassium in the *chloride* form is suitable for plants that are either not harmed by chloride, or which benefit from it, even in larger amounts. These are the "salt-loving" plants (halophytes) like beta beets, etc. (Chap. 3.4.1).

The varyingly large sodium component of the accompanying cations is beneficial to halophytes. In grassland it ensures a higher Na-content in the grass, which is to the benefit of grazing animals with their large sodium requirements. Sodium has a detrimental effect on the structure of medium and especially of heavy, agricultural soils, due to its clogging action and crust formation. *Magnesium* is also usually present in ordinary potassium fertilizers, although only in a few per cent if the content is not explicitly stated. Magnesium is an advantageous supplement. Substances with higher Mg-content are KMg-fertilizers.

The *salt load* due to minor constituents of water-soluble potassium fertilizers is not only important in the case of salty soils, but also with high fertilization intensities in horticulture and sometimes even in intensive-fertilization agriculture, e. g., in situations of plants germinating during dry periods. Fertilizers with low K-contents, in particular contain large amounts of *ballast salts* that should be avoided in intensive fertilization. In this respect, chloride is worse than sulfate, so that potassium in sulfate form should be preferred when there is a risk of higher salt concentrations (Chap. 6.3.4). Ballast-free potassium nitrate may be advisable for soils with increased natural salt content.

Effects on *soil reaction* are of minor importance. Chemically, potassium salts are neutral. They do have a weak acidifying action on the soil, but this effect is slight. Fertilization with KCl causes considerable chloride losses from the soil, because of the preferential absorption of potassium. In turn this causes the loss of an equivalent amount of calcium. Decreases in pH, after fertilization with potassium, are not usually observed, however, calcium losses may be slightly increased.

Practical Application of K-Fertilizers
K-fertilizers can be applied relatively simply. The correct fertilizer form is chosen (as described) mainly according to plant tolerance of, or sensitivity to, *chlorides*.

K-fertilizers can simply be placed on the surface of the soil, where they then penetrate slowly to the root region with water. Application of potassium salts as foliar nutrients is possible, but the danger of corrosion requires that the quantity applied be very limited in comparison with requirements. The quantity of potassium necessary is discussed in Chapter 5, and for various particular plants, in Chapters 7 and 8.

The *recovery rate* of K-fertilizers is about 50 to 60% in the first year. Allowing for residual effects, provided no losses have to be taken into account, complete utilization may be assumed in well supplied soils (Chap. 5.5.1).

Many plants have a tendency to *excessive consumption* when potassium supplies are abundant. However, even extremely high K-contents are not really harmful to plants although their quality for various purposes (e.g., as fodder, Chap. 9.3.2), is affected.

Certain K-losses through leaching in surface water and groundwater are unavoidable, but the magnitude of the losses makes them insignificant from the agricultural and "environmental" viewpoints. Significant losses only occur in very moist soils with low storage capacity, e.g., sandy and boggy soils.

2.4 Magnesium Fertilizers (Mg-Fertilizers)

Mg-fertilizers are chemical substances containing the nutrient element magnesium in the form of absorbable magnesium cations or yielding Mg^{2+}-cations after conversion.

In recent decades, magnesium has evolved from a "neglected" to an "equally-ranking" major nutrient, when considered from the fertilization point of view.

Some important designations are as follows:

Mg   chemical symbol of the element magnesium, reference basis for magnesium content;

Mg^{2+} magnesium ion, bivalent cation occurring in soils, plants, fertilizers;

MgO   magnesium oxide, also termed *magnesia* especially in fertilizers and largely used as a reference basis for Mg-content; this is suitable for Mg-limestones in accordance with their principal effect, but not for Mg in plant fertilizers.

Origin and Production

The raw materials of Mg-fertilizers are minerals occurring in nature. Water-soluble Mg-salts are associated with potassium salts in salt deposits and are separated from them by solution processes (Chap. 2.3.1). Mg-limestones are produced from Mg-containing (dolomitic) lime in a manner corresponding to that of limestone (Chap. 4.1.1). *Global reserves* of magnesium are large, since entire mountain ranges consist of Mg-containing rocks, i.e., dolomite $(CaCO_3 \cdot MgCO_3)$.

2.4.1 Mg-Fertilizers and their Properties

It is advisable to divide M-fertilizers into two groups with respect to their use, namely, easily water-soluble Mg-fertilizers and those barely soluble or insoluble in water (mainly Mg-limestones). (Table 2-9).

Water-Soluble Mg-Festilizers

Magnesium sulfate is used for fertilization in two forms. Kieserite has a higher Mg-content, but is much less water-soluble than epsom salt with which solutions of up to 60% concentration can be prepared. This solution is important only for applications in dissolved form (e. g., as foliar nutrient), whereas the solubility of kieserite is quite sufficient for application as soil dressing.

    Magnesium chloride is highly soluble in water and is marketed as a fertilizer solution for application as foliar nutrient. A saturated solution contains 9% Mg. A certain amount of calcium chloride ($CaCl_2$) is usually present, but the Ca-content is limited to 2%. Two-nutrient fertilizers containing magnesium in the form of sulfate or chloride should be evaluated according to these forms. They are almost exclusively applied as soil dressing, so that the two Mg-forms may be considered to be approximately equivalent, apart from the sensitivity of some plants to chloride.

    Magnesium nitrate has hardly any role to play as a fertilizer.

Table 2-9. Selection of magnesium fertilizers.

Fertilizer	Formula of Mg-content	Mg-content in %	Properties
magnesium sulfate kieserite epsom salt	$MgSO_4 \cdot H_2O$ $MgSO_4 \cdot 7H_2O$	17 10	less readily soluble readily soluble (up to 60%)
magnesium chloride (Mg-fertilizer solution)	$MgCl_2(\cdot 6H_2O)$	from 8	readily soluble in water (up to 75%)
magnesium oxide	MgO (>70%)	from 42	slightly water-soluble powder
magnesium rock powder	Mg-silicates	12	slightly soluble, very slow-acting
Mg-limestones limestones containing Mg	$MgCO_3$, etc.	.	*see* limestone fertilizers
NMg-fertilizers KMg-fertilizers MgCu-fertilizers	$MgSO_4$, $MgCO_3$ $MgSO_4$, $MgCl_2$ $MgSO_4$	5 3 to 8 13 to 16	*see* two-nutrient fertilizers (Chap. 3.2.1) *see* Cu-fertilizers (Chap. 3.1.4)

Slightly Soluble Mg-Fertilizers

The solubility of fertilizers in this group varies within wide limits.

Magnesium oxide (MgO) must be classified as only slightly water-soluble. However, it does slowly dissolve and therefore also seems to be suitable as a foliar nutrient. Thus, *Maneltra-Mg* fertilizer contains MgO combined with kaolin as a dispersing agent and a wetting agent to form a slow-acting foliar nutrient.

Magnesium limestones, which react basically, should be considered mainly as soil dressing. Their chief function, apart from the supply of magnesium, is to increase soil reaction. The following substances should be considered for this purpose:

• Magnesium oxide (MgO)   } relatively well converted in the soil,
• magnesium hydroxide (Mg(OH)$_2$) } acts moderately quickliy
• magnesium carbonate (MgCO$_3$):   acts slowly
• magnesium silicate in easily decomposable form: acts very slowly.

Because of the close relationship of Mg-limestones to other limestone, their basic action will be discussed later (Chap. 4.1.1).

Mg-rock powder consists of finely ground siliceous rocks rich in magnesium. The silicate need not be easily decomposable, but fine grinding is a precondition for obtaining any, even a slight, effect; fineness of grinding: 65% smaller than 0.032 mm. This fertilizer must be designated as acting particularly slowly.

There is another group of compounds that belong to the slow-acting Mg-fertilizers, but for certain reasons they still play practically no role, e.g., magnesium-ammonium phosphate ($MgNH_4PO_4$). Fertilizers with a Mg-content of less than 2.4% (= 4% MgO) are not considered to be Mg-fertilizers, but their Mg-content may be indicated separately. Thus, Thomas phosphate contains approximately 2% Mg in easily decomposed siliceous form.

2.4.2 Application of Mg-Fertilizers

The use of magnesium in fertilization has gained in importance, since magnesium deficiency has occurred more frequently in soils in humid climates with more intensive NPK-fertilization. This increased deficiency, which is partly acute (visible) and partly latent (hardly recognizable with the unaided eye) is due to increased removal, losses by leaching (magnesium deficiency especially in acid soils), reduced supply as minor constituents of other fertilizers, and antagonistic decrease in Mg-intake (e.g., reduced Mg-intake due to large potassium supply).

Choice of Correct Fertilizer Form

The optimal fertilizer form depends primarily on the lime requirements of the soil:

- *Soils requiring lime* are supplied with Mg-fertilizers in the cheapest way by means of Mg-containing limestones, as their application for liming requires no additional expenditure;
- *soils not requiring lime* can be fertilized with neutrally reacting, water-soluble magnesium salts (e.g., kieserite) or other Mg-containing fertilizers without lime effect. These fertilizers are approximately equivalent as regards Mg-supply.

The choice furthermore depends on

- *the speed of action:* spraying the leaves acts quickest (Chap. 6.1.4), but soil dressing with water-soluble Mg-salts also acts rapidly;
- the *possibilities of combination* with other nutrients, e.g., for potatoes the combination of potash and magnesium in sulfate form (potassium magnesia), or for grassland the combination of N, Mg, and Cu (Chap. 3.1.4).

Other side effects of Mg-fertilizers are of subordinate importance. Only the application of larger amounts of Mg-chloride to plants sensitive to *chloride* should be avoided.

Practical Application of Mg-Fertilizers

Mg-fertilizers are relatively simple to apply. Water-soluble forms can be placed on the soil surface, they then penetrate slowly with the water into the root region. Leaf fertilization is indicated in cases of acute deficiency, or may be indicated as a preventive measure. Maximum concentrations should be taken into account in this case (Chap. 6.1.4).

The required *Mg-quantity* is discussed in Chapter 5.4.4, and consideration of the fertilization of various plants in Chapters 7 and 8. There is little danger of overfertilizing with magnesium, and Mg-fertilizers are relatively cheap; magnesium may therefore be applied in excess. This greatly facilitates correct dosage. Extremely high, detrimental Mg-levels only rarely occur in plants, even on salty soils particularly rich in Mg.

The *utilization rate* of Mg-fertilizers approximately corresponds to that of K-fertilizers but decreases with increasing potassium supplies. As already discussed with reference to potassium, there are certain unavoidable losses by leaching in the case of magnesium too, but their effects are essentially unimportant.

2.5 Calcium and Sulfate Fertilizers

2.5.1 Calcium Fertilizers (Ca-Fertilizers)

Ca-fertilizers are chemical substances containing the nutrient element calcium in the nutrient form of absorbable calcium cations or yielding Ca^{2+}-cations after conversion.

Ca-fertilizers primarily serve for improving plant nutrition with calcium. However, they only play a minor role, since calcium is not deficient in most sites. Liming materials are also Ca-fertilizers but are seldom used as such, being mainly applied to increase soil reaction (Chap. 4.1.1).

Important designations are:

Ca chemical symbol of the element calcium; reference basis for calcium contents;

Ca^{2+} calcium ion; bivalent cation occurring in soils, plants, and fertilizers;

CaO calcium oxide (quicklime); suitable reference basis and active constituent for lime effect.

Origin and Production

The raw material of Ca-fertilizers is lime found in nature. It is either processed directly into Ca-fertilizers (Chap. 4.1.1) or yields Ca-fertilizers after conversion in chemical processes.

Global reserves of calcium are large, since whole mountain ranges consist of limestone ($CaCO_3$).

Ca-Fertilizers and their Properties

Water-soluble Ca-fertilizers are chiefly used for supplying additional calcium to plants. *Calcium chloride* in solid form, e.g., as $CaCl_2 \cdot 6H_2O$ contains 18% Ca according to the formula (at least 15% as fertilizer). It is highly water-soluble (maximum concentration in solution = 84%), and can therefore be easily dissolved for application as a foliar nutrient in solutions containing a maximum of 15% Ca.

Calcium-chloride solution containing at least 10% Ca is marketed ready for use. Application as a foliar nutrient or for spraying fruit (e.g., apples) for better preservation, requires a highly purified product, i.e., calcium chloride satisfying the rules laid down in the German Pharmacopoeia.

Calcium nitrate is another water-soluble Ca-fertilizer (20% Ca); however, it is justifiably classified as an N-fertilizer (Chap. 2.1.2).

The Ca-component is especially emphasized in some multiple nutrient fertilizers, e.g., type 2 Wuxal fertilizer suspension (Chap. 3.3.3).

Slightly water-soluble fertilizers are:

- *Calcium sulfate*, mostly used as gypsum ($CaSO_4 \cdot 2H_2O$, containing 23% Ca). This is actually a CaS two-nutrient fertilizer; its importance lies less in the replenishment of supplies for plants, than in supplying Ca-ions to the soil for improving its structure. Gypsum has a maximum concentration of 0.26% in aqueous solution.
- *Liming materials* of various types (Chap. 4.1.1), which improve Ca-supplies to the soil in addition to increasing the pH.
- Other Ca-containing fertilizers, e.g., phosphates.

Use of Ca-Fertilizers

Use of Ca-fertilizers for additional supplies of this major nutrient element to plants is far less important quantitatively than supplies of the nutrients already discussed. On the other hand, fertilization with Ca for traditional purposes is increasing, and is being extended to unexpected new areas.

Calcium is only slightly mobile in plants. Proper supplies to large reserve organs are particularly difficult to provide. The *brown-spot disease* of apples, caused by Ca-deficiency, is the best known example, but deficiency in head cabbage (Chinese cabbage) is gaining importance.

Ca-deficiency in plants is rarely caused by a shortage of available reserves in the soil, except for highly acid soils, and thus cannot usually be eliminated by replenishing supplies in the soil. *Leaf fertilization* therefore plays a decisive role as the only practical possibility of providing adequate additional Ca-supplies. Calcium chloride or nitrate may be applied as foliar nutrients. Nitrate offers the advantage of simultaneous N-supply. However, quality considerations exclude nitrate from use for spraying fruit (e. g., apples). The concentration of solutions used as foliar nutrient should not exceed 1 to 2%, to avoid damage to leaves (Chap. 6.1.4).

2.5.2 Sulfate Fertilizers (S-Fertilizers)

S-fertilizers are chemical substances containing the nutrient element sulfur in the nutrient form of absorbable sulfate anions or yielding such sulfate anions after conversion.

S-fertilizers serve to improve plant nutrition with sulfur. The sulfur requirements of plants are approximately two thirds of their phosphorus requirements. Nevertheless, S-fertilization is of far less practical importance than P-fertilization, because substantial S-supplies occur as minor constituents of various N-, P-, and K-fertilizers.

Important chemical designations are:

S   chemical symbol of the element sulfur, reference basis for sulfur contents;

SO_4^{2-} sulfate anion; anion of sulfuric acid, occurs in soils, plants, and fertilizers.

Origin and Production

Sulfates occur abundantly in salt deposits and mountain formations (gypsum). The processing of sulfates has already been discussed in connection with various fertilizers. Elemental (yellow) sulfur, of volcanic or other origin, also plays a role as a fertilizer.

S-Fertilizers and their Properties

S-fertilizers are predominantly sulfates; some of which are easily, and some only slightly, soluble in water [40]. Sulfate-containing N-, K-, and Mg-fertilizers have already been discussed in the relevant chapters. All these are readily soluble in water (Table 2–10).

Gypsum is a CaS-fertilizer (Chap. 2.5.1), which, because of its slight solubility in water is slow-acting. The same applies to superphosphate whose sulfate component largely consists of anhydrite (anhydrous calcium sulfate).

Aluminum sulfate is primarily used as a soil-improving agent in view of its soil-acidifying properties (Chap. 4.1.3). *Elemental sulfur* is also an S-fertilizer with a strong soil-acidifying action. It can be used directly for fertilization or as

Table 2-10. Selection of sulfate fertilizers.

Fertilizer	Formula	S-Content in %
ammonium sulfate	$(NH_4)_2SO_4$	24
ammonium sulfate nitrate	$(NH_4)_2SO_4 \cdot NH_4NO_3$	15
potassium sulfate	K_2SO_4	18
potash magnesia	$K_2SO_4 \cdot MgSO_4$	23
magnesium sulfate	$MgSO_4 \cdot 7 H_2O$	13
superphosphate	phosphate + $CaSO_4$	12
gypsum	$CaSO_4 \cdot 2 H_2O$	18
aluminum sulfate	$Al_2(SO_4)_3 \cdot 18 H_2O$	14

an additive to other solid fertilizers, e. g., sulfate-coated urea (Chap. 2.1.3) or fertilizer solution.

Use of Sulfate Fertilizers

Little attention is devoted to sulfate in practical fertilization, since the S-requirements of plants are largely provided from various sources without precise fertilization. Thus, sulfate is carried by the wind from the sea to fields near the coast. Fields near industrial zones are supplied with 10 to 30 kg/ha of S annually from sulfur-dioxide waste gases. Moreover, considerable amounts of sulfur can enter the soil through ordinary fertilization.

Fertilization with sulfur thus only becomes necessary in industrial countries with rising agricultural output (and thus increasing removal of soil constituents), especially in the case of plants with particularly high S-requirements (e. g., rape). S-deficiencies otherwise occur increasingly in countries where urea (as pure N-fertilizer) is replacing ammonium sulfate (NS-fertilizer), which had been used for a long time. No particular problems are involved in the use of sulfate fertilizers, however they have no real role to play as foliar nutrients.

3 Micronutrient Fertilizers, Multiple-Nutrient Fertilizers, etc.

This chapter deals with the properties and principles of application of several groups of fertilizers.

- Micronutrient fertilizers as single-micronutrient fertilizers;
- micronutrient fertilizers as multiple-micronutrient fertilizers (combinations);
- multiple-nutrient fertilizers with major nutrients;
- multiple-nutrient fertilizers with major- and micronutrients;
- fertilizers with beneficial nutrients;
- fertilizers with additional substances important to man and animal.

3.1 Micronutrient Fertilizers

Higher green plants are known to require seven *micronutrients,* but this number may have to be increased in future (Chap. 3.4.1). We distinguish between the two groups of cation-forming and anion-forming elements (Table 3-1). The unit of reference of nutrient contents is always the element basis in the case of trace elements (Mn, Cu, etc.). Chlorine hardly has any role to play as a micronutrient, and will therefore be discussed only in connection with the beneficial elements

Table 3-1. Essential trace elements (micronutrients).

Name	Symbol	Chemical classification	Absorption as nutrient	
iron	Fe	heavy metal	cation Fe^{2+}	
manganese	Mn	heavy metal	cation Mn^{2+}	
zinc	Zn	heavy metal	cation Zn^{2+}	or metal chelate
copper	Cu	heavy metal	cation Cu^{2+}	
chlorine	Cl	halogen	anion Cl^-	
boron	B	.	anion $H_2BO_3^-$, HBO_3^{2-}	
molybdenum	Mo	heavy metal	anion MoO_4^{2-}	

(Chap. 3.4.1). The following list of references summarizes fertilization with micronutrients [63, 68, 70, 71, 75].

The importance of fertilization with trace elements is increasing. Deficiencies in trace elements at medium yield levels, were formerly limited to poor soils that had often in addition been wrongly treated. However, in recent decades better soils have been increasingly seen to be deficient. It is precisely in intensive high yield agriculture that yield levels are increasingly limited by a, usually latent, deficiency in one micronutrient element. There are various reasons for this (Synopsis 3–1).

Synopsis 3–1. Causes of Increases in Trace-Element Deficiencies.

1. *Changes in plants:*
- Higher yields imply greater removal;
- the lower mobilization capacity of high-yield strains requires a higher nutrient mobility.

2. *Changes in soils:*
- More intensive liming ⎫
- more intensive drainage ⎬ Increased soil reaction and aeration cause greater immobilization of most heavy metals (except molybdenum).
- more intensive working ⎭

3. *Changes in fertilization:*
- Intensive NPK-fertilization has a diluting effect on other nutrients when reserves in the soil are inadequate;
- antagonistic action, due in part to excessive fertilization with major nutrients;
- less minor constituents in N-, P-, and K-fertilizers of higher concentration.

4. *Changes in overall growth conditions:*
- e.g., increase in stress situations owing to spraying with plant protectives.

3.1.1 Iron Fertilizers (Fe-Fertilizers)

The majority of Fe-fertilizers are water-soluble substances, being either salts or organic complexes (chelates). They are predominantly applied as foliar nutrients.

Ferrous sulfate ($FeSO_4 \cdot 7H_2O$, containing 20% Fe) is the simplest water-soluble fertilizer. However, today it plays only a subordinate role, in view of the greater tolerance to chelates in leaf spraying, as well as the smaller extent to which they are fixed when applied as soil dressing.

Fe-chelates (Synopsis 3–2) differ mainly in the stability of the Fe-complex. They are mostly suitable for application as foliar nutrients; however, the various chelates are not equally effective with different plants. Larger differences be-

Synopsis 3–2. Chelate Fertilizers.

Metal chelates are organometallic complexes in which the metal cation is bound on several sides by chelating agents (ligands) like scissors or clamps, and is thus surrounded (the term chelate is derived from the Greek word *chele* — crab's claw).

Metal chelates are used as chelate fertilizers to prevent fixation in the soil of the heavy-metal cations essential for plants, or to facilitate absorption through the leaves. Plants can absorb metal chelates as complete molecules and then metabolise the metal. An ideal chelate would, on one hand, be stable enough to be preserved in deficient soils, thus being protected against fixation, but on the other hand, could be decomposed in the plant after its absorption as a whole molecule, thus permitting further utilization of the iron.

The corrosion danger when they are applied as *foliar nutrients* is less than with salts. The chelates used must be harmless to health, in terms of quality of food for which the plant is used. Some chelates are natural constituents of plants.

Chelates without metal cations, i.e., in their fundamental form, may also have a fertilizing effect if applied as soil dressing. They form complexes with heavy metals from the reserve fraction, which is difficult to mobilize by soil and plants, and thus become micronutrient fertilizers absorbable by plants. This method, however, is not precise and is therefore unreliable.

$$Na_2 \left[\begin{array}{c} O \!\!\!\diagup \!\!\! {}^{\diagdown} C-H_2C \!\!\! \diagdown {}_{N} \overset{H_2C-CH_2}{\diagdown N} {}^{\diagup} CH_2-C \!\! \diagup {}^{O} \\ O \quad H_2C \diagdown {}_{C} Fe {}_{C} \diagup CH_2 \quad O \\ O\!\!=\!\!C {}^{\diagdown} O \quad O {}^{\diagup} C \!\!=\!\! O \end{array} \right]$$

Structural formula of Fe-EDTA (Na-salt)

tween chelates exist when they are applied as soil dressing. Thus, *Fe-EDDHA* remains more available in lime-containing soils than Fe-EDTA, which is effective in acid soils. Some Fe-chelates, apart from many still in the trial stage, are

• Fe-EDTA: iron *e*thylene*d*iamine*te*traa*c*etate (Na-salt), which contains bivalent iron (for the structural formula *see* Synopsis 3–2).
  Examples of commercial products are:
  Fetrilon (BASF) containing 5% Fe;
  Fe-Chelate-Jost containing 9% Fe.
  Fe-EDDHA: iron *e*thylene*d*iamine *d*i*h*droxyplenyl *a*cetate, in which the iron is in trivalent form. An example of a commercial product is *Sequestren* with 6% Fe.

  Fe-fertilizers, that are insoluble or only slightly soluble in water, may serve to replenish soil reserves (if this is considered necessary at all) or as slow-acting foliar nutrients. An example is Fe-oxalate ($Fe(COO)_2$); as *Maneltra-Iron* containing 22% Fe it is intended chiefly as a foliar nutrient.

Fe-oxides in principle are also Fe-fertilizers but must first be mobilized, like Mn-oxides (Chap. 3.1.2). *Metallic iron*, Fe-powder, might be suitable as fertilizer, since it is converted into Fe-compounds in the soil. A special sort of application is the insertion of iron nails into tree trunks to achieve additional Fe-nutrition of trees.

Iron is a minor constituent of several fertilizers, but its supply is seldom of practical importance. The admixture of Fe-chelates to other fertilizers is discussed later (Chap. 3.3).

Fertilization with Iron
Fe-fertilization is problematic, since the deficiency is usually not due to soil impoverishment, but to fixation despite very large reserves in the soil. (At contents of 1 to 4%, iron is a major constituent of soils.) *Fe-deficiency* is characteristic of soils with high soil reaction in arid zones, where *lime chloroses* caused by iron deficiency occur; sometimes there are other factors as well. It is remarkable that iron deficiency, or at least acute iron deficiency, is rare in lime-containing soils in humid climates. This indicates that Fe-reserves are reasonably easily mobilized. Fe-deficiencies are more frequent only in soils extremely poor in iron (e.g., high moors) or in artificial substrates (certain horticultural substrates, hydroponics), and should then be corrected by fertilization. On the other hand, Fe-deficiency may cause ornamental effects; pronounced green veins in yellow-green leaves of flowering plants and thus serve to promote sales.

In addition to supplying iron to deficient soils, it is also advisable *to mobilize* the iron in the soil itself through acid N-fertilization (Chap. 3.1.2). Many Fe-fertilizers are strongly fixed in deficient soils, so that *leaf spraying* plays an important role in Fe-supply. However, this requires repeated applications and large amounts of Fe-chelates are used in horticulture. The amounts required for leaf fertilization and their concentrations are discussed in Chap. 6.1.4.

Fe-removal amounts to a few kg/ha per year. However, the balance is of quite secondary importance for soils, since the mobility of iron in the soil itself determines availability.

3.1.2 Manganese Fertilizers (Mn-Fertilizers)

Mn-fertilizers are available in water-soluble and water-insoluble forms each of which serve different purposes. *Manganous sulfate* is the best known water-soluble fertilizer. This pink salt is used in two forms with different Mn-contents, depending on the amount of water of crystallization (Table 3-2). It dissolves in water, forming a solution of 30% maximum concentration (based on $MnSO_4$), and is suitable for leaf fertilization. It can also be used as soil dressing but is easily fixed in deficient soils when the pH is high.

Fertilization is also possible with other readily soluble salts, e. g., manganous nitrate ($Mn(NO_3)_2$) and manganous chloride ($MnCl_2$); but these offer hardly any advantages over the sulfate.

Mn-chelates (corresponding to Fe-chelates) are now playing an increasing role, especially outside Germany.

Table 3–2. Selection of manganese fertilizers.

Fertilizer	Formula	Mn-content in %
manganous sulfate	$MnSO_4 \cdot 4\,H_2O$	24
manganous sulfate (monohydrate)	$MnSO_4 \cdot H_2O$	32
Mn-chelate	Mn-EDTA	13
manganous oxide (e.g., Maneltra-Mn)	MnO	48
granulated Jost manganese fertilizer	Mn, MnO, Mn_2O_3, etc.	20

The largest number of *Mn-fertilizers not soluble in water* consist of various Mn-oxides (MnO, Mn_2O_3, MnO_2) and metallic manganese. An example is Jost Mn-fertilizer which contains about 0.5% Zn and Cu in addition to Mn. Other Mn-compounds, e. g., Mn-carbonate or Mn-phosphate, can also be used as fertilizers after suitable processing. Mn-fertilizers not soluble in water can be utilized by plants only after mobilization (reduction under acid conditions), e. g.,

$$MnO_2 \quad +\,4\,H^+ + 2\ \text{electrons} \longrightarrow \quad Mn^{2+} + 2\,H_2O$$

Mn-dioxide   acid   reducing   Mn-ions
conditions

Mn-oxides serve mainly as soil dressing to replenish reserves. Especially easily mobilizable forms may be applied as slow-acting foliar nutrients if suitably processed, e. g., *Maneltra-Mn*.

Manganese also occurs as a minor constituent of Thomas phosphate (2 to 4%) and blast furnace lime (1 to 5%). Manganese as an admixture is commonly added to single- and multiple-nutrient fertilizers in the form of manganous sulfate (Chap. 3.3).

Fertilization with Manganese

Mn-fertilization (like Fe-fertilization) is problematic since deficiency is usually not due to soil impoverishment, but to fixation of the frequently abundant reserves in the soil. *Mn-deficiency* is characteristic of many neutral to alkaline soils. It frequently occurs in acute visible form in over-limed sites in sandy soils and in lime-containing low moors (e. g., as grey-speck disease of oats). Mn-deficiency also occurs increasingly as a consequence of more intensive utilization, e. g., through increased liming at pH-values exceeding 6, and through improved soil aeration, which brings about higher oxidation instead of Mn-reducing conditions. On the other hand, intensified liming of medium to heavy soils is necessary, at least for reasons of the structural improvement of soils, and this unfortunately necessarily promotes a diminished Mn-supply.

Moreover, since drought increases manganese fixation, fertilization of deficient soils only leads to transient or minor results. The fertilizer is most often either fixed rapidly, or mobilized only to a small extent. Depot fertilization thus largely replenishes reserves, and only in special cases is it effective in deficient soils for several years.

Improvement of *Mn-supplies* from the soil through the use of soil-acidifying nitrogen fertilizer (Fig. 2.2, p. 59), compaction (rolling) of loose soils, prevention of excessive drying-out, or supply of easily decomposed organic substances, whose conversion creates reducing conditions and thus releases manganese, seems to be more important than fertilization with manganese. These measures alone sometimes permit elimination of at least slight deficiencies, but should accompany fertilization with manganese.

Mn-removal is 300 to 500 g/ha per year and reaches 1 kg/ha at very high yields. However, consideration of the Mn-balance is unimportant, since there are large Mn-reserves in nearly all soils. More important than comparisons of removal and supply is mobilization of the manganese in the soil.

The fertilizer quantities applied as soil dressing are 10 to 30 kg/ha in cases of deficiency. However, far smaller amounts are sufficient for continuous replenishment, (if required) i. e., approximately the quantity removed. Its application as a foliar nutrient is considered in Chap. 6.1.4.

3.1.3 Zinc Fertilizers (Zn-Fertilizers)

Zn-fertilizers in water-soluble or water-insoluble forms play an important role, especially in regions with frequent Zn-deficiency.

Zinc sulfate is the simplest form of the water-soluble Zn-fertilizers. This whitish salt has different zinc contents depending on the amount of water of crystallization (Table 3–3). It dissolves in water to form a solution of 35% maximum concentration (referred to $ZnSO_4$). Applied as a foliar nutrient, it easily causes corrosion damage, since it has a relatively acidic action. *Basic zinc sulfate* or *Zn-chelates* (corresponding to Fe-chelates) are therefore better than foliar nutrients.

Table 3-3. Selection of zinc fertilizers.

Fertilizer	Formula	Zn-content in %
zinc sulfate	$ZnSO_4 \cdot 7\,H_2O$	23
zinc sulfate	$ZnSO_4 \cdot H_2O$	36
basic zinc sulfate	$ZnSO_4 \cdot 4\,Zn(OH)_2$	55
Zn-chelate	Zn-EDTA	14
zinc oxide (e.g., Maneltra-Zn)	ZnO	70
Excello (Cu-fertilizer)	Zn, ZnO, etc.	5

Water-insoluble Zn-fertilizers above all, are Zn-oxide and Excello (copper fertilizer). Zn-oxide is only slightly soluble in water (2 mg ZnO in one liter of CO_2-containing water), and is therefore used as a slow-acting foliar nutrient *(Maneltra-Zinc)*. *Excello fertilizers* have a substantial zinc content (e.g., 5%), and are thus a combination of zinc and copper fertilizers (Chap. 3.1.4).

Significant amounts of zinc are practically absent as minor constituents of other mineral fertilizers. Zinc is added as an admixture to other fertilizers, mainly in the form of zinc sulfate (Chap. 3.3). Garbage and sewage-sludge products are organic fertilizers which frequently contain considerable amounts of zinc, and can thus be used as zinc fertilizers.

Fertilization with Zinc

Zn-deficiency (e.g., of maize or fruit trees) is characteristic of regions with intensive sunlight and lime-containing soil that frequently dries out. However, even slight deficiencies could increasingly play a role in the Central European climate with high grain yields. Nevertheless, zinc as a minimum factor is less important than manganese or copper.

Zn-supplies should primarily bridge critical periods, so that sufficient available zinc is present in the soil during dry periods. Improvement of natural Zn-supplies by the application of acid N-fertilizers is advisable in deficient soils.

Zn-removal is approximately 100 to 400 g/ha per year. However, balancing of deficiencies is of minor importance (as in the case of Fe and Mn). On the other hand, the balance should be taken into account from the aspect of Zn-toxicity in cases of large quantities (Chap. 6.2.1).

The *fertilizer quantities* applied as soil dressing are 10 to 20 kg/ha where there is serious deficiency. However, far smaller amounts are required for continuous replenishment, if required (approximately the quantity removed). Its application as a foliar nutrient is discussed in Chap. 6.1.4.

Pickling with 100 g zinc sulphate per dt of seeds may be indicated in case of slight zinc deficiency of young plants [63]. Another method is to immerse the roots, e.g., of vegetable plants, in zinc oxide paste. A special procedure to be mentioned is the occasional direct fertilization of trees with metallic zinc, i.e., insertion of a zinc nail into the trunk, so that plant saps can then convert the metal into soluble salts (caution against poisoning the tree!).

3.1.4 Copper Fertilizers (Cu-Fertilizers)

Cu-fertilizers in water-soluble and water-insoluble forms have been used for a long time to correct copper deficiencies.

Copper sulfate (e.g., blue vitriol) is the oldest Cu-fertilizer. Its Cu-content varies depending on the amount of water of crystallization (Table 3-4). It dissolves in water to form solutions of 17% maximum concentration (referred to $CuSO_4$). Copper sulfate can be applied as soil dressing or foliar nutrient. However, its acid side-effects are liable to cause corrosion damage when applied as foliar nutrient. Less caustic agents are therefore safer to use, i.e., *green copper* (copper calx) or Cu-chelates. Copper sulfate acts rapidly if applied as soil dressing but is difficult to distribute because of the small amounts needed and is therefore used mostly as an admixture.

Table 3-4. Selection of copper fertilizers (and fertilizers with Cu-additives).

Nutrient element	Fertilizer	Formula of Cu-component	Cu-content in %	Other nutrient elements in %
Cu	copper sulfate (blue)	$CuSO_4 \cdot 5\,H_2O$	25	
	copper sulfate	$CuSO_4 \cdot H_2O$	36	
	green copper (copper calx) (e.g., Cupravit OB 21)	$Cu_2Cl(OH)_3$	48	
	Excello (e.g., granulated) ⎫	Cu, CuO	2.7 or 5	Zn = 5
	Excello 25% ⎬	Cu-silicates	25	Zn = 15
	Urania copper fertilizer ⎫	Cu, CuO,	2.5 or 5	Co
	granulate ⎬	Cu-silicates		
	cupric oxide, e.g., Maneltra Cu	CuO	71	
Mg + Cu	Excello-magnesium	(like Excello) ⎱ Mg as	1.7	Mg = 16 Zn = 1
	Urania copper kieserite granulate	(like Urania) ⎰ kieserite	2.5	Mg = 13
N + Mg + Cu	nitrogen magnesia	$CuSO_4$	0.2	N = 20 Mg = 5
NPK + Cu	NPK-fertilizer	$CuSO_4$	0.1	

Water-insoluble copper fertilizers are Excello and Urania, as well as cupric oxide and (outside Germany) basic copper salts.

Excello copper fertilizer powder or granulate is a by-product of brass production and contains copper in metallic, oxide, and silicate form in differing amounts (metal-alloy fertilizer). As it has a relatively high Zn-content (5% Zn), it is also used as a Zn-fertilizer (Chap. 3.1.3).

Urania ground copper slag or fertilizer granulate is a by-product of copper smelting and also contains copper in metallic, oxide, and silicate form. Both fertilizers act slowly, since after mobilization by soil agents they must first be converted into absorbable nutrient forms by plants (e.g., like the Mn-oxides referred to in Chap. 3.1.2). Basic oxides and salts are thus formed from metallic copper, from which the nutrient Cu^{2+} enters into solution.

Decomposition is surface-dependent, so that these fertilizers are ground very finely (70% finer than 1.6 mm); they are sometimes granulated again to improve application (dust-free bonding).

Cupric oxide can be applied as soil dressing and, with suitable additives, as a slow-acting foliar nutrient, with depot action *(Maneltra-copper)*.

Copper hardly ever occurs in significant quantities as a minor constituent of mineral fertilizers. Used as an admixture to other fertilizers, copper is often added in the form of copper sulfate to NPK-fertilizers; it is sometimes added in water-insoluble form, as in MgCu-fertilizers.

Garbage and sewage-sludge products are organic fertilizers that frequently contain considerable Cu and can thus be used as copper fertilizers.

Considered from the long term, the various Cu-fertilizer forms are approximately equivalent as regards Cu-action, if worked into the soil. However, certain differences in effect are due to minor constituents. Thus, Excello has a large Zn-content; this is a considerable advantage in cases of combined Cu- and Zn-deficiency. Urania ground copper slag also contains 0.05% cobalt which can promote cobalt supplies to grassland (Chap. 3.4.2).

Fertilization with Copper

Cu-deficiency is characteristic of boggy soils and was formerly known as *heathbog disease* of oats. Slight, latent deficiency is prevalent today with high yields even on good mineral soils, sometimes limiting these as a minimum factor. Cu-deficiency is intensified at higher pH-values and during droughts, however, the pH dependence is less pronounced than that of manganese.

Fertilization with copper poses few problems, since depot fertilization for longer periods is quite effective in most soils. Copper deficiency is thus corrected far more easily than is manganese deficiency.

Copper removal is 30 to 100 g/ha per year. However, the balance is of little significance with respect to deficiencies. On the other hand, the balance should be taken into account from the aspect of Cu-toxicity in cases of large supplies, e.g., by plant protectives and organic fertilizers (Chap. 6.2.1).

The *fertilizer quantities* to be applied for depot fertilization are approximately 5 kg/ha, or 10 kg/ha for the correction of strong deficiencies. However, much smaller amounts are required for continuous replenishment, if necessary, i.e., approximately the amount removed, with a certain safety margin.

Utilization of Cu-fertilizers, like all micronutrient fertilizers, is very low, being about 0.5 to 5% per year. Its application as a foliar nutrient is discussed in Chap. 6.1.4.

Copper is highly immobile in the soil. The fertilizer should therefore be well mixed with the top soil of arable fields. Water-soluble forms penetrate fairly quickly into the upper few cm in grassland, whereas water-insoluble fertilizers penetrate deeper only after natural mixing with the soil.

3.1.5 Boron Fertilizers (B-Fertilizers)

Historically, borax was of considerable significance as a B-fertilizer. Some of the effectiveness of Chile saltpeter, formerly very popular in the cultivation of sugar beets, might be due to the natural admixture of borax (0.05 to 0.1% in the fertilizer).

Mention should therefore first be made of borax as water-soluble boron fertilizer (Table 3-5). This white, gritty salt can be applied as a single fertilizer in

Table 3-5. Selection of B-fertilizers (and fertilizers with B-additves).

Nutrient element	Fertilizer	Formula	B-content in %
B	borax (= Na-tetraborate)	$Na_2B_4O_7 \cdot 10\,H_2O$	11
	dehybor (= anhydrous borax)	$Na_2B_4O_7$	22
	boric acid	H_3BO_3	18
	solubor, polyborate	$Na_2B_8O_{13} \cdot 4\,H_2O$	21
	colemanite, pure	$Ca_2B_6O_{11} \cdot 5\,H_2O$	16
	colemanite fertilizer	colemanite, etc.	9–14
	fritted B-silicate	.	10–15
N + B	ammonium sulfate nitrate with boron (Rustica)	borax	0.2
P + B	boron superphosphate	borax	0.5
	boron-Novaphos	borax	0.5
P + K + B	Bor-Rhe-Ka-Phos	borax	0.25
NPK + B	NPK-fertilizers	borax	0.1

the form of soil dressing or foliar nutrient, but because of its better distribution possibilities is predominantly used as an additive to other fertilizers such as P-fertilizers or multiple-nutrient fertilizers. It dissolves in water to form solutions of only 2.6% maximum concentration, referred to $Na_2B_4O_7$.

Boric acid as a white crystalline powder, is also a useful B-fertilizer. Applied as a foliar nutrient, it offers the additional advantage of higher solubility, as solutions of up to almost 5% concentrations are possible. On the other hand, boric acid is relatively toxic (the lethal dose for human beings is approximately 8 g). Even higher B-concentrations in solutions can be obtained with *polyborates* and similar compounds; they are used as additives in nutrient solutions.

Water-insoluble or slightly water-soluble B-fertilizers, which can be mobilized in the soil, however, are important, because their application involves no risk of B-toxicity for plants with small B-requirements. On the other hand, a disadvantage is their poor initial activity. The substances used, especially in the USA, are the B-mineral *colemanite* and *fritted B-silicate* (finely ground glass containing boron). Colemanite also plays a role as a slow-acting, not too corrosive foliar nutrient.

Organic fertilizers containing substantial amounts of boron (100 ppm or more) such as garbage, composts with high ash content are thus boron fertilizers, but their use involves the risk of overdose.

Fertilization with Boron

B-deficiency occurs in acute form during dry periods, especially in light soils that have a high pH, due to liming. Little is known about latent deficiencies in high-yield production. However, boron might play a role as a minimum factor, e.g., in case of high rape yields.

The main problem in B-fertilization is the *different requirements* of crops. Some plants with large requirements are followed in crop rotation by plants with low requirements. In addition, the latter are highly sensitive to B-excess.

The groups of different B-requirements are

- plants with *large* B-requirements, e.g., beta beets, Swede turnip, cauliflower, celery, kohlrabi, rape, legumes;
- plants with 6 to 8 times *smaller* requirements, e.g., grain and all grasses.

Within a broad range, overdosage with the usual water-soluble B-fertilizers causes no harm to fertilized plants with large B-requirements, e.g., sugar beets, but may cause boron toxicity in wheat or barley, for example, as an after-effect in the following year. The purpose of fertilization with boron is therefore not to render the soil capable of supplying all plants abundantly but rather precisely to improve the supply to plants with large requirements.

B-removal is about 50g/ha per year in the case of plants with small requirements, and about 500g/ha in the case of plants with large requirements.

It is almost only plants with large requirements that are fertilized, when necessary, at amounts of 2 to 3 kg/ha B. Amounts of 4 kg or more involve the risk of after-effect toxicity, especially in acid soils. In contrast to other trace elements, continuous replenishment at the level of removal makes less sense than the recommended precise fertilization. The *utilization rate* of B-fertilizers varies between 2 and 20%. *Additions of boron* to other fertilizers such as P-fertilizers and multiple-nutrient fertilizers, are calculated such that the required quantities of boron are supplied with the customary N-doses, for example.

Slow-acting B-fertilizers have no great significance despite the occasional *damage* caused by toxic after-effects in case of overdosage (e. g., of borax), since water-soluble B-fertilizers permit effective, precise use.

Water-soluble B-fertilizers are converted in the soil into B-forms existing in the soil (e. g., Ca-borate) at varying speeds, more rapidly at higher pH-values. They are thus largely immobilized and partly fixed. This *conversion* of the bound form is the reason toxicity damage is not more widespread. However, this also explains why utilization in deficient soils is low, and renewed fertilization becomes necessary. Its application as a foliar nutrient is discussed in Chap. 6.1.4.

3.1.6 Molybdenum Fertilizers (Mo-Fertilizers)

Mo-fertilizers are required only in small amounts. However, they play an important role under certain production conditions (Table 3-6).

Table 3-6. Selection of Mo-fertilizers.

Fertilizer	Formula	Mo-content in %
sodium molybdate	$Na_2MoO_4 \cdot 2H_2O$	40
ammonium molybdate	$(NH_4)_6Mo_7O_{24} \cdot 4H_2O$	54
	$(=3(NH_4)_2MoO_4 \cdot 4MoO_3 \cdot 4H_2O)$	
molybdenum trioxide	MoO_3	66
calcium molybdate	$CaMoO_4$	48

The most important fertilizers are the *water-soluble* salts *sodium molybate* and *ammonium molybdate,* which contain molybdenum in directly utilizable form. They are suitable for application as soil dressing or foliar nutrients, as well as for treating seeds, if possible with adhesive agents. Only small quantities are required as soil dressing. Molybdenum is therefore mostly added to other fertilizers to permit better distribution. Thus, outside W. Germany, 0.05% Mo is added to superphosphate.

Water-insoluble Mo-fertilizers may be Mo-compounds convertible into soluble molybdates, e.g., *Ca-molybdate* or *Mo-oxide* . These fertilizers act slowly, but until now have hardly had any practical significance.

Fertilization with Molybdenum

Mo-deficiency is characteristic of acid soils rich in iron, in which plants are intensively fertilized with nitrate and have little capacity for absorbing molybdenum. Cauliflower is considered to be particularly sensitive. Deficiencies occasionally occur only temporarily in a critical phase, and can be avoided by proper supply to the seed or young plants in the nursery bed.

Mo-removal is low at 5 to 20 g/ha per year. The *amount of fertilizer* to be supplied is about 0.5 kg/ha Mo when there is a slight deficiency or for replenishment, and about 1 kg/ha in cases of pronounced deficiency. These quantities must be doubled on sites with greater Mo-fixation (2 kg/ha). Fertilization of nursery beds (e.g., cauliflower) requires 0.5 g Mo per m^2 bed area or 2 g per m^3 nursery soil, in order to provide abundant initial supplies. Slight deficiencies in grain are corrected, e.g., in tropical regions, by soaking the seeds for 8 hours in a 0.5% solution of Na-molybdate. Its application as a foliar nutrient is discussed in Chap. 6.1.4.

Mo-deficiency can also be corrected by indirect measures intended to increase Mo-availability, e.g., liming or intensified loosening and drainage of the soil, since Mo-compounds are best available in the most highly oxidized form.

3.1.7 Combinations of Micronutrients

Frequently, plants lack not only one, but several micronutrients. It therefore seems that combinations of all, or at least all important, trace elements be developed. These are termed multiple-micronutrient fertilizers.

Their advantage is their broad spectrum of action. They are intended not only to improve supplies of the minimum factor in question but also to fill other supply gaps that might subsequently occur.

Trace element combinations are useful aids to fertilization, as long as the state of supplies in a field is not known (Chap. 5. 3) , or if multiple deficiencies are to be eliminated. A method, which in principle is superior to the application of combination fertilizers, is the precise supply of the nutrient element representing the minimum factor in the case considered. This precise supply can above all be abundant and might thus be more effective.

Some combinations have already been discussed under the individual micronutrient fertilizers, since they are usually designated according to the trace elements most important for fertilization. An example is Excello, which according to its composition is a copper-zinc fertilizer.

This section deals only with "pure" micronutrient combinations. Additional combinations with major nutrient elements, e.g., magnesium, will be considered in Chap. 3.3.

Micronutrient Combinations as Depot Fertilizers

In theory it should be possible to improve a soil so as to form a nutrient substrate by the supply of water-insoluble but easily mobilizable micronutrients. This nutrient substrate would supply sufficient nutrients to the plants without risk of damage through excesses. This is in fact realized in the best soils.

However, this would incur very high expenditure for most agricultural lands. Moreover, since supplies of only one micronutrient are generally deficient, precise application of individual micronutrient fertilizers is therefore indicated so that the minimum factor in question can be improved. Exceptions are the use of larger quantities of town wastes, e.g., garbage compost, which may also be considered to be a multiple-micronutrient fertilizer (Chap. 4.3.3).

Depot Fertilizers for Horticulture

The situation is different with regard to supplying horticultural substrates. For ornamental plants especially, it may be advisable to add long-acting micronutrient fertilizers to the nutrient substrate, which sometimes is very poor in all nutrients, so that only one single fertilization during planting is required. It is then obviously necessary to prevent toxic excesses. This can be achieved by using depot fertilizers with slow action, as has already been discussed with reference to nitrogen in Chap. 2.1.4.

An example of such a fertilizer (based on Excello-Copper) is *Radigen*, manufactured by Jost. Its composition is as follows: Mg and Cu 1.5% respectively; Fe 2%; Mn, B, Mo 0.6 to 0.8% respectively; Zn 0.2%. The recommended quantity to be applied is 1 g Radigen per 10 l (peat) nutrient substrate.

Trace-Element Combinations for Seed Treatment

Rapid initiation of growth forms the basis of high yields. Proper nutrient supplies should therefore be ensured, especially to young plants, in particular supplies of trace elements. In this way young plants overcome the initial difficult growth phase more rapidly and can precisely during this period better resist harmful effects (Chap. 6.5).

A germinating plant obtains certain supplies from the seed grain itself: this amounts to about 1 g/ha for copper, i.e., only 1 to 2% of the requirement. It is possible to make additional supplies available to young plants by the placement of fertilizer around the seed grain, if the soil does not have sufficient supplies of available trace elements. The procedure is *to pickle* the seed with agents containing the micronutrient, e.g., with *Dynamal*, produced by Albert. Boron, copper, and manganese compounds are added to the seed dressing and then adhere to the seed grain, and are thus in easy reach of the radicle. Up to 80 g/ha of Cu can thus be supplied to grain (this represents a considerable supply), but only about 20 g boron to beets, which is relatively little. Manganese supplies might be less important since mobilization conditions in the soil are decisive. Dynamal thus offers only limited overall possibilities for additional supplies of trace elements, with the best results obtained in cases of copper deficiency. Several forms should be available for selection according to the purpose of application.

The addition of molybdenum appears to be promising, since small amounts are sufficient to cover the very slight requirements.

3.2 Multiple-Nutrient Fertilizers with Major Nutrients

3.2.1 Two-nutrient fertilizers

By the appropriate combination of six major nutrient elements, 15 two-nutrient fertilizers can theoretically be realized. However, calcium and sulfate are of minor importance as single-nutrient fertilizers. It therefore appears more purposeful to use only the four major nutrient elements that on a practical basis are the most important. This yields six two-nutrient fertilizers, in the narrow sense (Synopsis 3-3).

Synopsis 3-3. 15 possible Two-Nutrient Fertilizers formed from the Six Major Nutrients.

NP-fertilizer	PK-fertilizer	KMg-fertilizer	MgCa-fertilizer
NK-fertilizer	PMg-fertilizer	KCa-fertilizer	MgS-fertilizer
NMg-fertilizer	PCa-fertilizer	KS-fertilizer	CaS-fertilizer
NCa-fertilizer	PS-fertilizer		
NS-fertilizer			

The *fertilizer list* only distinguishes between three forms, since potassium and magnesium fertilizers are combined there. This can be attributed to historical reasons, since magnesium became of practical important much later than N, P, and K. All the combinations listed in Synopsis 3-3 are available. However, some have been preferentially classified as single-nutrient fertilizers according to the predominant importance of one nutrient, e.g., calcium nitrate as N-fertilizer and not as NCa-fertilizer.

Multiple-nutrient fertilizers, and thus also two-nutrient fertilizers, are offi-cially designated according to the $(N + P_2O_5 + K_2O)$-contents (Chap. 3.2.3).

N-Containing Two-Nutrient Fertilizers

As indicated above, only three types will be considered in detail, namely, NP-, NK-, and NMg-fertilizers.

NP-Fertilizers

Nitrogen-phosphorus fertilizers are significant from various points of view:

- As starting materials for many NPK-fertilizers;
- for fertilizing soils rich in potassium or supplementing separate fertilization with potassium;
- as fertilizer solutions for liquid fertilization;
- as fertilizers with extremely large nutrient contents *(ultrahigh-analysis fertiliz-ers)*.

For historical reasons, we shall first consider the *guano* fertilizers from natural deposits. They primarily contain nitrogen and phosphate, predominantly in mineral form. However, they are classified as organic fertilizers because of their origin (Chap. 4.3.3).

Mineral NP-fertilizers may be classified into different groups, i.e., solid and liquid fertilizers, or fertilizers with medium or particularly large contents.

The official *indication of contents* is the same as for NPK-fertilizers, i.e., $N + P_2O_5 + 0$. The zero represents the missing potassium; it will be omitted for simplicity in the formulas given below.

The following classification of NP-fertilizers is intended to provide a bet-ter picture of their multiplicity:

a) *Superphosphate mixed fertilizers* or ammonium-containing superphosphate. The composition is either superphosphate with ammonium sulfate, or am-monium phosphate with Ca-phosphates, the P-component being water-sol-uble to at least 50%. One well known fertilizer of this type is $9 + 9 (N + P_2O_5)$.

b) *Ammoniated products* of the acid decomposition of crude phosphates.

In pure form they consist of ammonium phosphates and are thus largely wa-ter-soluble.

- *Monoammonium phosphate* $(NH_4H_2PO_4)$ reacts as an acid. Such a fertilizer type is $11 + 52$.

- *Diammonium phosphate* (($NH_4)_2HPO_4$), whose basic reaction may cause ammonia to split off, depending on the soil conditions (Chap. 2.1.5). One fertilizer of this type is $16+46$.

Combinations of ammonium phosphates and calcium phosphates are the *nitrogen phosphates* ($20+20$). The P-component is water-soluble to at least 30%. Trade names are Cederan-, Complesal-, Enpeka-, Kampka-, Rustica-Nitrogen Phosphate.

c) *Mixed products* consisting of carbamide and soft-earth rock phosphate, e.g., of type $17+17$.

d) *NP-fertilizer solutions* consist of ammonium phosphates and polyphosphates. The P-component present in the form of polyphosphate is limited to 60%. The specific gravity of the solution is 1.4, hence a content of 10% N by weight thus corresponds to 14 kg N per 100 l. The solution remains liquid down to $-17\,°C$ and can be applied both as a soil dressing (100 to 300 l/ha) and foliar nutrient. The following formulation of the fertilizers is important:

- *Enpesol, Praysol:* $10+34$.

A different type ($18+18$) is produced by adding a solution of ammonium nitrate and urea.

e) *Ultrahigh-analysis fertilizers.* Fertilizers with especially large nutrient contents are being developed [66, 73] to save costs in transportation and application and also to permit better P-utilization. Referred to the oxide basis, these fertilizers have total nutrient contents exceeding 100%. Three groups are mainly involved:

- *Linear polyamides,* e.g., phosphoric triamide with the basic formula $(NH_2)_3PO$ and various formulations, e.g., $43+74$ ($N+P_2O_5$).
- *Phosphorus nitriles,* e.g., phosphorus nitrilamide with the cyclic basic formula $(NH_2)_6P_3N_3$ and different formulations, e.g., $48+98$.
- *Cyclic metaphosphimates* with the basic formula $(NH_4)_3P_3N_3O_6H_3$. One type is $27+60$.

NK-Fertilizers

Nitrogen-potassium fertilizers are intended for soil rich in phosphates or for supplementation of separate P-fertilization. All are granulated. An ideal NK-fertilizer would appear to be *potassium nitrate* (13% N, 38% K). It consists of the two chief nutrients absorbed by plants. However, besides its inherent dangers (it is a constituent of explosives), the nutrient ratio of $N:K=1:3$ is not suitable for most fertilization purposes.

Some common NK-fertilizers are listed in Table 3-7. The official indication of contents is the same as for NPK-fertilizers, i.e., $N + 0 + K_2O$. However, the zero indicating the missing P-content has been omitted in Table 3-7.

NK-fertilizers with additives of other nutrients are suitable for liquid fertilization in view of their solubility in water, e.g., type-1 Wuxal suspension.

Table 3-7. Selection of NK-fertilizers.

Fertilizer	Composition	N-K_2O content in %
Nitroka (BASF)	NH_4NO_3, $(NH_4)_2SO_4$	16 + 24
Rustica-nitrogen potassium	KCl, sometimes K_2SO_4	20 + 20
NK-Hakaphos	carbamide, $(NH_4)_2SO_4$, KNO_3	20 + 20
suspension fertilizer	NH_4NO_3, carbamide, K_2SO_4	10 + 20
suspension fertilizer	NH_4NO_3, carbamide, KNO_3	20 + 15

NMg-Fertilizers
Nitrogen-magnesium fertilizers are classified as N-fertilizers in the type list, since they supply Mg within the framework of fertilization with N. The contents are indicated in the type list as "MgO" (8%); this, however, is only sometimes correct; i.e., if there is a lime action. The general type designation *nitrogen magnesia* includes two different fertilizers:

a) *Nitrogen-magnesium sulfate* with the composition $(NH_4)_2SO_4$, NH_4NO_3 ($2/3\,NH_4 - N$), and $MgSO_4$. The contents are 20%N and 5% Mg. This fertilizer is water-soluble and soil-acidifying.

b) *Nitrogen-magnesium carbonate* with the composition $NH_4NO_3(1/2NH_4 - N)$ and dolomite ($CaCO_3 \cdot MgCO_3$). The contents are also 20% N and 5% Mg. The Mg-component of this fertilizer dissolves in acids. This fertilizer corresponds to lime-ammonium nitrate (Chap. 2.1.2) with magnesium (e.g., HOECHST *nitrogen magnesia*)

c) *Mixed nitrogen magnesia*, a blend of a and b. NMg-fertilizers corresponding to N-fertilizers are produced by several firms. Nitrogen magnesia is a typical grassland fertilizer and is treated with 0.2% copper for this purpose.

PK-, PMg-, and KMg-Fertilizers

PK-Fertilizers
Phosphorus-potassium fertilizers are suitable for basic and depot fertilization within the framework of crop rotation, since these can be applied largely irres-

pective of time and quantities. Fertilization with N, which in any case is variable, can then be separately matched to the prevailing conditions.

The type list contains about 25 PK-fertilizers produced by mixing phosphate and potassium fertilizers. They may be grouped in the order of P-form solubility (Chap. 2.2.1). The potassium component consists of either K chloride (KCl) or K sulfate (K_2SO_4), and more recently also potassium filter dust. Some PK-fertilizers are also marketed with the addition of magnesium: all are granulated. The official indication of contents is as for NPK-fertilizers, i.e., $0 + P_2O_5 + K_2O$. However, the zero representing the missing N-content has been omitted in Table 3-8.

Table 3-8. Selection of PK-fertilizers.

Trade name	P-Fertilizer	K-fertilizer	$P_2O_5 + K_2O$ contents
Peka phosphate potash	superphosphate	KCl	20 + 30
phosphate potash R (= Reform)	Novaphos	KCl	15 + 20
phosphate potash R, blue		K_2SO_4	14 + 22
Rhe-Ka-Phos	Rhenania phosphate	KCl	15 + 25
			18 + 20
Thomas potash	Thomas phosphate	KCl	10 + 20
			12 + 18
Hyperphos potash	soft-earth rock phosphate	KCl	15 + 25
			20 + 20

PMg-Fertilizers

P-fertilizers with significant magnesium contents still play no important role in Mg-supply. An example is the type of *soft-earth rock phosphate with magnesium* or the fertilizer *Hyperphos-Magnesia 23/7* with 23% P_2O_5 and 4% Mg in the form of magnesium sulfate (7% "MgO"). Some P-fertilizers contain Mg as a minor constituent, e.g., Thomas phosphate with 1 to 2% Mg.

KMG-Fertilizer

Potassium-magnesium fertilizers are included in the type list under the K-fertilizers, since they are intended for Mg-supply within the framework of fertilization with K. Magnesium salts are contained in many crude potassium salts. They are either left in the crude salt, e.g., *magnesia kainite* or added to the potassium fertilizers. The Mg-content is given as "MgO" in the type list. This is incorrect, since magnesium is present in the form of neutral salts (chloride or sulfate) and has no lime action. KMg-fertilizers are listed in Table 3-9.

Table 3-9. Selection of KMg-fertilizers.

Fertilizer Type	Composition	Contents in % $K_2O + $ "MgO"	$K + Mg$
magnesia-kainite	KCl, $MgCl_2$, etc.	12 + 6	10 + 3.6
granulated potash with magnesia	KCl, $MgCl_2$, etc.	40 + 5	33 + 3
potash magnesia (patent potash)	$K_2SO_4 \cdot MgSO_4$	30 + 10	25 + 6

3.2.2 Three- to Six-Nutrient Fertilizers

Plants require large amounts of three major nutrient elements, and altogether six major nutrient elements.

Three-nutrient fertilizers in the usual (narrow) sense are fertilizers containing the three nutrient elements most important for practical fertilization on many sites (nitrogen, phosphorus, potassium). Three-nutrient fertilizers in the wider sense are fertilizers containing any three major nutrient elements, with 20 possible combinations (Synopsis 3–4). NPK-fertilizers, being the most important three-nutrient fertilizers, are discussed separately (Chap. 3.2.3).

Four-nutrient fertilizers, obtained by the addition of magnesium, are in accordance with the type list best discussed in conjunction with NPK-fertilizers.

Various *three-nutrient fertilizers* listed in the Synopsis have already been discussed under single-nutrient, two-nutrient, or lime fertilizers, e.g.,

- NPS-fertilizers under NP-fertilizers (ultrahigh-analysis fertilizers);
- NKS-fertilizers under NK-fertilizers with sulfate;
- NMgS-fertilizers under nitrogen magnesia;
- PKCa-fertilizers under PK-fertilizers;
- PKS-fertilizers under PK-fertilizers with superphosphate;
- PMgCa-fertilizers under magnesium lime with phosphate;
- PCaS-fertilizers under superphosphate;
- KMgS-fertilizers under potassium magnesia;
- KMgCa-fertilizers under potassium quicklime.

All the possibilities indicated in Synopsis 3–4 have been realized in fertilizers. This, however, cannot be discussed here in detail.

A *six-nutrient fertilizer*, i.e., a fertilizer containing all major nutrient elements, could be obtained by combining nitrogen magnesia with a PK-fertilizer. This would then be a *NPKMgCaS-fertilizer*. However, combining all major nutrient elements in a single fertilizer is of practical importance only in combination with micronutrients and is discussed in Chap. 3.3.3.

Synopsis 3-4. Three-Nutrient Fertilizers as 20 Possible Combinations of 6 Major Nutrients.

NPK	NMgCa	PKMg	KMgCa
NPMg	NMgS	PKCa	KMgS
NPCa	NCaS	PKS	KCaS
NPS		PMgCa	MgCaS
NKMg		PMgS	
NKCa		PCaS	
NKS			

Plant ash is a *five-nutrient fertilizer* containing trace elements. It contains all vital mineral elements except nitrogen.

When the six most important micronutrients in addition to the six major nutrient elements are included, this yields 220 qualitatively different three-nutrient fertilizers.

3.2.3 NPK-Fertilizers

The most important three-nutrient fertilizers are undoubtedly the NPK-fertilizers (allowing for magnesium in addition). The type list (1976) contains about 80 kinds of solid and liquid NPK-fertilizers with different contents. Their number continues to increase.

The *composition* of fertilizers is indicated by content as follows:

$$N + P_2O_5 + K_2O + MgO \text{ (if present)}$$

e.g., 12 + 12 + 17 + 2.
These content combinations are sometimes also indicated as follows, e.g.: 9·9·9 or 15/15/15.

The following *content ranges* have been considered until now:

3 to 25% N
4 to 22% P_2O_5
5 to 32% K_2O.

Future standards envisage the combination 3+5+5 as a lower content limit (also the EEC minimum standard).

Allowing for all possible combinations in steps of 1% would yield a total number of *2197 (!) different possibilities* (i.e., types) ranging from a low content of 5+5+5 to the maximum attainable content of 17+17+17. This number would be still greater if magnesium were to be included. This multiplicity is certainly not desirable. Practical fertilization does not require all possible quantity combinations, but a clear choice of multiple-nutrient fertilizers with contents differing by large steps.

EEC-standards do include all NPK-fertilizers as a single type with the above-mentioned minimum contents. This simplifies legal problems of fertilizer licensing; however, the number of marketed fertilizers can and will be large even with this regulation.

Solid NPK-fertilizers

Most NPK-fertilizers are marketed in the form of solid fertilizer salts. Sixty-five types were licensed in 1976, and their compositions are given in Table 3-10: fertilizer solutions and suspension fertilizers are discussed separately. The following remarks refer to the value-determining constituents:

The *N-component* frequently consists of ammonium and nitrate in equal parts, unless some other N-compound such as urea is present.

The *P-compound* of many NPK-fertilizers consists of 30% water-soluble and 70% citrate-soluble phosphate. However, the water-soluble component exceeds 50%, and sometimes even 90%, in fertilizers based on superphosphate.

The *K-component* is nearly always water-soluble, since it usually consists of K chloride, or K sulfate for plants sensitive to chloride. Some fertilizers are not only *poor in chloride*, but, practically *free of chloride*.

The *Mg-component* frequently consists of Mg-sulfate, however, other compounds are also used. The contents are indicated as MgO, in analogy to K_2O. Both these designations are unsuitable.

NPK-fertilizers are grey granulates (mostly of 2 to 3 mm size), and are sometimes *sealed dustfree*. Some are coated with plastic (foils) to slow down their action, e. g., "Osmocote" $(15 + 15 + 15)$.

Some types were formerly *dyed* for recognition. However, the added dyes are disadvantageous in application; identification is therefore now ensured by colored printing on the fertilizer bag or colored pieces of paper added to soluble fertilizers in bulk.

Some *standard types* stand out from the multiplicity of solid NPK-fertilizers (Table 3-10). More details on them appear in Table 3-11. The composition of NPK-fertilizers is seen from Table 3-12.

Production of NPK-Fertilizers

NPK-fertilizers are sometimes produced by mixing the components (mixed fertilizers). However, the predominant method is *acid decomposition* of rock phosphates, followed by the addition of potassium (complex fertilizer) [67]. There are various processes for acid decomposition. They are compiled in a simplified manner in Synopsis 3-5. Several firms produce NPK-fertilizers and market them under trade names (Synopsis 3-6).

Table 3–10. 65 Types of solid mineral NPK-fertilizers (1976).
(K-contents printed in bold type indicate that potassium occurs as sulfate or sometimes as nitrate. Only contents exceeding 4% are indicated in column for MgO, referred to type mentioned.)

Contents in % of fertilizer

N	P₂O₅ a	b	c	d	e	f	g	K₂O a	b	c	d	e	f	g	MgO
3	10							15							
4	12	20						16	**32**						
5	5	10	10					**20**	15	20					5 (a)
6	8	10	12	17				**10**	18	18	6				
7	8	40						**10**	8						12 (b)
8	8	12	14	15	16			8	16	**18**	20	24			
9	9	9	11					9	15	6					
10	5	8	10	10	10	12	15	**20**	18	5	10	15	**18**	20	6 (a)
12	6	11	12	12				12	16	17	**20**				
13	7	13	13					13	13	21					
14	7	10	10	12	12			14	14	24	**14**	17			
15	5	9	9	11	12	15		20	5	15	**15**	24	15		5 (b)
16	8	8	10	12				12	**16**	16	18				
17	13	17						20	17						
18	6	6	8	9	18	22		12	**18**	16	18	9	12		
19	6							**6**							
20	5	5	8	10	10			8	**10**	12	10	**15**			6 (e)
22	8							12							
24	8							8							

Examples: 5 + 5 + **20** + 5; 5 + 10 + 15; 5 + 10 + 20

Table 3–11. Selection of some standard NPK-fertilizers.

Colour marking	Content N + P₂O₅ + K₂O + MgO	K-Form	Remarks
violet red yellow	10 + 15 + 20 13 + 13 + 21 15 + 15 + 15	alternative KCl	for plants tolerant to chloride
blue	12 + 12 + 20 12 + 12 + 17 + 2	K₂SO₄	sometimes with trace elements as "blue extra" or "blue trace"

Table 3-12. Examples of compositions of NPK-fertilizers (contents in % according to [31]).

Constituent	red 13+13+21	Nitrophoska yellow 15+15+15	blue 12+12+20
ammonium phosphate	10	11	9
dicalcium phosphate	14	16	13
ammonium nitrate	1.7	11	—
ammonium chloride	20	18	—
ammonium-sulfate nitrate	—	—	27
potassium nitrate	38	35	19
potassium sulfate	—	—	22
potassium chloride	4.9	—	—
tricalcium phosphate	0.3	0.8	0.3
calcium sulfate	2.4	1.5	—
magnesium chloride	0.4	0.4	—
magnesium sulfate	—	—	2.0
calcium fluoride	2.3	2.9	2.3
sodium chloride	2.3	1.2	2.3
sodium sulfate	—	—	0.8
substances insoluble in acids	1.3	0.9	1.1
substances insoluble in water	0.8	0.3	0.7

NPK-Solutions and Suspensions

Technical and economic reasons, and labor considerations impose the need for testing new fertilization procedures such as the development of liquid fertilizers, both solutions and suspensions [53].

Fertilizer solutions of the components of most solid fertilizers can be prepared only in relatively low concentrations, since some salts are of limited solubility (Ca-phosphate, potassium salts). However, the use of ammonium and potassium phosphates allows the total contents to be raised to approximately 25 to 30% $(N + P_2O_5 + K_2O)$ (Table 3-13).

Only the preparation of stable suspensions permits further increases in concentrations. Suspensions contain fine fertilizer crystals dispersed in saturated solutions (suspended), their dispersion being stabilized by additives like clay minerals, etc. This prevents precipitation and formation of sediments even after prolonged storage. Solubility of the salts is further promoted by the use of polyphosphates (Chap. 2.2.2). Suspension fertilizers may have total-nutrient contents of up to 50% (Table 3-13).

Direct *application* of a suspension requires special nozzles on the sprayers or dilution until the solid salt particles dissolve. These fertilizers are economical

Synopsis 3–5. Processes for Producing NPK-Fertilizers by Phosphate Decomposition.

1. *Decomposition with sulfuric acid and ammonation* (of superphosphate)

$$\text{rock phosphate} + H_2SO_4 \rightarrow Ca(H_2PO_4)_2 + \begin{pmatrix} NH_3 \\ NH_4NO_3 \end{pmatrix} \rightarrow \boxed{\begin{array}{l} NH_4H_2PO_4 \\ CaHPO_4 \\ \text{etc.} \end{array}} + \text{potassium salt}$$

$$\underbrace{}_{\substack{\text{ammonation} \\ \text{solution}}} \quad \underbrace{\text{NP-fertilizer}}$$
$$\underbrace{}_{\text{NPK-fertilizer}}$$

2. *Decomposition with nitric acid and ammonation* (e.g., ODDA-process)

$$\text{rock phosphate} + HNO_3 \rightarrow \left.\begin{array}{l} CaHPO_4 \\ H_3PO_4 \\ HNO_3 \end{array}\right\} + NH_3 \rightarrow \boxed{\begin{array}{l} NH_4\text{-phosphates} \\ CaHPO_4 \\ NH_4NO_3 \\ \text{etc.} \end{array}} + \text{potassium salt}$$

$$\underbrace{\begin{array}{c} Ca(NO_3)_2 \\ \text{separated and processed} \\ \text{to lime-ammonium} \\ \text{nitrate} \end{array}} \quad \underbrace{\text{NP-fertilizer}}$$
$$\underbrace{}_{\text{NPK-fertilizer}}$$

3. *Decomposition with mixed acids, and ammonation*
(Process acording to PEC, Potasse et Engrais Chimique, or TVA, Tennessee Valley Authority)

$$\text{rock phospate} + \left.\begin{array}{l} HNO_3 \\ H_3PO_4 \\ (H_2SO_4) \end{array}\right\} \rightarrow \left.\begin{array}{l} CaHPO_4 \\ H_3PO_4 \\ HNO_3 \\ Ca(NO_3)_2 \\ CaSO_4 \end{array}\right\} + NH_3 \rightarrow \boxed{\begin{array}{l} NH_4\text{-phosphates} \\ CaHPO_4 \\ NH_4NO_3 \\ Ca(NO_3)_2 \\ CaSO_4 \\ \text{etc.} \end{array}} + \text{potassium salt}$$

$$\underbrace{\text{NP-fertilizer}}$$
$$\underbrace{}_{\text{NPK-fertilizer}}$$

Synopsis 3–6. Some Trade Names of NPK-Fertilizers.
(Used primarily in agriculture, listed in alphabetical order.)

Trade name	Manufacturer
Am-Sup-Ka	Superphosphat-Industrie, Hamburg
Cederan	Chemische Düngerfabrik, Rendsburg
Complesal	Farbwerke Hoechst AG, Frankfurt
Enpeka	Guano-Werke AG, Hamburg
Kampka	Chemische Fabrik Kalk GmbH, Köln
Nitrophoska	BASF-AG, Ludwigshafen
Rustica	Ruhr-Stickstoff AG, Bochum

Table 3–13. NPK-fertilizer solutions and suspensions (1976 fertilizer types).

	contents in % by weight $N + P_2O_5 + K_2O$	nutrient form		
		N	P	K
solution (specific gravity = 1.2–1.3)	5 + 8 + 10	carbamide	ammonium phosphates	alternative (KCl, K_2SO_4, KNO_3)
	8 + 8 + 6	NH_4, NO_3		
	9 + 9 + 7	NH_4, NO_3	potassium phosphates	
	10 + 4 + 7	carbamide NO_3		
	12 + 4 + 6			no Cl
	12 + 8 + 11			
	13 + 8 + 7	carbamide		alternative
	15 + 6 + 9	carbamide NO_3		
suspension (specific gravity = 1.5)	3 + 10 + 30	NH_4	polyphosphates (up to 60% of P-component) ammonium phosphates	KCl
	12 + 12 + 18	carbamide		
	13 + 13 + 13	NH_4, NO_3		
	6 + 10 + 16	carbamide NO_3	polyphosphates	alternative
	20 + 4 + 8	carbamide		no Cl
	16 + 16 + 12	NH_4, NO_3	ammonium phosphates	alternative
	25 + 6 + 20			

to transport, in view of their high specific weight, and can also be applied as foliar nutrients after dilution, i.e., complete solution.

NPK-solutions and -suspensions are often treated with micronutrient additives (Chap. 3.3).

Considerations of Optimum NPK-Ratios

Different *nutrient ratios* will be required depending on the nutrient supply in the soil and the needs of the plants, if fertilization is to be carried out entirely with NPK-fertilizers.

However, proceeding on the basis of average supplies in the soil and average requirements of the most important plants in crop rotation, it is the problem of optimum NPK-ratios that has to be considered in this simplified case. These considerations are derived from removal ratios, to provide tentative values for standardizing fertilization (Table 3–14). This derivation has become questionable. Thus, removal ratios should be adapted to modern high yields and the utili-

Table 3-14. Derivation of optimal nutrient ratios (according to [65]).

	N	:	P_2O_5	:	K_2O
A. nutrient removal in one crop rotation	1	:	0.5	:	1.2
B. utilization rate in % (over several years)	80	:	35–50	:	60–70
C. computed ratio $\dfrac{(A \cdot 100)}{B}$	1.25	:	1.0–1.4	:	1.8–2
corrected ratio	1	:	0.8–1.1	:	1.4–1.6
D. optimal ratio determined from fertilization trials	1	:	0.8–1	:	1.5–2
E. example of NPK-fertilizer (13 + 13 + 21)	1	:	1	:	1.6

zation rate of P-fertilizers should be assumed to be higher for well supplied soils (Chap. 5.5.1).

Moreover, trials undertaken for the sake of experimental corroboration provided an inadequate basis and were carried out without checking the state of supplies to the plants, so that derivation requires revision. Tentative data on modern developments are provided by the trends in nutrient ratios for the total *fertilizer consumption* (Table 3–15). The tendency is a narrowing of nutrient ratios towards 1:1:1; an increasing emphasis on nitrogen is also evident (Chap. 1.3.4). However, it should be remembered that, especially as regards global statistics, certain fertilization patterns are obtained from which direct evaluation of an optimal nutrient ratio (e.g., for conditions in Germany) is not necessarily possible.

Table 3-15. Nutrient ratios in fertilizer statistics (according to [1, 20]).

Region	Year	N	:	P_2O_5	:	K_2O
Germany or Federal Republic of Germany	1936/38	1	:	1.3	:	1.9
	1954/57	1	:	1.1	:	1.8
	1970	1	:	0.8	:	1.0
	1975	1	:	0.7	:	1.0
global	1951	1	:	1.4	:	1.1
	1956	1	:	1.2	:	0.9
	1965	1	:	0.9	:	0.6
	1975	1	:	0.6	:	0.5

What is needed, in the final analysis, is not a fertilizer with optimal nutrient ratios for the average case, nor hundreds of fertilizers with different nutrient ratios, but a small choice of the most frequently applicable combinations.

3.2.4 Single- or Multiple-Nutrient Fertilizers

The problem is posed frequently in simplified form as a question of *single* or *complete fertilizer* and is worth some thought. NPK-fertilizers, also called "complete fertilizers", were developed for the purpose of simplifying fertilization. Ideally, it would be desirable, for the sake of simplicity, if a only one fertilizer could ensure nutrient supplies for at least a simple scheme of crop rotation.

It would hardly be possible to achieve the ideal of a "universal standard fertilizer" in view of the multiplicity of production conditions. However, a certain standardization of fertilizers through pre-prepared nutrient combinations is desirable, since it would simplify fertilization and thus reduce calculation and labour. The application of single fertilizers mixed on one's own farm, as opposed to standardized NPK-fertilizers has advantages and disadvantages that have different significance, depending on the production conditions involved (Synopsis 3-7).

The problem of the *correct nutrient supply* is of particular significance (from the "agronomics" aspect), and will therefore be discussed in detail:

a) Plants must be nourished *harmoniously*, i. e., they should obtain all necessary nutrients in, approximately, correct ratios in the soil;

b) this "harmony" of nutrients is advantageously pre-prepared in the fertilizer granulates, though this is not absolutely necessary;

c) balanced supplies to plants are often possible even with separately applied single fertilizers, since a fertile soil balances supplies, and in any case soils form nutrient mosaics in microregions.

d) *Harmonious* fertilization may cause *disharmonious* supplies to the plant, under certain conditions, for example, if the soil is already abundantly supplied with one nutrient, and this supply is further increased by fertilization.

e) Application of standardized NPK-fertilizers *simplifies* fertilization, since supplies of P and K are approximately ensured together with the necessary amounts of N. This prevents gross mistakes in fertilization.

f) Precise adaptation of fertilization to requirements in each case, i. e., *fine tuning* of fertilization, is often best done by a combination of single-nutrient fertilizers. This, however, presupposes a reliable diagnosis of fertilizer requirements, and is therefore worthwhile above all in intensive agriculture.

g) On the other hand, an optimally combined supply of nutrients is of minor importance, compased with other factors, if yield levels are average.

Synopsis 3–7. Specific Advantages of Single-Nutrient Fertilizers and Standard NPK-Fertilizers [8].

Single-nutrient fertilizers	Standard NPK-fertilizers
Agronomic advantages	
Nutrient supplies	
good matching of nutrient quantities to plant requirements on different soils	simplified fertilization through all-in formulation (standardization)
Fertilizer form	
mostly choice between different solubility and dispersal forms	fixed combination of several solubility forms and uniform, good dispersal form
Minor constituents	
frequently minor constituents with favourable effects on soil and plant	mostly no ballast substances
Economic advantages	
Nutrient price	
frequently low price per nutrient unit	purchase of large amounts of one type possible (rebate for large quantities)
Mixing costs	
saving of factory mixing costs	no mixing work by farmer
Transportation costs	
occasionally smaller total amounts of fertilizer (with highly concentrated types)	mostly smaller total quantities of fertilizer

Considerations of fertilizer forms are concentrated on the *phosphate form*. The use of single-nutrient fertilizers permits selection of the respectively optimal P-form or cheapest P-form, if proper P-supplies to the soil require only replenishment of reserves (Chap. 2.2.4). However, standardized NPK-fertilizers provide fixed combinations of water-soluble and citrate-soluble phosphates. This is often desirable but not always necessary, so that this may be an unnecessary expense. This affects *price comparison* between the two fertilization systems, which may therefore be carried out by two different methods:

- Comparison on the basis of *equivalent P-forms:* This standard method consists in comparing the water-soluble P-component of NPK-fertilizers with a water-soluble phosphate in single-nutrient fertilizers.
- Comparison on the basis of the *cheapest P-form:* This simplified method consists in comparing the different P-forms of NPK-fertilizers with the cheapest P-form in single-nutrient fertilizers, that would have the same effect. Obviously, this procedure is acceptable only under production conditions in which no differences in the action of the P-forms are to be expected.

Minor constituents are characteristic of many single-nutrient fertilizers. These substances may be important for fertilization, without being charged for separately, or may represent unnecessary ballast, or may even be harmful. They play different roles in comparative evaluation, depending on conditions on the particular farms.

The approximately 10% higher price, based on equivalent P-forms, may have maximum significance in *economic comparisons.* It is often compensated for by labor-economic and other advantages. The costs of mixing fertilizers on the farm may be assessed differently, depending on the time. Suitable fertilizer dispersers make the simultaneous application of several single fertilizers without premixing possible. The fertilizer quantities to be transported are usually larger when single fertilizers are used, but this disadvantage can be overcome by the application of highly concentrated single fertilizers.

To summarize, we can say that a general decision on the alternatives "single-nutrient" or "NPK-fertilizers" is for various reasons not possible. However, the farmer can frequently make a decision suitable for his farm, taking the above-mentioned aspects into account. The inclusion of two-nutrient fertilizers in the considerations, often leads to the best and most practical solution in the form of a combination of single-nutrient and multiple-nutrient fertilizers. This is also seen from fertilizer sales figures, in which multiple-nutrient fertilizers account for about 50%.

3.2.5 Use of Multiple-Nutrient Fertilizers

Some fundamental considerations on their use (besides those already offered for single fertilizers), will be listed to facilitate selection from the many multiple-nutrient fertilizers on the market.

Nitrogen represents the most expensive and most important component. Hence, N-containing multiple-nutrient fertilizers must be applied in such a way that above all the nitrogen is optimally effective. The rules of N-application are thus largely generally valid (Chap. 2.1.5).

The *phosphate form* essentially determines the mode of application of multiple-nutrient fertilizers without nitrogen, e. g., PK-fertilizers. These fertilizers, in particular, pose no problems in application and have the advantage that the nitrogen is "kept in hand", i. e., can be applied separately and precisely. This also applies to NPK-fertilizers having small N-components.

Considering NPK-fertilizers themselves, one must first decide whether *solid* or *liquid* forms are to be applied. This is in essence a problem of labor, economy and price. Certain uses obviously require a form which can be applied with spraying devices.

The choice of multiple-nutrient fertilizers is certainly larger than necessary, as far as *nutrient ratios* are concerned. There is certainly no need for a special fertilizer formulation for every purpose. The enormous choice of fertilizers is partly due to the competition between producers to "cash-in" with their own fertilizer formulations. Actually, 5 to 15 NPK-fertilizers would be sufficient to cover nearly all fertilization requirements.

Nutrient supplies in the soil which is to be fertilized, decisively influence selection of the multiple-nutrient fertilizer. Thus, only NK-fertilizers or NPK-fertilizers with small P-contents can be considered for soils abundantly supplied with phosphates. A similar reasoning applies to soils rich in potassium. Magnesium as an additional component is particularly important for poor soils.

Other *soil properties* should be taken into account, particularly with regard to phosphate action. A large water-soluble P-component acts optimally in slightly acid to neutral soils. However, this is of minor importance when the soil is well supplied with P. On the other hand, the latter point raises the question of whether the relatively expensive, easily soluble P-component is necessary.

Plant requirements play an important role in the selection of NPK-fertilizers. Many formulations were developed to accurately satisfy plant requirements through fertilization. However, a rough match of the fertilizer to requirements generally suffices when plants are grown in natural soils. The storage capacity of the soil and the selective power of the plants then ensure fine tuning. Selection of NPK-fertilizers should take the following *plant requirements* into account:

- Plants with abundant vegetative growth, e. g., green forage, foliage vegetables, require large amounts of N and K, and thus these two components should be accentuated in NPK-fertilizers;
- sugar beets require particularly large amounts of potassium, so that a large K-component is indicated (e. g., 13 + 13 + 21);
- brewers' barley requires relatively little nitrogen, so that a formulation with little N should be chosen;
- pastures require relatively little phosphate and potassium, so that a combination rich in N is indicated;

- potatoes require large amounts of magnesium as well as potassium, so that a NPK-fertilizer containing Mg should be chosen;
- sugar beets, being "halophytes", should be given fertilizers containing Cl;
- potatoes and tomatoes, being sensitive to Cl, prefer fertilizers containing little Cl, or in special cases no Cl;
- maize has a large phosphate requirement in the initial growth stage, and on insufficiently supplied soils needs large amounts of water-soluble phosphate, which, if possible, should be applied by placement.

Further details of the nutrient requirements of plants are discussed for individual plants in Chapters 7 and 8.

The correct time of application of multiple-nutrient fertilizers chiefly depends on the N- and P-forms. Many NPK-fertilizers are applied at the time of sowing, i.e., neither a long time in advance (to avoid losses through leaching or immobilization) nor a long time afterwards, since introduction into the soil is then difficult. Multiple-nutrient fertilizers are best scattered or sprayed on the crop-free soil, and then worked into it. All solid multiple-nutrient fertilizers are granulated, so that dispersion with the centrifugal fertilizer distributor is simple.

Influence of Multiple-Nutrient Fertilizers on Soil Reaction

The pH-altering influence of multiple-nutrient fertilizers is due to the sum of the single components. Quantitatively, the influence of the nitrogen component is greatest (apart from lime that might be present). The reader can thus be referred, to a large extent, to the comments on the influence of N-fertilizers on soil reaction (Chap. 2.1.5). However, there are complications caused by the somewhat different N-forms in multiple-nutrient fertilizers (Table 3–12). The various N-compounds should in reality be evaluated separately, but this would be laborious and still gives only an approximation.

Making some simplifying assumptions, a comprehensive formula was developed to determine the acidifying action of a fertilizer, particularly a multiple-nutrient fertilizer [72]. The lime loss or the opposite, i.e., the supply of lime, is summarized as the *influence* I, on the lime balance. For arable fields it is given by the formula

$$I \text{ (in kg CaO)} = 1.0 \cdot CaO + 1.4 \cdot MgO + 0.6 \cdot K_2O + 0.9 \cdot Na_2O$$
$$- 0.4 \cdot P_2O_5 - 0.7 \cdot SO_3 - 0.8 \cdot Cl - 1.0 \cdot N$$

into which the corresponding percentage fertilizer contents have to be inserted, and yields either acidification, as a negative value (loss of lime in kg CaO) or lime gain as a positive value. For grassland the coefficient 0.8 should be substituted before nitrogen, N, allowing for the increased ammonium intake in this case. But it should be remembered that the above formula represents only a rough approximation; the principal sources of error are the insufficient allowance made for the N-form, and the full allowance made for the calculated lime value of phosphates (Chap. 2.2.3). Most NPK-fertilizers have a soil-acidifying action.

3.3 Fertilizers with Major- and Micronutrients

Plants require many nutrients. It would therefore seem appropriate to develop fertilizers containing all the necessary nutrients. These would be *complete fertilizers* in the *real meaning of the term*, whose nutrient supply could completely nourish the plant. However, different nutrients have different significance for various fertilization purposes; it therefore seems advisable to also market incomplete combinations of several nutrients.

The many combinations marketed as fertilizers will be separated into three groups to provide a clearer picture:

- Micronutrient combinations with magnesium;
- nitrogen fertilizers with magnesium and micronutrients;
- multiple-nutrient fertilizers with micronutrients.

Micronutrients may be optionally added to other fertilizers as desired, as long as the minimum contents specified in the type list are satisfied (Table 3-16).

It certainly may be appropriate to reduce these values in some cases. Extremely low contents should not be considered as the criterion of value, but in view of the fertilization of well supplied soils it would be sufficient to replace

Table 3-16. Minimum contents of trace-element additives to other fertilizers [24].

(Solid) fertilizers according to type list		Minimum contents in %			
		Mn	Cu	Zn	B
mineral single-nutrient fertilizers		1.0	0.2	—	0.2
mineral multiple-nutrient fertilizers	general (to 1977) for horticultural fertilizer and general from 1978 onward	0.1 0.05	0.04 0.01	0.02 0.01	0.05 0.02
organic and organic-mineral fertilizers	from organic wastes in case of large organic components (e.g., peat and peat mixed fertilizers)	0.05 0.01	0.01 0.003	0.01 0.002	0.02 0.01

the micronutrients removed by annual application. It might therefore be advisable to specify the low contents valid in horticulture for all mineral multiple-nutrient fertilizers. It should, however, be stressed that this proposal is not aimed at a general reduction of the contents of trace elements in fertilizers but should provide additional possibilities, particularly with regard to undesirable accumulations of heavy metals.

It is not always advisable to use complete combinations of all micronutrients together with other fertilizers. It might frequently be better to limit additions to trace elements of practical importance, e. g., manganese, copper, boron, for a given fertilization purpose.

3.3.1 Micronutrient Combinations with Magnesium

Water-soluble combinations of solid fertilizers have been developed mainly for application as *foliar nutrients*. They contain practically all nutrients "beyond NPK" that are often neglected in fertilization, e. g., they include the major nutrient magnesium.

A typical example is *Fetrilon-Combi* which was developed from the iron fertilizer Fetrilon (Chap. 3.1.1), and is thus based on chelates. The composition of Fetrilon-Combi and the amounts supplied to plants with this fertilizer are indicated in Table 3-17. Fetrilon-Combi is used to bridge deficiency caused by inadequate supplies from the soil. Spraying is performed once or repeatedly, e. g.,

- at 4 kg/ha when the plants are young,
- at 1 kg/ha from the appearance of ears till blossom time.

As expected, yield increases achieved in trials vary within wide limits according to the deficiency situation. Of special interest is the possible yield increase on high yielding fields having grain yield levels of 60 to 70 dt/ha. The in-

Table 3-17. Composition of Fetrilon-Combi and standard supplies.

Element	Content in %	Supply in g/ha by application of	
		1 kg/ha	4 kg/ha
magnesium (Mg)	2.4*	24	96
iron (Fe), manganese (Mn)	1.5	15	60
zinc (Zn), copper (Cu)	0.5	5	20
boron (B)	0.3	3	12
molybdenum (Mo), cobalt (Co)	(trace)	.	.

* Officially specified as 4% MgO

crease is often 2 to 3 dt/ha after a single spraying at the time of ear appearance, and is therefore extremely worthwhile in such cases. However, the entire problem requires more detailed consideration. Plants supplied optimally with all nutrients throughout the vegetation period should generally have the maximum possible yields if losses are prevented through proper plant protection. Yield increases of 3 dt/ha or even 10 dt/ha seem modest in comparison with these possible increases. Thus, the question is why spraying leaves with all the nutrients which might be lacking does not ensure larger yield increases.

Leaf fertilization can never provide more than additional supplies, at least under agricultural conditions. Results depend on the general state of trace-element supplies to the plant, the appearance of the principal deficiency phase, the time of leaf spraying, etc. The real situation will thus lie between the following two extremes:

- The plants have been supplied optimally with trace elements and magnesium, from the youth stage onward; leaf spraying can then provide no additional gains.
- The plants lack one trace element during the youth stage, and possibly several trace elements later; timely, adequate leaf spraying in such cases can ensure large yield increases or even maximum yields, provided that other conditions are optimal.

Application trials of a foliar nutrient are mostly carried out on fields with latent deficiency (whose onset cannot be recognized). Therefore, in practice, leaf spraying will begin only when plants have already suffered deficiencies for varying periods. Additional supplies can therefore only mitigate damage that can no longer be completely repaired. The later fertilization is undertaken, the smaller will be the yield increase expected. This reasoning does not invalidate later fertilization as such, since it may be of value as a final additional supplement.

The *quantities* supplied also play a major role in assessing effects on the yield. This will be explained by means of the following rough calculation: Let supplies from the soil amount to three quarters of requirements. This would mean that about 200 g/ha of manganese would have to be supplied at the high removal rate of 800 g/ha. This supply would be theoretically possible when Fetrilon-Combi is applied as follows:

- 3 sprayings at 4 kg/ha = 180 g/ha;
- 1 spraying at 1 kg/ha = 15 g/ha.

However, utilization is not complete even with leaf fertilization, so that approximately twice as many sprayings would be required in practice. The supply/demand ratio is more favourable in the case of copper, but is particularly unfavourable for magnesium. A single spraying will thus supplement insufficient supplies only for a short time or to a small extent at best. In the particular case of high performance areas most micronutrients must be supplied via the soil, which must therefore be fertilized accordingly. Nevertheless, additional sprayings represent important and often profitable auxiliary measures.

3.3.2 Nitrogen Fertilizers with Magnesium and Micronutrients

These liquid fertilizers were developed primarily for application as *foliar nutrients* for additional supplies to growing plants. They are intended to fill gaps in requirements, nearly always present in the case of nitrogen, and frequently so with other nutrients.

Plants on high-performance fields are generally well supplied with phosphate and potassium from the soil; moreover, calcium and sulfate are rarely minimum factors. The necessary supplement of major nutrients thus chiefly involves nitrogen and magnesium. Only *manganese and copper* are added as trace elements, since these are significant as minimum factors for common cultivated plants.

These are thus NMg-fertilizers with some micronutrients of practical importance. These foliar nutrients were developed from the N-fertilizer solution (Chap. 2.1.3). They are also called *Grade 34 Fertilizer*, because their content is 34% N by volume. Examples are:

- *BASF grade 34 foliar nutrient;*
- *Hoechst 34 N-fluid,* and
- *Rustica fluid N 34*, made by Ruhr-Stickstoff.

All contain nitrogen as ammonium nitrate and urea, magnesium as Mg-sulfate, as well as manganese and copper in chelate form (Table 3–18).

Their *field of application* is mainly a large number of agricultural crops, provided the latter have no special boron requirements or that these are covered by other means, as well as many other plants. The amount applied is 8 to 12 l/ha

Table 3–18. **Composition of "Grade 34 Liquid Fertilizer" and standard supplies** (specific gravity: 1.28).

Nutrient element	Content in %		Supply by application of 10 l/ha
	by weight	by volume	
N	27	34	3.4 kg
Mg	0.3*	0.4	40 g
Mn	0.1	0.13	13 g
Cu	0.1	0.13	13 g

* Officially specified as 0.5% MgO

(e. g., for grain). This corresponds to a solution of 2 to 3% concentration at an application of 400 l/ha. Solutions of only about 0.5% concentration should be used in fruit growing (Chap. 6.1.4).

The possibilities of adequate supplies of trace elements, e. g., to grain, leaf fertilization correspond to those indicated with reference to micronutrient combinations discussed in the previous section.

3.3.3 Multiple-Nutrient Fertilizers with Micronutrients

Fertilizers containing all nutrient elements in the ratios required by the plants can be produced easily, at least in solid form, but are really only needed for restricted purposes. On the other hand, largely complete combinations are required much more frequently. It has already been stated that calcium and sulfur are often of minor significance for fertilization; iron and molybdenum are rarely required, at least outdoors.

Multiple-nutrient fertilizers and complete fertilizers as combinations of many or all nutrient elements have been developed for diverse agricultural, and especially, horticultural purposes:

1. Solid multiple-nutrient fertilizers with important micronutrients:
   a) Soil dressings based on ordinary NPK-fertilizers;
   b) soil dressings with depot nitrogen;
   c) foliar nutrients and soil dressings in one (water-soluble).

2. Suspension multiple-nutrient fertilizers with some major nutrients and all micronutrients.

3. Liquid complete fertilizers with all, or almost all, nutrient elements.

It should be recalled that, apart from commercial mineral fertilizers, other *multiple-nutrient fertilizers with micronutrients* exist, e. g., many natural organic fertilizers or plant *ash* (which contains all necessary and beneficial elements in matched composition, apart from nitrogen).

Solid Multiple-Nutrient Fertilizers based on Ordinary NPK-Fertilizers

These fertilizers are NPKMg-fertilizers with the addition of micronutrients of practical importance. They can be developed from all Mg-containing NPK-fertilizers. However, the *NPK-fertilizer type 12 + 12 + 17 + 2* with guaranteed contents of manganese, zinc and boron is of greatest significance (Table 3–19).

One third of the phosphate is water-soluble, corresponding to the basic formulation (Chap. 3.2.3). This fertilizer contains little chloride, and can thus also be used for fertilizing plants sensitive to chloride. However, copper is absent from this combination and is thus classified in Table 3–19 as a fertilizer

Table 3-19. Composition of some solid and liquid complete fertilizers.

Contents in % by weight

	N	P_2O_5	K_2O	Mg	Ca	S	Fe	Mn	Zn	Cu	B	Mo	Remark
solid NPK-fertilizer with 3 micro-nutrients	12	12	17	1.2	×	×	—	0.1	0.02	—	0.05	—	contains little chloride
solid NPK-fertilizer with depot-N and trace elements	15	9	15	1.2	×	×	0.1	0.1	0.02	—	0.35	—	e.g., Nitrophoska permanent
solid NPK-fertilizer with 4 micro-nutrients	14	10	14	0.4	•	•	—	0.1	0.02	0.04	0.05	—	e.g., Poly Crescal

Contents in % by volume

	N	P_2O_5	K_2O	Mg	Ca	S	Fe	Mn	Zn	Cu	B	Mo	Remark
suspension fertilizer with trace elements	15	—	—	1.8	16	•	0.075	0.15	0.03	0.06	0.075	0.0015	specific gravity = 1.5 e.g., Wuxal type 2 (Co = 0.0004%)
liquid complete foliar nutrient	11	8	6	—	•	×	0.02	0.016	0.006	0.008	0.01	0.001	specific gravity = 1.2 e.g., Bayfolan (also Co)
liquid complete foliar nutrient 12+4+6	14	5	7	0.14	•	0.8	0.012	0.012	0.006	0.012	0.024	0.006	specific gravity = 1.2 contains no chloride (also Co)

Explanations: × present; — practically absent; • no data given.

with three micronutrients. Cobalt is added in very small amounts, as an element beneficial to some plants.

This multiple-nutrient fertilizer represents a largely complete combination. It is marketed by several firms for use in agriculture and horticulture (*Complesal blue, Enpeka, Kampka SE, Nitrophoska special, Rustica*).

Solid Multiple-Nutrient Fertilizers with Depot-N

A multiple-nutrient fertilizer containing nitrogen either in a particularly slow-acting form as depot-N (Chap. 2.1.4), or rendered slow-acting through suitable coating (Chap. 2.1.5) is used in horticulture with substrates which are relatively poor in nutrients. This permits (complete) depot fertilization for longer periods, e. g., of potted plants.

There are various possibilities of combining such fertilizers. Examples are:

- *Nitrophoska permanent* (with 40% of the N as Isodur) without chloride (Table 3-19);
- *Plantosan 4 D* (based on K-Mg-phosphate): 20 + 10 + 15, etc.

Dosage of these fertilizers for potted plants is basically according to N-demand. The quantities supplied vary between 0.2 and 1 g per liter of peat (substrate).

Some of these fertilizers should be classified as *fertilizers for ornamental plants* or *lawns* and thus need not correspond to any licensed type; even when applied improperly, there is no risk to human health, since these plants do not serve as food. Nevertheless, much care is devoted to the production of *fertilizers for flowers* and *lawns* mainly for the following reasons:

- Fertilizers for flowers are often used by persons with little knowledge of fertilization, so that the risk of overfertilization must be avoided (application of slow-acting fertilizers);
- the nutritional requirements of flowers differ, so that the use of various formulations appears worthwhile;
- fertilizers for flowers should be simple to apply and be provided with precise dosage instructions;
- the fertilizer price plays a minor role in hobby gardening, so that expensive fertilizer components may be used if they best meet the needs.

Water-Soluble Solid Multiple-Nutrient Fertilizers

Various completely water-soluble NPK-fertilizers with magnesium and trace elements have been developed especially for *horticultural purposes*. Their contents of major nutrients vary, but the Mg-content is often 0.4% indicated as 0.7% MgO. The contents of four micronutrients are guaranteed (Table 3-19). Molybdenum is added in extremely small amounts, as is cobalt as an element beneficial to some plants (Chap. 3.4.1). Since these fertilizers are intended for horticultural use, they obviously must contain little chloride. An example given in Table 3–19 is *Poly-Crescal*, made by Philips-Duphar. Similar products by the same manufacturer but with different formulations are *Poly Fertisal* (8 + 14 + 18) and *Alkrisal* (18 + 6 + 12).

These fertilizers are applied, e.g., once a week at about 5 liters per m^2 of nursery bed:

- 0.1 to 0.2% solution for young plants;
- 0.2 to 0.3% solution for older plants (sometimes up to 0.5%).

Multiple-Nutrient Suspensions

Suspension fertilizers of different compositions are marketed for various horticultural as well as agricultural purposes, especially for application as *foliar nutrients*. These are suspension fertilizers developed further through additives (Chap. 3.2.3). They are transported and applied in "liquid" form, so that nutrient contents are best indicated in % by volume. These liquids have relatively high specific gravities (approximately 1.5); percentages by weight, multiplied by 1.5, thus yield the (higher) contents in % by volume. Examples are some of the seven *Wuxal fertilizers* made by Philips-Duphar (the important constituents are indicated first):

Type 1: NKMg   + all micronutrients;
Type 2: N—Ca   + Mg + all micronutrients (Table 3–19);
Type 3: NK—Fe   + Mg + all micronutrients;
Type 4: NPK—Zn + Mg + all micronutrients;
Type 5: NPK (without Mg) + all micronutrients.

These fertilizer types also differ in their quantitative composition. Thus, Type 2 contains, besides 10% N, 16% Ca (indicated as 22.5% CaO), and is therefore primarily a Ca-fertilizer. Most types contain 0.05% iron, but Type 3 has the relatively high content of 1.5% Fe by volume. The same applies to Type 4 with respect to zinc. As a foliar nutrient, the material is applied by coarse or fine spraying (Chap. 6.1.4) in concentrations of 0.1 to 0.3%. The instructions for various plants should be followed.

Liquid Complete Fertilizers

Complete fertilizers in the true sense contain all the mineral nutrients required by plants. However, plant requirements differ, so that complete fertilizers with different nutrient ratios are needed.

Complete fertilizers might be said to form the link with *nutrient solutions* (Chap. 4.4.2), since they can be used as nutrient solutions in suitably diluted form. However, their chief importance is as *foliar nutrient*. A complete combination of all major nutrients with the entire "range" of micronutrients is not always required. Certain simplifications from the fertilization aspect are therefore quite admissible. There is also a technical problem of keeping the dissolved salts in solution. Occasionally this imposes compromises in the composition, e.g., with regard to magnesium. Some micronutrients are usually added in the form of chelates.

Several liquid complete fertilizers have been developed especially for horticultural purposes, but they may also be used for agricultural purposes. A "complete fertilizer" without magnesium is Bayfolan (made by Bayer and obtainable only outside the FRG). The greenish-blue solution contains little chloride but does contain vitamin B_1 and growth hormones.

Liquid *complete fertilizers with the composition 12+4+6* are produced by several firms under the respective trade names

- *BASF complete foliar nutrient* (BASF);
- *Complesal fluid* (Hoechst);
- *Rustica fluid NPK* (Ruhr-Stickstoff).

The respective contents are indicated in Table 3–19. Liquid complete fertilizers are used in agriculture as solutions of 1 to 2.5% concentration (4 to 10 l/ha), and for horticultural plants under glass in solutions of 0.1 to 1% concentration. The respective guidelines should be adhered to according to the different sensitivities of plants to foliar nutrients (Chap. 6.1.4).

Kamasol green (10+4+7+0.5 MgO), which contains 0.36% Mg by volume, is a liquid complete fertilizer that contains magnesium. The liquid is colored with a green dye. It contains no chloride but may contain polyphosphate. Other liquid complete fertilizers that should be mentioned have the formulation 8+8+6, e.g. *Complesal, Kamasol* (with magnesium), or *Wuxal super* (without magnesium).

The quantitative effects of spraying with highly dilute solutions have already been discussed in Chap. 3.3.1. Supplies of magnesium and trace elements are sometimes extremely small and can at most serve for continuous replenishment but hardly for alleviating any deficiency.

3.4 Fertilizers with Other Nutrients

Only fertilizers containing substances *beneficial* for plant growth and in some way significant as important constituents of food for man and animal will be considered here. *Carbon dioxide,* being a fertilizer of a special kind, also belongs in this section.

3.4.1 Fertilizers with Beneficial Nutrients

Plant growth, especially in unfavorable site-conditions, can be *promoted* by various substances, which are not vital under conditions of optimal growth. This may be due to various reasons, e.g., partial substitution of nutrient elements (sodium instead of potassium), or increased resistance of grain stalks to lodging (silicon). Sometimes the effects might be even more difficult to determine.

Frow the point of view of correct plant nutrition, it follows from this that beneficial substances should also be considered in fertilization unless natural supplies from the soil are sufficient. The practical significance of beneficial elements primarily concerns certain groups of plants, and its importance is increased by the fact that these substances are also vital for the nutrition of man and animal (Chap. 9.4). These elements need not actually be supplied to herbivorous animals through their plant food, but good supplies to the plants simultaneously ensure high-quality animal fodder.

For plants the chief *beneficial elements* are sodium, chlorine (as chloride), silicon, cobalt, and aluminum. Some of these may in fact be vital in very small amounts. Cobalt will be discussed in the next section, in view of its predominant importance for animals.

Sodium Fertilizers (Na-Fertilizers)

Sodium improves the growth, above all, of the so-called "Na-loving" plants, i.e., beta beet, spinach, cabbage, barley, etc. The Na-contents in their leaves should lie between 10 and 30‰ [61]. Other plants are sensitive to larger Na-supplies (as little as 1 to 2‰ Na in the leaves lowers the yield of wheat).

The common-salt requirements of cows make Na-contents of between 1 and 2‰ desirable for pasture grass (Chap. 7.5.2). Moreover, sodium lowers the potassium requirements of many plants, since it improves the water balance in a similar way. A particularly impressive example is *Rhodes grass (chloris gayana)*: it is claimed to require only about 4‰ K in the leaves when Na-supply is abundant.

Various fertilizers provide a sodium supply; their Na-contents have already been indicated as important minor constituents in the appropriate sections.

Important *Na-containing fertilizers* are e.g.:

- potassium fertilizers with various NaCl-contents,
- sodium nitrate or Chile saltpeter (about 25% Na),
- Rhenania phosphate (about 12% Na),
- multiple-nutrient fertilizers with Na, e.g., pasture fertilizers with 15% N, etc., and 3% Na (4% Na_2O).

Sodium chloride (NaCl = common salt, 40% Na) is in practice used for fertilization only as a constituent of potassium salts. The quantity of Na supplied for fertilization usually depends on the amount of the major nutrient supplied with the fertilizer concerned. Very large supplies of sodium (sodium chloride) are undesirable because of likely salt damage.

Chloride Fertilizers (Cl-Fertilizers)

The element chlorine is vital to all plants, but the amounts required are so small that fertilization specifically for this purpose would be unnecessary. However, chloride is a beneficial nutrient for many plants, especially the "salt-" or "chloride-loving" plants (beta beet, spinach, cabbage, celery, etc.). Cl-contents of 1 to 5‰, on a dry basis, in the leaves of these plants promote their growth.

Even plants with special Cl-requirements can often obtain these from the substrate. Soils, particularly near the sea, obtain continuous supplies from the atmosphere. Moreover, chloride is supplied primarily with potassium fertilizers, often in larger amounts then required.

When additional chloride supplies are desirable, *potassium chloride* can be used if there are also K-requirements. Fertilization with chloride in other forms hardly seems necessary.

In this connection, the sensitivity of the so-called *sulfate-plants* to *chloride* should be mentioned. Chloride is in fact also vital to these plants, but is detrimental in relatively low concentrations (e. g., for starch displacement in the potato plant).

Silicate Fertilizers (Si-Fertilizers)

Silicate or silicic acid is beneficial for many cultivated plants because of its positive influence in resistance to lodging. Silicic acid strengthens the supporting tissue especially of cereals thus increasing the resistance of the stalks to *lodging*. Silicic acid also seems to increase resistance to fungal diseases, and it is possible that it fulfills other functions, as well [64]. The silicon contents of leaves, on a dry basis, vary between 1 and greater than 10‰.

Silicon represents up to one third of soil content, so that it would be thought that supplies to plants should be adequate. However, the silicon in the soil itself is obviously not always mobilizable to a sufficient extent, so that additional supplies of soluble or readily mobilizable silicate may be required. Fertilization with silicate is especially important for the cultivation of rice in certain regions.

The following *Si-fertilizers* should be considered:

- Si-containing phosphate fertilizers (Chap. 2.2.2),
- soluble silicic acid or soluble silicates.

Si-containing rock powder probably contributes little to better Si-supplies because of its low reactivity and should at most be considered as an extremely slow flowing Si-source, like the silicates in the soil itself.

The quantities applied as Si-fertilizers vary within wide limits. Determination of the *available silicate* in the soil permits quantitative estimation of requirements. Large contents of silicic acid in pasture grasses are harmful to cows, since needle-shaped residues may cause injuries to the digestive tract.

It has neither been proved nor demonstrated to be likely that the application of very small amounts of silicic acid may produce special "effects", by activating growth-promoting forces, beyond the known and perhaps still unknown material effects. This problem will be discussed in Chapter 9.2.2.

Aluminum Fertilizers (Al-Fertilizers)

Aluminum appears to be beneficial to some plants. Tea leaves contain 2 to 3‰ Al, on a dry basis and appears to promote the growth of the plant, although little is known about its mechanism of action.

On the other hand, aluminum is one of the major constituents of the soil and its mobility increases with decreasing pH, so that any deficiency might be due to fixation at higher pH-values. In this rare set of circumstances, supply of aluminum is expedient. Aluminum sulfate ($Al_2(SO_4)_3 \cdot 18\,H_2O$) is used for this purpose (Chap. 2.5.2).

However, Al-sulfate acts primarily as a soil-acidifying fertilizer. The favorable effect of this fertilizer on some acid-loving plants (e.g., *blueberries* in horticulture) may be due not to better Al-supplies, but to the mobilization of micronutrients as a result of acidification. For most cultivated plants, Al is a toxic element responsible for much acid damage.

3.4.2 Fertilizers with Essential Substances for Man and Animals

If we proceed from the basic assumption that complete nutrition of animal and man implies adequate quantities of everything needed for full-value nutrition, this raises the problem of whether fertilization with other elements may be needed, beyond the requirements of and benefits to the plant.

Plants, animals, and man have, in many respects, *similar* qualitative and quantitative *requirements*. However, there are important quantitative differences in the substances vital to all organisms. Beyond this, animals and man require a greater range of elements (Chap. 9.4.1). In certain cases these elements must therefore also be considered in fertilization, although they are unimportant for plant growth.

The most important case is the supply of cobalt to grazing animals, and thus indirectly to human beings via milk. Herbivorous animals need not obtain all necessary minerals via plant feed, but for various reasons this method may be required in some cases.

Thus, iodine deficiency is common to animals in some regions. However, it appears to be simpler to alleviate this deficiency by adding iodine salts to the feed, rather than by accumulating iodine in plants through fertilization. The needs of animals and man for certain elements has been only recently demonstrated, so that no statements can yet be made on the possible consequences for fertilization.

Fertilization with Cobalt (Co-Fertilization)

As far as is known at present, plants do not require cobalt, however, it is essential for the activity of nitrogen-fixing bacteria and blue algae. It therefore represents a *beneficial* element for legumes and other N-fixing plants and plays an important role in agriculture.

Cobalt is also important for the complete nutrition of grazing animals. Cows require the cobalt-containing vitamin B_{12}, which they take in either with fodder or synthesize in their rumens from the cobalt in plants, using their own microflora. Pasture fodder should therefore contain at least 0.1 ppm Co, on a dry basis.

Legumes can usually obtain their very small Co-requirements from the soil, so that Co-fertilization should produce results only on very poor soils. On the other hand, the Co-content of grasses and herbs on pastures does not always correspond to the large requirements of cows (Chap. 7.5.2).

The main fertilizers to be considered are simple *cobalt salts,* e.g., *cobalt sulfate* ($CoSO_4 \cdot 7 H_2O$) with 21% Co. The amounts of Co required are very small; Co-fertilizer is thus, e.g., in Australia, applied as an additive to superphosphate (50 to 80 g/ha of Co). Depot fertilization with Co should be at the rate of 0.5 to 1 kg/ha.

3.4.3 Carbon Dioxide (CO_2)

The normal content of *carbon dioxide* in the air is 0.03% by volume (300 ppm). This CO_2-content is usually sufficient outdoors, under ordinary conditions of light and temperature, for a high photosynthesis performance, provided that some motion of the air (light breeze) continuously replenishes supplies from the atmospheric layers above the plant. Fertilization with CO_2, in practice, thus plays a role only in (largely closed) greenhouses, since replenishment from the free atmosphere is restricted under these conditions.

The CO_2-concentration easily decreases below the normal value in *greenhouse cultivation.* This happens especially in hydroponic installations, where no CO_2 is released from the substrate. Moreover, a larger than normal content can be utilized productively by plants under optimal light and temperature conditions. It therefore appears to be of value to check the CO_2-content of the air in greenhouses continuously. The *optimal content* is considered to be 1 000 to 2 000 ppm, depending on other growth conditions, provided that this level can be maintained continuously near the plants. On the other hand, if CO_2 is applied only intermittently, replenishment to a concentration of 3 000 ppm for 1 to 2 hrs, three times a day, is recommended. Concentrations of 10 000 ppm (1%), which are sometimes proposed, may have detrimental effects if maintained for long periods. Fertilization with CO_2 (fumigation) is possible in various ways [62, 69]:

- Direct supply of CO_2 (1 kg in liquid or solid form yields about 50 l of gaseous CO_2);
- combustion of briquettes, fuel oil, propane, ect. (1 kg propane yields 1 500 l CO_2).

Combustion of fuel has the advantage of simultaneous heat release. However, care must be taken as far as possible, to burn sulfur-free fuel since damage from sulfur dioxide is otherwise possible.

4 Fertilizers for Soil Improvement and General Growth Support

Only very few soils are "by nature" ideal substrates for plant growth (Chap. 6.3). Since antiquity, therefore, much effort has been devoted to improving imperfect soils by different methods. Unfavorable soil properties can be changed more easily, the more variable they are. Generally, chemical properties are easier to improve than physical, and some properties, at least with regard to variations over large areas, can be considered to be almost invariant, i.e., soil type, range of grain size. However, more and more methods of soil improvement become profitable as the intensity of exploitation increases (Synopsis 4–1). An essential precondition of improved structural and chemical soil properties is the pronounced activity of soil organisms. Therefore, many measures for soil improvement are directed to this end, for example regulation of pH, supply of humus, etc. The reader is referred to the bibliography for special problems of soil science [76, 78, 88, 89, 94, 96, 97].

4.1 Fertilizers for Improvement of Soil Reaction

Optimal soil reaction is a precondition for the success of many other fertilization measures. Improvement of soil reaction requires:

- in soils with too low pH-values: raising the pH by fertilization with lime (substances with an alkaline reaction);
- in soils that have an excessively high pH: lowering the pH with acidifying fertilizers.

4.1.1 Lime Fertilizers and their Properties

Lime fertilizers are alkaline-acting substances produced from natural carbonates (limes) by grinding or chemical decomposition; they are used mainly for increasing soil reaction.

Their nutrient content (Ca, Mg, etc.) (Chap. 2.4 and 2.5.1) is only of secondary significance. The chemical components are listed in Synopsis 4–2.

Synopsis 4–1. Possibilities for Soil Improvement.

A. Improvement of *chemical* soil properties; aim: optimum supply of nutrients and elimination of toxic substances.

1. Regulation of *soil reaction*, to improve the supply of available nutrients (liming or acidification of the soil).

2. Improvement of *nutrient supplies* by providing nutrients or improving the storage capacity.

3. Elimination of detrimental or toxic *excesses*, e.g., of natural salts (Chap. 6.3.4) or heavy metals (Chap. 6.5.3).

B. Improvement of *physical* soil properties.

1. Improvement of *soil structure*; aim: stable crumb structure as a precondition for optimum water and air balance.
   a) Improvement by tillage, regulation of soil reaction, etc.;
   b) supply of substances that form granulates directly (Chap. 4.2.1) or indirectly, e.g., lime;
   c) elimination of substances that damage the structure (Chap. 6.3.4).

2. Improvement of soil texture (range of grain size). The aim is a granulation of medium composition (loamy sand to sandy loam), with an adequate humus component.
   a) Reduction of grain size: supply of clayey material, clearing of stones;
   b) increase of grain size: supply of sandy and gravelly material (e.g., rock powder);
   c) increase of mineral component: supply of minerals to bog soil;
   d) increase of humus component: supply of organic matter to mineral soil.

3. Increase of useful *profile depth*; aim: optimum depth of root penetration.
   a) Removal of impermeable strata: mechanical breaking-up and stabilization of loosening by fertilization in depth;
   b) establishment of water-retaining strata in the subsoil.

Synopsis 4–2. Important Formulae of Limes.

Ca	chemical symbol of element calcium
Ca^{2+}	calcium ion, bivalent cation of calcium salts
CaO	a) chemical formula of quicklime (calcium oxide)
	b) reference basis for lime fertilizer components that have a basic reaction (Synopsis 4–3)
$CaCO_3$	calcium carbonate ⎱ mixture designated
$MgCO_3$	magnesium carbonate ⎰ as "lime"
$CaCO_3 \cdot MgCO_3$	dolomite (dolomitic lime if Mg-component is small)

Origin and Reserves

The raw materials for lime fertilizers are limestone deposits. They were formed in the seas during earlier geological periods, especially the Cretaceous and Jurassic, and to a lesser extent, in other periods as well. The calcium and magnesium carbonates of seawater were either precipitated chemically (this still occurs today with spring limes), or they have accumulated as lime-containing shells or other parts of small animals and plants. This yields a lime sludge from which solid, hard or loose, soft limestone is formed according to the prevailing pressure and temperature. Lime-containing substrates with larger silicate contents are also quarried as marl.

Reserves are immense if the limestone mountain ranges are taken into account, however, there may be a regional lack of raw materials, since some agricultural regions in humid tropical zones are distant from lime deposits, and transport is expensive.

The production of lime fertilizers is detailed in Synopsis 4–3.

Lime Fertilizers

Lime fertilizers include a large group of fertilizers, with respect to origin and properties, namely:

A. Lime-containing soil material, e.g., marl;
B. Natural limes;

- ground limestone: calcium carbonate;
- ground lime-containing plant substance, e.g., algae carbonate;
- chemically processed, converted limes: quicklime, slaked lime, also lump lime and mixed lime.

C. Industrial limes, e.g., blast furnace limes, converter limes, residue limes, etc.
D. Mixtures of lime and organic fertilizers.
E. Some other alkaline-acting fertilizers also exhibit a lime effect (Chap. 2.1.5).

The basis for evaluating lime fertilizers is their content of basic active substances, referred to collectively as "CaO" reference basis (Synopsis 4–4). Important lime fertilizers are listed in Table 4–1 [83, 100].

A special kind of soil material was used more than 2 000 years ago in ancient Greece and Rome for improving the soil. This was *marl*, called *marga* (lime earth) to distinguish it from ordinary clayey soil. It was a lime-containing substance ensuring good yield-increases, particularly in humid and cool regions, i.e., on acid soils. The use of marl was introduced to Central Europe around 1700, but was also rediscovered by chance in some regions (e.g., around 1770 in

Synopsis 4–3. Production of Lime Fertilizers.

A. Processing of Natural Limes

The composition of the carbonate component of limestone varies:

Ca-carbonate:	85 to 90% $CaCO_3$, 0.5 to 2% $MgCO_3$
carbonates with a larger Mg-component:	55 to 85% $CaCO_3$, 5 to 40% $MgCO_3$
pure dolomite:	54% $CaCO_3$, 46% $MgCO_3$

1. *Physical* processing:
Limestone, also in the form of soft chalk, is worked in open-pit mining and then ground. The degree of grinding for the purpose of increasing the surface asea must be in relation to the hardness of the rock: the harder the rock the finer the grinding (Table 4–1).

2. *Chemical* processing:
Burning (calcining) of the carbonate yields quicklime:

$$CaCO_3 \xrightarrow{900\,°C} CaO + CO_2$$

Ca carbonate   quicklime   carbon dioxide

Slaking (addition of water) yields *slaked lime* from quicklime:

$$CaO + H_2O \rightarrow Ca(OH)_2$$

quicklime   water   slaked lime

B. Industrial Limes

Many chemical processes require the addition of natural lime. Lime-containing substances or substances with liming action are obtained as by-products or waste. Examples of these *industrial limes* are:

- *Blast furnace lime:* ground blast-furnace slag from the smelting of iron ores; the addition of lime to iron ore binds the silicates, undesirable at high temperatures, in the form of calcium silicate.

- *Converter lime:* ground converter slag obtained in steel production from pig iron. The addition of lime to the pig iron binds the undesirable minor constituents, at high temperatures, in the form of Ca-silicates and phosphates. The by-product obtained resembles Thomas phosphate.

- *Residue limes:* various lime-containing substances from industrial production, e. g., separation slime from sugar refining.

the Probstei area of North Germany, at the Baltic Sea). Marl is a mixture of sand, silt, clay, and 10 to 30% lime. Many soils were formed in the glacial age from boulder clay and therefore contain this material at a certain depth. Marl was applied extensively in Northern Germany in the nineteen-twenties and thir-

Synopsis 4-4. Basic Active Substances.

The purpose of liming is primarily the neutralization of acid (H^+-ions), i. e., raising the pH. This can finally only be obtained with hydroxides (bases, alkalis, i. e., OH^--ions). Technical hydroxides (potassium hydroxide, sodium hydroxide) are excluded from fertilization for cost and other reasons. Alkaline-earth hydroxides (calcium and magnesium hydroxides) or substances yielding these alkalis after conversion, i. e., which act as bases in the soil are therefore used as fertilizers.

The following eight basic active substances are used as fertilizers:

- $Ca(OH)_2$ or $Mg(OH)_2$: act directly in aqueous media;
- CaO or MgO: converted into $Ca(OH)_2$ or $Mg(OH)_2$ after the addition of water;
- $CaCO_3$ or $MgCO_3$: e. g., converted via $Ca(HCO_3)_2$ into $Ca(OH)_2$ after the addition of CO_2 and H_2O;
- Ca silicate or Mg silicate: hydrolysis to $Ca(OH)_2$ or $Mg(OH)_2$; however, only easily decomposable silicates, which dissolve at least with N/2 HCl at high temperature, can be used.

Reference base "CaO"

Since hydroxides are finally effective, these might therefore be used as a uniform reference basis. However, it is simpler and better to use their anhydrides, i. e., CaO instead of $Ca(OH)_2$ and MgO instead of $Mg(OH)_2$ respectively. Further simplification is possible by using "CaO", the sum of CaO and MgO as a *reference basis*, indicating the total basic action as a number. Computation of MgO as "CaO" is possible for practical purposes, since the Mg-component is always low, often even very low. In theory, however, a higher value should be assigned to MgO, since its neutralization power is 40% higher than that of CaO (20 g Mg have the same acid-neutralizing effect as 28 g CaO).

The suggestion to use Ca as a reference basis, in analogy to the nutrient fertilizers, makes little sense since Ca as such plays no role in the lime effect. Thus, gypsum is not a lime fertilizer, but a Ca-fertilizer (Chap. 2.5.1).

The content of basic-acting substances, and thus the pH-increasing effect of the various lime fertilizers is assessed by means of the uniform reference basis "CaO". It is computed as the sum of CaO and MgO.

ties; and there are still many marl pits to remind us of this fertilization activity at that time. The primary importance of marl is its lime content, but it also supplies nutrients.

To be effective as fertilizers in the soil, *calcium carbonates* (marly limestone) and *magnesium limes* must be finely ground, depending on the hardness of the rock, because of their slight solubility in water (Table 4-2). The magnesium content is an important quality criterion, in addition to the lime effect. The fertilizer may have the additional designation "with magnesium" if the $MgCO_3$-content is 5 to 15%. Smaller amounts, i. e., below 1.5% Mg, are not taken into account. The carbonate-free residues may be significant because of their quantities

Table 4-1. Selection of lime fertilizers.

Fertilizer	Formula of lime component (content in fertilizer)	Reference basis for lime effect in % CaO	Minor constituents	Properties
calcium carbonate (marly limestone)	$CaCO_3$ (75 to 95%)	42 to 53%	silicates etc., above 5% $MgCO_3$ designated "with Mg"	whitish, sometimes dark grey, slow-acting, grinding fineness for marly limestone: hard limestone 97%<1 mm, 70%<0.3 mm; soft limestone: 97%<2.5 mm, 50%<0.8 mm; chalk: 97%<4 mm, 70%<2 mm
magnesium carbonate (magnesium marl)	$CaCO_3$ (60 to 80%) $MgCO_3$ (15 to 40%)			
quicklime	CaO (65 to 95%)	65 to 95%	above 5% MgO designated "with Mg"	whitish, hot and corrosive after the addition of water, quick-acting, grinding fineness: 97%<6.3 mm, particle size: 0.4 to 6.3 mm (at most 5% finer)
magnesium quicklime	CaO (50 to 80%) MgO (15 to 22%)			
slaked lime	$Ca(OH)_2$ (80 to 93%)	60 to 70%	$Mg(OH)_2$, above 5% MgO designated "with Mg"	whitish, corrosive, quick-acting, grinding fineness: 97%<4 mm, 80%<2 mm, also granular
magnesium slaked lime	$Ca(OH)_2$ $Mg(OH)_2$ (>15% MgO)			
blast furnace lime	(about: Ca_2SiO_4) (75%)	40 to 50%	oxides, etc., 2 to 3% Mn, above 3% MgO designated "with Mg"	grey, acts very slowly, grinding fineness: 97%<1 mm, 80%<0.3 mm

Table 4-2. Specific surface area of 1 g limestone (specific gravity = 2.7) (according to [100]).

Diameter (mm)	Number of particles	Specific surface area (cm²/g)
1	$7 \cdot 10^2$	22
0.1	$7 \cdot 10^5$	225
0.01	$7 \cdot 10^8$	2 250
0.001	$7 \cdot 10^{11}$	22 500

(1 to 15% of sand, clay, or organic matter). These supplies mostly have positive effects, e. g., clay minerals on sandy soil. The larger contents of organic matter or dark minerals sometimes cause the white limes to have a dark colour, but this does not alter the lime effect. Trace elements also play a certain role in the supply of nutrients with limes. However, there are only occasionally significant amounts of manganese, up to 700 ppm Mn instead of low concentrations of 20 to 50 ppm.

Lime from marine algae is particularly soft. Thus, *algae lime* is obtained by drying and grinding red algae (*lithothamnion* that occurs in the sea at depths of 30 to 50 m) (at least 70% $CaCO_3$). The boron content, which is larger than in other limes, about 50 ppm, is significant. However, the content of common salt should be limited to below 3%.

Peat marl is produced from lime with 5% peat added; it is sometimes treated with *azotobacter* spores for soil inoculation.

Carbonates act slowly, since they are only slightly soluble in water (Table 4-3) and must first be converted in the soil. The solubility increases with the CO_2-content of the water in the soil, especially when the acid action is still stronger. Timely application is necessary because of the slow action, on the other hand, the risk of overliming is therefore smaller. Marly limestone can be stored better than quicklime or slaked lime. Ca carbonates yield carbon dioxide upon decomposition and are thus also CO_2 fertilizers.

Table 4-3. Solubility of limes in water.

Lime fertilizer	Solvent	Dissolved quantity (mg/l)
dolomite	water	8
$CaCO_3$	water	15
$CaCO_3$	water with 0.03% CO_2 } from air	50
$CaCO_3$	water with 2% CO_2 } from air	200
$CaCO_3$	water, CO_2-saturated	1 100
$Ca(OH)_2$	water	1 600

Quicklime and slaked lime have similar properties and effects. They act quickly; fine grinding of the fertilizers is important for uniform distribution in the soil, because of the concentrated lime effect. But in the interests of economy of labour, granulation might offer advantages for application. This involves only a slight reduction in effect if the lime content of the soil is approximately correct. Quicklime is also marketed as lump lime without granulation prescriptions. *Mixed lime* consists of quicklime or slaked lime with the addition of carbonates.

Quicklime is strongly corrosive and should therefore not be used for "top liming" (fertilizing the soil while crops are growing). Slaked lime has a greater volume and is slightly more expensive than quicklime.

Blast furnace lime is a silicate-based, relatively slow-acting lime fertilizer with a sustained effect. Its properties render it suitable for "tender" liming of light soils. The manganese content of 2 to 3% also helps to prevent overliming damage, which frequently implies manganese deficiency.

Converter lime may be considered to be a mixture of quicklime and silicate lime. It must contain at least 35% CaO (evaluation basis) and may be designated "with magnesium" if it contains more than 3% MgO, or "with phosphate" if it contains more than 3% P_2O_5. The P-component corresponds to Thomas phosphate in its action. Grinding fineness is the same as for blast furnace lime.

Residue limes contain diverse lime forms and minor constituents. The so-called "separation slime" from sugar refineries consists of dried carbonate with approximately 45% CaO (reference basis). Lime dust from cement production consists of different lime forms. One lime mixture with organic fertilizers that is on the market is *poultry-manure lime.* it consists of slaked lime, sometimes converted into carbonate, and poultry manure, and contains at least 30% "CaO" (reference basis).

Gypsum, used for soil improvement, is *not a lime fertilizer* (Chap. 4.2.1).

4.1.2 Application of Lime Fertilizers

The effects of liming are varied (see scheme). Correct fertilization with lime thus requires selection of the proper lime fertilizer, use of the correct amounts, and expert application.

The Correct Lime Form

An important criterion for selecting the lime fertilizer is the *speed of action of the lime.* Marly limestone in particular, is only slightly soluble in water, and it is therefore very important that it be finely ground (Table 4–2, p. 145).

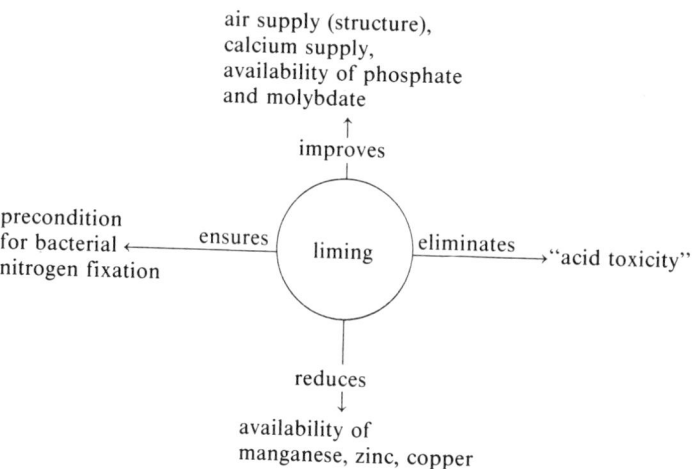

air supply (structure),
calcium supply,
availability of phosphate
and molybdate

↑
improves

precondition
for bacterial ←———ensures——— liming ┤eliminates———→"acid toxicity"
nitrogen fixation

reduces
↓
availability of
manganese, zinc, copper

Effects of liming a soil on nutrient supplies to plants.

Quick-acting limes are *quicklime* and *slaked lime,* and they quickly eliminate excess soil acidity. Because of their relatively high solubility in water, the basic action spreads rapidly in the topsoil, after proper mixing. Some of the lime effect also involves the subsoil. Small zones of excessively high pH-values are formed around very small lime grains, but the buffering capacity of the soils should be sufficient to compensate for this. Quick-acting limes are therefore indicated for medium and heavy soils in which no overliming effects occur even when the lime distribution is not ideal.

Slow-acting limes are *calcium carbonate* and in particular *blast furnace lime.* They must first be converted in the soil; this takes some time, especially in the case of blast furnace lime. The slow action of these limes is a great advantage in light soils, since the low buffering capacity of these soils could otherwise easily lead to overliming, especially when there is nonuniform lime distribution. However, overliming damage is also possible with slow-acting limes if the required quantity of lime (per unit area) is greatly exceeded.

In principle, all lime fertilizers may be used on all soils, provided that the quantity is dosed precisely and the lime is mixed optimally with the topsoil. However, allowing for certain imperfections in application makes it advisable to use lime fertilizers according to the following scheme:

- *Restoration liming of medium and heavy soils,* i.e., making-up of supplies, should be done primarily with quick-acting limes; (current) maintenance liming may be performed with either quick- or slow-acting limes, depending on their price and availability.

- Slow-acting limes are indicated for *light soils* ranging from sand to loamy sand), especially for maintenance liming. Quick-acting limes should be considered for first liming only when there are large make-up requirements.

The Significance of Minor Constituents

Marly limestone with a relatively small CaO-content often contains clay minerals that should be considered suitable for light soils. Some natural limes acquire a dark colour from organic matter, and thus provide a supply of humus, even if only a small one. For the replenishment of soil reserves, the contents of certain micronutrients in natural limes should also be taken into account. This applies in particular to the large Mn-content of blast furnace lime. However, the most important minor constituent is magnesium (Chap. 2.4). Magnesium lime should always be used as fertilizer when magnesium is required, since this is the cheapest method of Mg-supply.

The Correct Quantity of Lime

Correct dosage of lime is very important. The purpose of liming is not to eliminate the entire "soil acidity" but only undesirable acidity. Determination of the correct lime quantity is discussed in Chap. 5.1.2.

Maintenance liming requires an average quantity of 10 to 15 dt/ha over a period of three years. This value should be used if, for any reason, lime requirements cannot be determined. However, this is a very rough estimate that is not recommended, since determination of lime requirements is very cheap.

The Correct Application of Lime

The purpose of liming is a uniform increase of soil reaction. This requires proper distribution of the lime over the area and proper mixing of the lime with the soil stratum to be limed, usually the A-horizon and occasionally also the subsoil when there is need of melioration liming.

Consequently, it would seem advisable to lime the crop-free soil after the harvest, before it is worked more intensively, for example, *stubble liming* in the case of grain. Lime may also be applied at other times when the above-mentioned conditions exist. Thus, liming of potatoes a long time before planting is often inadvisable, since as a result of a smaller supply of manganese this can cause a higher incidence of scab. "Top dressing" is indicated in such cases, i.e., lime is spread into the growing crop, so that part of it unintentionally also reaches the leaves. Either slaked lime or calcium carbonate should therefore be used to reduce corrosion damage. If possible, the material should be applied in dry weather. Quicklime would greatly corrode the leaves, since it becomes hot when water is added (this further increases the basic action).

Liming of *grassland* is more problematic than liming of fields, since in the latter case the lime cannot be worked in. Correct liming during sowing of the grass is therefore particularly important. Lime penetrates into the A-horizon of permanent grassland through natural displacement. This, however, is a slow process, so that more frequent application on grassland is indicated.

The problem of correct *application* also influences selection of the proper *lime form.* Fertilization with lime often involves the transportation and storage of large quantities of fertilizer. If the material must be stored outdoors, hardening of the lime should be prevented. Calcium carbonate is much more suitable in this respect than quicklime or slaked lime. On the other hand, far more marly limestone than quicklime is required for the same lime effect.

The Timing of Lime Application
Liming is not necessary every year and is therefore particularly indicated at certain stages of crop rotation, e. g.,

- liming of plants preferring higher pH-values, e. g., sugar beet, rape, etc.;
- liming of potatoes, since this root crop permits good mixing of lime and soil;
- before more intensive soil working, e. g., after turning-up grassland.

Liming of grassland is indicated when the grass is short, e. g., in winter or after cutting, so that the lime will penetrate the turf as far as possible and reach the soil.

Neither on fields nor on grassland should the lime turn into a hard crust on the soil surface, when large quantities are applied, since the lime distribution in the soil would then be very uneven.

4.1.3 Fertilizers for Soil Acidification

The soil reaction can be reduced by acidifying fertilizers if it is above the optimum. We should in this case distinguish between

- fertilizers for lowering the pH (acidification), i. e., fertilizers applied precisely for this purpose;
- fertilizers that lower the pH as a side-effect; this is usually undesirable, but may occasionally also have positive results (Chap. 2.1.5).

Intentional *fertilization to lower the pH* may be required on

- soils erroneously overlimed, e. g., to increase the manganese supply;
- highly alkaline soils, e. g., soda soils, to eliminate the highly to extremely alkaline-acting soda content (Chap. 6.3.4).

Fertilizers for intentional acidification should be either acids or yield acids after reaction. The most important substance of this kind is *sulfuric acid.* The reaction between sulfuric acid (1%) and sodium carbonate (soda) is as follows:

$$H_2SO_4 \;+ Na_2CO_3 \longrightarrow \quad Na_2SO_4 \;+ \quad CO_2 \quad + H_2O$$

sulfuric acid   soda   sodium sulfate   carbon dioxide   water

This reaction, sometimes used for the melioration of soda-salt soils, involves the mutual neutralization of two approximately equal masses. However, the use of pure sulfuric acid is technically difficult. Therefore, it is better to use substances that yield sulfuric acid after reaction.

The following reactions play a role in this case:

$$2\,S \;+\; 3\,O_2 \;+ 2\,H_2O \longrightarrow \quad 2\,H_2SO_4$$

sulfur   oxygen   water   sulfuric acid

$$FeSO_4 \quad + 2\,H_2O \longrightarrow \quad H_2SO_4 \;+ \quad Fe(OH)_2$$

ferrous sulfate   water   sulfuric acid   ferrous hydroxide

$$(NH_4)_2SO_4 \quad + H_2O \longrightarrow \quad H_2SO_4 \;+\; 2\,NH_3 \;+ H_2O$$

ammonium sulfate   water   sulfuric acid   ammonia   water

The quantities of *sulfur* required for soil acidification depend on the lime content and other soil properties. The quantity applied, for tea, for example, is about 10 to 20 dt/ha if the pH of the entire A-horizon is to be reduced by at least 0.5, in the absence of lime. Erroneous overliming (e.g., of a light soil) can also be corrected by fertilizing with sulfur. However, it is usually better to acidify with acidifying N-fertilizers (Chap. 2.1.5).

4.2 Fertilizers for Improving Soil Structure and Texture

A precondition for optimal plant growth, besides nutrient supplies, is an optimal water supply and aeration of the soil, as well as root penetrability into the subsoil.

An important measure for improving the structure and opening-up the subsoil is correct tillage of the soil. This ensures only temporary improvement but does create favorable conditions for the structure-forming forces in the soil, whose activation through a precise supply of substances is thus an indirect measure for improving the structure.

4.2.1 Fertilizers for Structure Improvement

Several fertilizers have recently been developed specifically for improvement of the structure. They are in part included in the type list as *soil conditioners*.

The main problems of structure improvement, in the broader sense, are:

- Increasing the *water capacity* and *resistance to erosion* of light soils;
- creating a crumb structure in heavy soils, chiefly for *aeration*.

Many fertilizers improve the soil by their side effects (limes, organic fertilizers). This chapter, however, deals with fertilizers for direct, precise application (Synopsis 4–1, p. 140).

Improvement of Soil Granulation

A *crumb structure* is formed automatically in fertile soils. Mineral particles combine to form stable aggregates mainly through the activity of soil organisms. This creates large pores for aeration and drainage, and small pores for water retention.

The following substances contribute to the bonding of the particles:

- *Inorganic matter*: oxide-, lime-, silicate coatings, and gypsum;
- *organic matter*: slimy "glues" (polysaccharides, especially polyuronides) produced by microbes, the hyphae of fungi ("contexture produced by the life process"), and humic substances derived from the formation of clay-humus complexes; conditions for this are especially favorable in the intestines of animals that live in the soil, particularly earthworms.

In some soils it may be necessary to improve or supplement natural crumb formation. An important precondition for a proper crumb structure is the far-reaching saturation of exchange complexes with calcium; so that a supply of calcium to acid soils is thus indicated. Suitable lime fertilizers should be applied, or *gypsum* if liming is not possible.

Gypsum ($CaSO_4 \cdot 2H_2O$) increases calcium reserves in the soil without raising the pH. It thus promotes the aggregation of clay particles (flocculation) and the combination of colloidal particles through adsorption by calcium. Fertilization with gypsum may be indicated particularly for heavy soils; the quantities required are considerable (20 to 100 dt/ha). Fertilization with gypsum plays a special role in the melioration of alkaline soils as it displaces sodium from the exchange complexes (Chap. 6.3.4).

Special Soil Conditioners as Fertilizers or Soil Consolidating Agents

These fertilizers are used for loosening heavy soils containing little humus and for stabilizing light soils.

The soil conditioner *ammonium iron sulfate* $(NH_4)_2SO_4 \cdot Fe_2(SO_4)_3 \cdot 6H_2O$ contains up to 85% of this compound plus 10% organic matter (e. g., peat). The material acts by forming bonding coatings of iron oxide.

Silicate colloids consist of sodium hydrosilicate and sometimes phosphate (e. g., *Agrosil LR* with 36% SiO_2 and 7% P). It forms a silicate sol and reversible gels with water, and increases the capacity of the soil to store water and nutrients. Agrosil can be applied in the dry state or as a suspension in water at a ratio of 1:20. The quantity applied may be 15 dt/ha.

Organic soil conditioners imitate the natural bonding through filamentary molecules. They must resist rapid decomposition, so that their effect may be sustained for several years. Various *polymer dispersions* and powders with polymers, long chain and filamentary molecules, are marketed in and outside Germany. One of the first of these soil conditioners was *Krilium* in the USA. It is based on polyacrylic acid. Polyacrylic acid esters are similar (e. g., *Acronal*). *Alginure* is obtained from algae; it is based on polyuronide. Other products were derived from polyvinyl acid, e. g., *VAMA* (polyvinyl acetate and maleic acid anhydride) or *polyvinyl propionate*. These substances are sprayed on the mechanically loosened soil in the form of *dispersions*, mixtures with water, or scattered as *powders* and "rained in" with water. The quantities applied vary between 1 and 20 dt/ha. The effect is sustained for several years. However, the considerable cost per unit area restricts this method chiefly to horticulture [79].

Improvement of Soil Aeration
Heavy soils can be improved by fertilizers that loosen the soil. A proven product is *Styromull* consisting of flocs of polystyrene foam (*Styropor*). The foamed material is chemically inert, i. e., it does not react with the soil, resists rot, and does not become moist internally, since it consists of cells filled with air. The addition of these 4 to 12 mm-size flocs considerably increases permeability to water and aeration. About 10% by volume or 1 to 2 m^3 per 100 m^2 are required.

Improved soil aeration can also be achieved through coarser granulation (Chap. 4.2.2), e. g., by adding rock powder.

Improvement of the Water-Retaining Capacity of Light Soils
Light soils often dry out, but this can be prevented by adding water-storing substances. Thus, Agrosil as an inorganic product has a favourable effect, whereas *Hygromull* is very important as an organic fertilizer. It consists of flocs of urea-formaldehyde resin. This foamed plastic has fine open pores in the 4 to 12 mm-size flocs, where water in stored in amounts up to 60 to 70% of the volume. Only about 5% is decomposed annually with a corresponding part of the N-component (30%) being mineralized. *Hygropor* consists of 70% Hygromull and 30% Styromull. The quantities applied outdoors are 2 to 4 m^3 per 100 m^2.

Activation of Soil Life

The addition of *activators*, sometimes with microbial spores, is advocated for promoting the formation of a proper structure by soil organisms. However, significant results can be expected only under suitable conditions (Chap. 4.5.2).

4.2.2 Mineral Substances for Texture Improvement

The structure of a soil is more easily improved the better the basic preconditions for a medium particle-size distribution (granulation).

Light soils (sandy soils) lack fine particles (clay particles), whereas heavy soils (clay soils) lack coarse particles. The consequences of extremely coarse or fine particle sizes is a low potential for natural structure formation and thus for a good structure. This potential can only be supported, but not replaced, by fertilization measures.

The obvious measures for *altering the particle size* are:

• Light soils require a supply of clay particles;
• heavy soils require a supply of sand particles.

But the problem is what quantities should be supplied. Addition of 1% of a mineral component represents a quantity of 30000 kg/ha (30 t/ha) at a topsoil weight of 3000 t/ha (of the 0–20 cm layer). Thus, increasing the clay content of a sandy soil from 4 to 10% to convert it into loamy sand requires 6×30 t = 180 t/ha of clay material. However, usually the available material only consists of up to one half clay, so that some 400 t/ha should be allowed for. This treatment with clay material would involve substantial transportation costs even if such material were available free of charge in the vincinity of the field. Similar considerations apply to the supply of sand.

This meliorative fertilization is simplest on sites where suitable soil material is available in the subsoil and must only be brought to the surface. This is the case with *melioration* of heavy, marshy soils with *blue sand*, which has the further advantage of containing lime.

The texture of a soil must generally be considered constant under agricultural production conditions. It cannot be changed practically within the scope of economic possibilities. The situation is different in horticulture or in transitional forms of particularly intensive agriculture. Favourable climatic conditions permitting several harvests a year, or favourable economic production conditions may make it advisable to improve the texture by suitable fertilization. A useful fertilizer for this purpose would be rock powder, which, however, could at best be considered a coarse-grain additive if suitably ground.

There are some exceptions to the general rule that a medium texture is optimal for the soil structure. Thus, many *vertisols*, dark clayey soils in subtropical regions, with clay contents of 50 to 60%, which have to be classified as extremely heavy soils, would not necessarily benefit from the addition of sand (Chap. 6.3.3). On the contrary, productivity studies indicate that growth conditions, with irrigation, improve rather than deteriorate with increasing clay content.

4.3 Organic Fertilizers

Organic fertilizers include a large and diverse group of substances that serve various fertilization purposes. They may be generally subdivided into the following categories:

- Organic *farm manures* of widely varying compositions, obtained in agricultural and horticultural enterprises or produced in them;
- organic *commercial fertilizers* whose production is subject to certain standards and whose marketing is subject to the fertilizer law.

Organic fertilizers are primarily evaluated according to the content of organic matter (%). It would be possible but not practical to indicate the carbon content (%C) as a measure of organic matter. The content of organic matter is determined either as a loss on ignition (thus yielding approximations) or by exact methods.

Organic farm manures, more precisely *fertilizers produced on the farm itself*, are important primarily because of their organic content. All soils require the supply of organic matter as a carrier of utilizable energy and nutrients for the soil organisms, as well as for many other reasons (Chap. 4.3.4). All soils are supplied with organic matter, even if only harvest residues left in the soil and on the field. This supply, which is provided in any case, has been supplemented since antiquity with organic wastes obtained on the farm itself or supplied from the surroundings (e.g., peat). The use of wastes as fertilizer offers not only the advantage of removing noxious waste products, but also permits their useful *recycling* by natural circulation.

4.3.1 Farmyard Manure, Liquid Manure, Semi-Liquid Manure

The excrement of domestic animals, sometimes with additives, provides the farmyard fertilizers known as farmyard manure, liquid manure, and semi-liquid manure.

Farmyard manure
Farmyard manure [91] was formerly the most important farmyard fertilizer and on many farms still is. It is a "universal" fertilizer in view of its many effects, but

its properties vary widely. Farmyard manure consists mainly of faeces, sometimes with urine and litter (straw, etc.). It is mainly produced by cattle, pigs, and poultry. Particulars of its production and properties are given in Synopsis 4–5.

The action of farmyard manure on plant yields is the sum of its soil-improving and nutrient effects. Obviously, it varies within wide limits according to site conditions. Yield effects to be expected might be best estimated from the nitrogen content.

Synopsis 4–5. Farmyard Manure.

1. *Amounts of fresh manure produced:*
   About 100 dt per *head of cattle* per stabling period (30 kg faeces per day);
   about 20 dt per *pig* per year (100 dt per livestock unit);
   about 6 kg per *hen* per year.

2. *Storage* either in the deep stable (this is advantageous as regards economy of labor and yields a higher N-concentration (0.8%)), or on the manure heap as
   a) *manure produced by cold fermentation:* far-reaching exclusion of air causes anaerobic fermentation at up to 30 °C; losses of organic matter = 10 to 15%;
   b) *manure produced by warm fermentation:* compact storage ensures digestion at up to 40 °C; losses of organic matter = 30 to 40%;
   c) *hot fermentation process* (fermented manure): loose storage ensures digestion at up to 60 °C; losses of organic matter = 40 to 50%.

   Losses of organic matter are not altogether detrimental, but ensure a relative enrichment in mineral nutrients.

3. The *nutrient contents* of digested manure vary within wide limits. The following contents may be assumed for cattle and pig manure (about 20 to 25% organic matter with a C:N-ratio of 20:1), see Table 4–4.

   Pig manure contains relatively more P than cattle manure; chicken manure contains significantly more nutrients, in part, because of the larger content of dry matter; it thus has about a five times higher concentration of N.

4. *Quantities applied:* 100 to 400 dt/ha.

5. *N-losses* during application: 2 to 30% (particularly large losses at high temperatures and winds); losses can be reduced by the addition of superphosphate, which binds ammonia.

6. *Nutrient utilization rate* [52]: for

   N: 20 to 30%   in the first year, after-effects barely one half of this in the following year (about 50% average rate for first years);

   P: 15 to 20% } as for mineral fertilizers.
   K: 50 to 60% }

Table 4-4. Nutrient content of farmyard manure.

Contents (%)	Contents (ppm)
N = 0.2 -0.6% (∅0.5)	manganese = 30 to 50 ppm
P = 0.04-0.3% (∅0.1)	zinc = 10 to 20 ppm
K = 0.1 -0.8% (∅0.5)	borum = 3 to 5 ppm
Ca = 0.07-1.0%	copper = 1 to 3 ppm
Mg = 0.06-0.3%	molybdenum = 0.1 to 0.2 ppm

The most important soil-improving effect is the increased humus content of the soil. Permanent humus forms one quarter to one third of the organic matter. Much of it remains stable in the soil for longer periods, apart from the formation of permanent humus from the decomposition of farmyard manure. Soil organisms also benefit from farmyard manure, since its *nutrient-humus* content is an important source of nutrients and particularly of energy. Farmyard manure has particularly favorable effects on root crops, where its soil-loosening and structure-improving action is of primary importance. Application should be in the autumn in the case of sugar beet, and possibly in the spring with potatoes.

Liquid Manure

Liquid manure [91] is a liquid, more precisely a suspension, consisting mainly of the urine of domestic animals, decomposed by microbes, necessarily admixed with faecal particles, and sometimes with seepage from solid manure and rainwater. The urine content of liquid manure is often about 50%. The amount of urine produced per day per animal is about 15 kg for cattle and 4 kg for pigs. This represents 4 to 5 m^3 liquid manure per livestock unit per year. The dry matter in liquid manure (only about 1 to 3%) consists predominantly of *urea* which is largely fermented to ammonium salts. Hippuric acid, which is converted to benzoic acid, also contains N. The considerable content of soluble potassium salts makes liquid manure a *NK-fertilizer*. The nutrients most often occur in easily available form, so that liquid manure is a quick-acting fertilizer.

The *nutrient content* varies by a factor of five or ten and should therefore be indicated only by rough average values: 0.2% N, 0.01% P, 0.5% K. The average *nutrient supply* in 10 m^3 liquid manure is thus 20 kg N and 50 kg K. The *quantities to be applied* largely depend on the N-content and may be 20 to 50 m^3/ha (40 to 100 kg N).

Large ammonia losses may occur during *application*, but they can be reduced by dilution with water or rapid introduction into the soil.

The effects on the yield chiefly depend on the supply of nutrients (N and K).

Semi-Liquid Manure

Labor-economic considerations have often necessitated transition from solid/liquid-manure management to *semi-liquid-manure* management. The combined excrements of domestic animals are generally termed semi-liquid manure, but the nomenclature is not uniform (this cannot be expected in view of the different methods of hydraulic manure removal).

Semi-liquid manure (Gülle), in the general sense, is a mixture of faeces and urine, sometimes with a small litter component, diluted with varying amounts of water, for the purpose of mechanical transportation.

Various forms can be distinguished, but in principle there are two groups:

- Urine, semi-liquid manure, thin semi-liquid manure, or hydraulic manure, sometimes diluted more than five times with water;
- semi-liquid manure rich in faeces, thick semi-liquid manure, diluted with very little water.

An important problem of semi-liquid-manure management is the compromise between the semi-liquid manure pumpability, which requires a dry-matter content of less than 12%, and savings in storage space, which require minimum dilution with water.

The following are approximate quantities of semi-liquid manure obtained as raw material:

per head of cattle: 50 kg per day, containing 10% dry matter;   } semi-liquid
per pig: 4 kg per day, containing 15% dry matter;   } manure
per hen: 50 g per day, containing 80% dry matter
                   (dry faeces).

The nutrient contents are indicated in Table 4–5.

Table 4–5. Nutrient content of semi-liquid manure.

Nutrient	% nutrients in semi-liquid manure, referred to 7.5% dry matter		15% dry matter
	cattle	pigs	chicken
N	0.4	0.6	1.0
P	0.1	0.2	0.3
K	0.4	0.2	0.4
Mg	0.04	0.05	1.0
organic matter	5.5	6.0	10.5

(Trace-element contents approximately correspond to those in farmyard manure.)

Raw semi-liquid manure ferments through intensive microbial decomposition. For example, nitrogen is largely converted from organic to mineral form. About 50% of the nitrogen then occurs as ammonium-N and acts relatively quickly. The $C:N$ ratio varies between $5:1$ and $8:1$. Some gaseous products of fermentation (hydrogen sulfide, ammonia, etc.) may cause annoyance because of their smell. Technical problems of semi-liquid manure preparation including disinfection, are discussed in [86, 99]. Semi-liquid manure should be "decomposed" as much as possible.

The use of semi-liquid manure for fertilization is similar to that of solid and liquid manure. The major effect of semi-liquid manure on the field is through the nutrients, whereas the humus action is insignificant. The criterion for determining the quantities necessary for effects on yields is the nitrogen content; one half to three quarters of the necessary N-supply can be provided by semi-liquid manure. The utilization rate of N is 30 to 50% in the first year, when semi-liquid manure is applied several times. Immediate introduction of semi-liquid manure into the soil prevents ammonia losses.

It is important to know the *maximum quantities* of semi-liquid manure to be applied (Table 4-6) in view of possible overfertilization (salt damage, etc.) and nutrient losses from the soil, which harm the environment. Nitrate losses may be considerable and are undesirable from the agricultural as well as the environmental point of view.

Table 4-6. Maximum quantities of semi-liquid manure to be applied (referred to liquid).

Semi-liquid manure from	m^3 per hectare and year for fields and meadows	pastures	Maximal number of animals per hectare of area fertilized with semi-liquid manure
cattle	40–80 ⎫		2– 4 heads of cattle
pigs	30–45 ⎬	15–22	10– 15 pigs
chicken	15–25 ⎭		200–300 chicken

A great benefit, but also a risk inherent in the extensive use of semi-liquid manure, is the considerable accumulation of available nutrients in the soil. The amounts may be up to five or ten times the normal content in the case of phosphate, for example. The problems associated with this are discussed in Chap. 6.2.1.

4.3.2 Straw, Compost, Green Manure

Farmyard fertilizers from plant matter are straw, compost, and green manure.

Straw

Straw is becoming less necessary in the stable as semi-liquid-manure management increases and is sometimes left on the field as waste product from the harvest. About one quarter of all the straw is "problem straw" in the FRG. Destroying it by burning only makes sense from the production aspect when there is an oversupply. It is useful as an organic fertilizer, especially on farms without livestock [77].

Although straw contains few nutrients, it yields decomposable *organic matter*, e.g., cellulose. Fertilization with straw thus supplies energy (1 kg straw = 12 kJ), but mainly improves the structure of medium and heavy soils. However, loosening of soil by straw should not be excessive, since this would interfere with the supply of water.

Nutrient contents vary but are small in any case. Rough averages for winter cereals are:

0.5% N, 0.1% P and 1% K.

These values apply equally to fresh and dry matter when the content of dry matter is approximately 85%. The C : N ratio is very large, about 100 : 1. This may cause inhibition of decomposition, because the microbes lack N. Compensating supplies of nitrogen (about 1% of the quantity of straw) is thus required.

Fertilization with straw should be carried out as early as possible before sowing, in view of the *decomposition* that is necessary (this is in any case preferred from the labor point of view). One third to one half of the straw should, if possible, already be decomposed before winter. Possible drawbacks of fertilization with straw, such as the proliferation of weeds and plant diseases, should be taken into account.

Fertilization with straw is utilized best by root crops and legumes. The humus content can hardly be increased by straw, apart from the undesirable accumulation due to poor decomposition conditions. Quantities of 50 to 100 dt/ha of straw are generally well utilized by the soil. Possible side effects of fertilization with straw are discussed in Chap. 4.3.4.

Compost

Dry or fresh plant- and animal wastes can be introduced directly into or on the soil. They are then decomposed at varying speeds, but have only slight, and sometimes even detrimental, effects and bring about scarcely any sustained soil improvement. Processes known since antiquity make it possible to convert such wastes into useful agents for soil improvement, i.e., composts. The word is derived from the Latin *componere* — put together.

Compost is a product of the decomposition of plant and animal wastes with various additives. The compost group has the largest range of variation of all organic fertilizers. It ranges from worthless matter on neglected garbage dumps to carefully composed and treated substrates with high fertilizing effects. The multiplicity of raw materials and purposes corresponds to the large number of recipes for optimal compost preparation [98]. Thus, the recipe for "Indore"-compost according to *Howard* is from India.

The following basic rules are important for the *production* of good-quality compost:

a) The purpose of composting is to *convert* organic matter *into soil* and growth-promoting substances for sustained soil improvement.

b) The organic matter is partly decomposed and converted by microbes. These require proper growth conditions for their activity, i. e., humidity (about 50% water capacity), and aeration. This is achieved through stacking and occasional *turning-over*. Microbes also need sufficient nitrogen for synthetizing their body mass (the optimum C:N ratio is 20 : 1 to 30 : 1), etc.

The addition of "mature" compost or *inoculation* with bacterial cultures may be required in order to provide a good start for the development of bacteria (Chap. 4.5.2).

c) *Soil animals* contribute considerably to the decomposition of organic matter through their comminuting activities. The larger of these animals also provide optimal conditions in their digestive tracts for the synthesis of valuable permanent humus and stable soil crumbs. A typical compost earthworm is *Eisenia foetida*.

d) Certain *additives* can speed up conversion and improve the final product:

- Minerals such as earth, lime, and other mineral fertilizers, possibly also rock powder. The addition of nitrogen (0.1 to 1%) is important in case of large C:N ratios, as is the addition of lime (0.3 to 0.5%) if sufficient lime is not present.
- Organic agents, e. g., herb extracts (however, their effects are difficult to determine and measure).

The *preparation of compost* takes several months to three years. Compost acts as fertilizer in various ways (Chap. 5.3.4). Its composition varies within wide limits. The following average values are rough approximations:

30 to 50% dry matter, 10 to 15% organic matter, 0.3% N, 0.1% P, 0.3% K.

The quantity of good-quality compost to be applied has no upper limit, since it is itself a good nutrient substrate. About 10 kg per m^2 (3 to 10 l/m^2) are required for the application of a 1 cm-thick compost layer. Compost as commercial fertilizers are discussed in Chap. 4.3.3.

Green Manure

Green manure consists either of whole green plants used as fertilizer or only of stubble and root residues. The latter represents about one half of the plant, while the green parts, above the surface, are used as forage (Chap. 7.5.1) [84, 91].

The formerly important nitrogen fixation (Chap. 7.3.4) now plays the principal role only in regions where fertilization is extensive.

Green manure is applied for nitrogen fixation and to improve the soil structure, and intensive root penetration of the growing plants already contributes to the process. In addition, green manure should exert as many effects of organic fertilization as possible (Chap. 4.3.4).

Green-manure plants apart from legumes (formerly, mainly yellow lupins and serradella on sandy soils) consist of other plants (cereals, oil plants). Rapid, proper start of growth is important, and here abundant nutrient supplies make an important contribution. From the agricultural point of view there are the alternatives of autumn intercrops, as stubble seed or underseed, or winter intercrops.

Yields depend on the duration of growth and other factors, and are

- 100 to 200 dt/ha of green mass that can also be used as forage ($=$ 20 to 40 dt/ha of dry matter);
- 7 to 30 dt/ha of dry matter in the form of root residues and stalks;
- about 100 kg N in the case of *legumes,* of which one quarter to one half occurs in the residues, with the utilization rate varying widely.

Fertilization with *beet leaves* as a special form of green manure plays a certain role in farms that have no livestock. Sugar beet yields 300 to 500 dt/ha of beet leaves. Nutrient contents are: 0.5% N, 0.05% P, and 0.5% K.

Green manure supplies nutrient humus and provides nitrogen. However, only legumes provide a real N-gain, if one disregards the fact that plants "rescue" nutrients from leaching. The increased decomposition of nutrients beneath growing crops may play a certain role, by causing greater mobilization of phosphate and trace elements. Green manure obviously supplies no additional nutrients to the soil. The humus content of the soil is not increased permanently by green manure. Certain losses of permanent humus are even possible, owing to the increased stimulation of decomposition (priming effect).

4.3.3 Commercial Organic Fertilizers

Commercial organic fertilizers up to now have been divided in the type list into

- *organic fertilizers* (fertilizers from organic matter: about 20 types);
- *organic-mineral fertilizers* (fertilizers from organic matter, with the addition of mineral fertilizers: about 35 types).

The two groups will be taken together because of their similarity (Table 4-7). This will also be done in future in the fertilizer law.

Marketing requires careful *processing*. This essential precondition is specified in the regulations. The raw materials, animal and plant wastes, sewage sludge, etc., must be processed to products that are harmless to health, through hot fermentation or some other process. The fertilizers must contain no pathological agents or substances that are hygienically questionable or harmful to plants. This requirement is difficult to satisfy strictly or to check. A supply of magnesium is desirable, so that up to 2.5% Mg or 4% MgO may be added.

Table 4-7. Organic commercial fertilizers (fertilizer types).

	Fertilizer type	Organic raw material	Value-determining constituents
1. organic fertilizers	organic fertilizer	peat, organic residues	30 to 50 organic matter
	organic N-fertilizer	{ blood powder, horn powder, etc.	5 to 14% N
	organic P-fertilizer	bone meal	13% P
	organic NP-fertilizer	mixtures of horn and bone powder, etc.	4 to 10% N, 2 to 10% P
	organic NPK-fertilizer	guano, animal residues	4 to 6% N, 4 to 5% P, 2% K
2. organic-mineral fertilizers	organic-mineral mixed fertilizer	town wastes, sewage sludge, animal and plant wastes	25 to 40% organic matter (about 1 to 3% N, P, K)
	peat mixed fertilizer	brown or black peat	more than 35% organic matter (about 1 to 3% N, P, K)
	organic-mineral N-fertilizer	lignin	15 to 19% N
	organic-mineral NP-fertilizer	animal and plant wastes and residues	6% N, 3 to 4% P
	organic-mineral NPK-fertilizer		4 to 14% N, P, 3 to 14% K

Only a few of the many organic and organic-mineral fertilizers on the market can be considered here [85]. Organic fertilizers will be discussed briefly in more detail below.

Fertilizers from Peat

Peat has been known from antiquity as an organic fertilizer and has been marketed for a long time in various forms as *fertilizer peat*. It was formerly marketed mainly as *brown peat,* but because of its scarcity *black peat* is also sold now. Peat consists of 40 to 50%, each, of organic matter and water, and contains 1 to 2% minerals ("ash"). It is marketed in bales weighing about 50 kg each. The *volume of the loose peat* (volume of loosely filled peat) is about 300 l. One to two bales are applied per 100 m² area, e.g., about 1 kg/m². Peat primarily improves the structure and is decomposed only slowly in the soil (example of a trade name: *Floratorf*).

Peat mixed fertilizers contain additives of mineral fertilizers that at least partially compensate for the absence of natural nutrients in pure peat. They are applied like pure peat (examples of trade names: *Huminal, Nettolin, Super-Manural*).

Fertilizers from Town Wastes

Dewatered sewage sludge is a useful fertilizer after suitable processing and can also be enriched by the addition of nutrients. It consists of up to 35% of dry matter, about one half of which is organic. The quantities applied are of the order of 30 to 40 dt/ha. Drying can increase the dry-matter content to 90%.

Refuse composts and refuse-sewage composts are increasingly being produced from town wastes and can also be converted into useful fertilizers by suitable processing. They are used primarily for improving the soil structure, but they also supplement the supplies of certain nutrients, especially some micronutrients. Processed composts contain about 0.7% N in the dry matter. The *quantities to be applied* vary between 50 and 200 dt/ha but may be even greater. However, there is an upper limit because on one hand, there is a risk of salt damage, which can be largely avoided by application a long time before sowing, and on the other hand, over a longer period in cases of continuous intensive application, there is a risk of excessive accumulations of heavy metals and other harmful substances. The "normal contents" of certain "problematic" elements in soils should not be greatly exceeded (Chap. 6.2.1).

Fertilizers from Animal Wastes

The fertilizing action of animal wastes, such as blood, has been known since antiquity. Fertilizer production began about 1800 with the recycling of bones.

Bone meal is produced from bones obtained from abattoirs. The bones are crushed, degreased, and cleaned, yielding bone grist. Finely ground, it serves as an organic NP-fertilizer. Removing the proteins from the grist by delamination yields *delaminated bone meal*. This is a P-fertilizer of organic origin (Ca-phosphate), used more frequently for fertilization.

Horn material yields *horn powder, horn grits,* or *horn chips,* depending on the degree of crushing. This is a slow-acting N-fertilizer and is sometimes also treated with mineral fertilizers.

Horn-and-bone-meal mixed fertilizers yield organic NP-fertilizers, depending on their composition (N-component from horn, P-component from bones).

Blood powder is a very effective fertilizer. Its principal component is N (up to 14% in slow-acting form). The adage "blood is a special sap" indicates that the contents of all important elements and organic agents also play a role. Other *animal wastes* of all kinds, hair, intestines, cartilage, fish waste, etc., form the basis of other organic fertilizers, like processed animal excrement, e. g., of cattle and poultry (Chap. 4.3).

Guano plays a special role as a fertilizer of animal origin. (Raw) guano is a product of sea bird excrement, converted over long periods and occurring in natural deposits. These birds live on islands with no rain or vegetation along the coasts of Peru and Chile and feed on the abundant fish in the sea. The name originated in Peru and denotes "manure" *(huano).* The deposits are up to 60 m thick; however, only the central layer has a large N-content. Natural conversion produces a predominantly inorganic substance from the originally organic matter; it contains 8 to 15% N and 2 to 3% P. The chemical constituents are mainly ammonium oxalate and ammonium phosphate, as well as calcium phosphates. There are important admixtures besides 2 to 4% K. Raw guano is sometimes processed into guano fertilizer by acid decomposition. Peru-Guano $6+12+2$ $(N+P_2O_5+K_2O)$ is one type on the market. Guano also occurs elsewhere, sometimes as *cave fertilizer* produced by bats.

Fertilizers from Plant Wastes

Various residues are left over from the processing of plant products. These can be used directly as fertilizer after processing or are marketed as composts after special preparation and treatment with additives. The many possibilities can only be summarized and represented here by some examples.

Grape-kernel oil residues or *grape residues* are processed into composts after the addition of farmyard manure and bacterial cultures for activation of the microflora (e. g., *Cofuna* or *Regenor* with about 50% dry matter and 1% N).

Residues of castor seeds, tobacco, cocoa, malt sprouts, molasses, seaweeds, etc., are also processed into fertilizers. Lignin is a residue of the production of cellulose from wood and can be processed into *N-lignin* fertilizer by the

addition of ammonia under pressure. However, this process is not very important at present. Pure *coal* (lignite) may be considered to be a "borderline case"
of organic fertilizers.

In *conclusion* it may be stated that in the final analysis nearly all plant and
animal wastes can be processed into fertilizer, unless they contain harmful substances.

Composts for Horticulture

Much organic matter is often required in commercial and private gardening. Organic commercial fertilizers are frequently used if the compost production of the
garden is insufficient. A choice of various composts is especially popular. *Processing agents* suitable for compost production are discussed in Chap. 4.4.2.

Composts (Chap. 4.3.2) exert various soil-improving effects (Chap. 4.3.4).
They therefore frequently increase yields and improve quality. Occasionally,
harmful substances occur, largely due to improper preparation. The quantities
applied vary within wide limits, e.g., 1 to 2 kg/m^2 (much like peat).

The age-old familiarity of gardeners with compost as a reliable fertilizer,
which almost always has positive effects and hardly ever involves any risks of
overdosage or undesirable side effects, makes composts very popular and manufacturers have not neglected this in their publicity material. An "ideal" compost
fertilizer might be advertised as follows:

"The Ideal Compost Fertilizer" for the Garden

has a large *content* of organic matter and has been treated with farmyard manure, rock powder, marine algae, herb extracts, and beneficial bacterial cultures;
contains all plant nutrients, especially the all-important trace elements, matched
harmoniously in a natural slow-acting form, also contains enzymes and hormones with considerable growth-substance effects, as well as antibiotics against
plant diseases; *improves* the water-retaining capacity and structure of the soil,
renders it crumbly and easy to work, promotes the activity of soil organisms including earthworms, reduces weed growth, contains no artificial additives, is
completely nontoxic, and is simple to apply and reliable.

In brief, "Ideal Compost Fertilizer" is a fertilizer ensuring ideal, naturally
pure soil fertility, which turns the garden into a flourishing paradise ... The consumer would then need to know only one thing, the *price.*

4.3.4 Effects of Organic Fertilizers

Even the effects of simply composed mineral fertilizers are more complex than
is generally assumed. This holds true to a far greater extent for organic fertilizers.

The many possible effects exerted by organic fertilizers on plant growth (the problems of *humus fertilization*) can only briefly be discussed here [95, 80, 93]. These possibilities are known qualitatively but comprehensive assessment of the quantitative effects of the various influences is still very difficult, if not virtually impossible. In any case, they can be elucidated, if necessary, only by extensive investigations of the individual site conditions concerned.

Organic fertilizers are important primarily as *humus fertilizers*. Differentiation between *nutrient* (unstable) *humus* and *permanent humus* [95] is important for a rough classification of humic substances. Nutrient humus is decomposed more or less rapidly by soil organisms, providing them with nutrients and energy. Permanent humus is stable for longer periods and improves soil fertility. The manifold effects of organic fertilizers on cultivated plants are partly direct, partly indirect, e. g., via the soil organisms whose activity is therefore of special significance. The effects on plants and soil organisms may be classified according to the action on the physical and chemical properties of the soil.

Effects of Organic Fertilizers on Plant Growth
A. Effects of humic substances on physical soil properties.

1. Humic substances improve the *soil structure,*
a) directly by loosening heavy soils with large humus particles (however, excessive loosening may be detrimental);
b) indirectly by better "contexture produced by the organism" in the soil, i. e., through increased production of bacterial "adhesives" (fine particles combined to form crumbs) or by means of aggregate-stabilizing humates (salts of humic acids).

2. Humic substances increase the *water-capacity* of the soil
a) directly by binding water to organic matter;
b) indirectly by improving the structure.

3. Humic substances improve *aeration* (a proper crumb structure implies more and larger pores), and thus
a) the supply of oxygen to the roots;
b) the escape of carbon dioxide from the root space.

4. Humic substances increase the *soil temperature*
a) directly by their dark color, which increases heat absorption by the soil;
b) indirectly by the improved structure, since, for example, more rapid elimination of excess water in spring causes a more rapid temperature increase.

B. Effects of humic substances on chemical soil properties.

1. Humic substances store nutrients at their surfaces in exchangeable form; this is important mainly for soils containing little clay;

2. humic substances *provide nutrients,* and energy, through the decomposition of nutrient humus, namely
a) *carbon dioxide* for photosynthesis;
b) *mineral nutrients,* especially with the elements nitrogen, phosphorus, and sulfur, but also with others, including trace elements;
c) *organic nutrients,* such as sugars, and amino acids formed as intermediate products during decomposition, which can be utilized by plants. This possibility, however, is of minor importance.

3. The decomposition of humus mobilizes mineral nutrients from inorganic reserves:
a) directly through the "humate effect", in which, e.g., microbial excretion of acids and complex-formers releases nutrients from the reserves and thus makes them available;
b) indirectly by the attack of acids, formed during the decomposition of humus, on the nutrient reserves or by lowering the redox potential, in which case some nutrients are mobilized under reducing conditions, e.g., iron and manganese, and indirectly, phosphate and molybdate, too.

4. The decomposition of humus promotes the *fixation of atmospheric nitrogen,* but only if special humic substances containing little N are offered to the N-fixing microbes.

5. *Immobilization* of nutrients by humic substances. There are two possibilities for this essentially detrimental effect.
a) Short-term *nutrient blockade,* in which microbes temporarily store available mineral nutrients in their bodies and thus "block" their availability to the plant root. Nitrogen blockade during the decomposition of organic matter with C : N ratios above 25 is especially important, but P-, S-, or even oxygen blockades may also have negative effects.
b) Long-term *nutrient fixation* in humic substances, e.g., through insertion into macromolecules of permanent humus or highly rigid complex fixation in immobile and thus non-absorbable complexes. Examples are the Cu-fixation by humic substances in half-bog soils, and the intentional reduction of the content of available copper by increased humus fertilization of rubber plantations, since a high Cu-concentration is detrimental to raw rubber (Chap. 7.4.2).

6. Organic fertilization and *active agents* in the soil.
Organic fertilizers, like soils, may contain many active agents or stimulate microbial production. Active agents are growth factors of varying importance for plants:

a) *Growth substances* or "substances with growth-substance character" (e. g., vitamins, quinones) may be beneficial mainly for germinating plants which do not yet produce enough themselves:
b) *growth inhibitors,* which in very small amounts retard plant growth, occasionally cause serious damage when present in larger concentrations, especially with certain monocultures;
c) *resistance improving substances (antibiotics),* occurring in small amounts in many composts and also in fertile soils, e. g., streptomycine (5 ppm), terramycine (0.1 ppm). These antibiotics are absorbed by plants and might be of importance in increasing their resistance to some bacterial diseases, but details must still be elucidated (Chap. 6.5.2).

Substitutes for Natural Organic Fertilizers

The soil-improving and growth-promoting effects of organic fertilizers are complex and nonspecific. Many side effects, even if positive, are frequently unnecessary. The use of organic fertilizers is often also expensive.

Attempts are therefore made to obtain a specific *substitution* of certain effects of natural organic fertilizers. Such synthetic substitutes often permit a desired effect to be achieved more simply, to a larger extent, and for a longer time. Many of these synthetic substances have already been discussed under the corresponding fertilizer groups. Synopsis 4–6 contains a summary of substitution possibilities.

4.4 Growth Substrates

All plants grow on nutrient substrates (Chap. 2.1.3). However, this chapter deals with *horticultural nutrient substrates* in particular. When marketed, they need only be identified (but not licensed).

4.4.1 Solid Growth Substrates

Various *horticultural substrates* were formerly prepared for potted plants and plant beds in gardening, according to certain individual recipes from compost, peat, mineral soil constituents, mineral fertilizers, etc. However, this procedure is involved and unnecessary. Each kind of plant does not require a specific nu-

Synopsis 4-6. Possible Substitution of Effects of Natural Organic Fertilizers by Synthetic Substances.

Desirable property of organic fertilizer for soil	Effect	Example of synthetic substitute
dark colour	rapid heating of soil	soot, coal
binding of water	greater water capacity	Hygromull
enlargement of pores	greater air capacity	Styromull
soil crumb formation	better structure	synthetic "adhesives" (crumb formers)
adsorbing surface	nutrient storage	exchange resins (clay minerals)
complex formation	a) mobilization of nutrients, especially heavy metals	complex formers, e.g., EDTA
	b) immobilization of nutrients (e.g., against surpluses)	special synthetic resins
reduction capacity	mobilization of Fe, Mn	hydroquinone, ascorbic acid
nutrient reserves	slow-flowing nutrient supply	depot fertilizers

trient substrate; for the sake of simplicity it therefore seems better to use some *standard substrates* instead of a multiplicity of *special substrates*. Horticultural substrates (growth substrates), should ensure optimal supplies of water, air, and minerals to the plants and be free of harmful substances. The basic substance of these substrates is chiefly *brown peat* with admixtures of lime and mineral nutrients, with or without the addition of clay.

The optimal pH of such a substrate for many ornamental and vegetable plants is in the acid region (pH = 5.5 to 6). Lime is best added in the form of magnesium lime or algae lime. The mineral fertilizer used is a complete fertilizer with trace elements, sometimes also with depot action. The nutrient ratios should be approximately 1 : 1 : 1.5 ($N : P_2O_5 : K_2O$). Separate addition of iron in the form of Fe-chelate is advisable.

Standard soil and *peat growth substrates* have become popular in horticulture [98].

Standard Soil according to *Fruhstorfer*

Peat is mixed with montmorillonitic (i.e., swellable) clay substance and enriched with phosphate. Nitrogen is added in amounts of 0.8 g per liter, e.g., in the form of ammonium sulfate-ammonium nitrate, together with 1 g K, in the form of potassium magnesia. Two to five grammes of lime are added for some plants, e.g., cabbage and celery. The pH is 5.5 (without lime). The standard soil should always be kept moist, since peat absorbs water only with difficulty after drying-out.

Standard soil is marketed in two forms:

- Type P: transplanting earth, with less nutrients;
- type T: pot earth, with more nutrients.

Peat Growth Substrate (TKS) according to *Penningsfeld*
Peat is mixed with lime and irrigated with a solution of complete fertilizer, which contains no chloride (12-12-17-2) but trace elements, to ensure uniform distribution. Fe-chelate is added in order to ensure the often problematic iron supply. TKS is commonly marketed in two forms (contents indicated per liter of moist peat):

- TKS 1: for young plants, with 3 g lime, 1.5 g complete fertilizer, 0.15 g Fe-chelate, and 0.03 g Na-molybdate;
- TKS 2: for potted plants, of the same composition but containing 3 g complete fertilizer.

TKS for bog-bed plants contains only one half of the above amount of lime.

4.4.2 Liquid Growth Substrates

Nutrient solutions form the substrates for hydroponics. Plants can grow and flourish on liquid substrates if the nutrient solutions are suitably prepared and the correct composition maintained. Nutrient solutions can also be used for liquid fertilization, e.g., on peat growth substrates. However, fertilization should then be in the form of a continuous discharge, to prevent the accumulation of salts.

Many suggestions have been made about the *composition* of nutrient solutions, corresponding to the variety of plants [90, 92, 82]. However, many plants grow properly in some standard solutions. Nutrient supplies in the solution need not precisely match requirements quantitatively, since the plant can choose through its selection capacity. A concentration range of 1 to 2‰ (sometimes up to 5‰) of total salts should be maintained in any case, corresponding to an osmotic pressure of 0.5 to 1.5 atm. Three examples of nutrient solutions are given in Table 4–8. The first, "No. 2 *Hoagland*", is known internationally (*Hoagland* et al., in California have contributed considerably to the further development of the nutrient solutions due to *Knop* and *Sachs* (Chap. 1.3.2)).

Preparation of the solutions can be simplified by first preparing stock solutions of the trace elements, and adding the following amounts per liter of nutrient solution:

- *1 ppm iron* in the nutrient solution is obtained by adding 1 ml of a stock solution containing 5 g Fe-citrate (1 g Fe) per liter.
- *0.02 ppm copper* in the nutrient solution is obtained by adding 1 ml of a stock solution obtained by 1 : 100 dilution of a stock solution containing 8 g/l of copper sulfate ($CuSO_4 \cdot 5H_2O$).

Table 4-8. Preparation and composition of nutrient solutions (3 examples).

Salt or element	Formula	*Hoagland* 1938 (No. 2)	*Pennings-feld* 1960 (B)	modified after *Gericke* (*Salzer*, 1965)
A. preparation of nutrient solution		distilled water containing no salt mg nutrient salt per litre		tap water
potassium nitrate	KNO_3	500	426	460
calcium nitrate	$Ca(NO_3)_2 \cdot 4H_2O$	118	868	.
K-dihydrogen phosphate	KH_2PO_4	136	284	105
ammonium nitrate	NH_4NO_3	.	.	75
ammonium sulfate	$(NH_4)_2SO_4$	.	10	.
magnesium sulfate	$MgSO_4 \cdot 7H_2O$	493	378	216
magnesium nitrate	$Mg(NO_3)_2 \cdot 6H_2O$	.	.	25
tap-water lime	$CaCO_3$	.	.	(85)
Fe-citrate/Fe-EDTA	Fe-citrate with 20% Fe	5	.	.
ferrous sulfate	$FeSO_4 \cdot 7H_2O$	.	20	15
manganous sulfate	$MnSO_4 \cdot 4H_2O$	2	5	2
zinc sulfate	$ZnSO_4 \cdot 7H_2O$	0.2	0.04	0.8
copper sulfate	$CuSO_4 \cdot 5H_2O$	0.08	0.04	0.6
borax	$Na_2B_4O_7 \cdot 10H_2O$	2	10	1.7
ammonium molybdate	$(NH_4)_6Mo_7O_{24}$	0.1	.	.

B. composition of nutrient solution (ppm = mg/l)

pH-value (initial value)		4.5	5.5	5.5
salt concentration (%)		1.7	2	0.9
remarks	N	212	192	93
• for plants and for	P	32	64	24
acclimatization at	S	64	50	28
first in solution of	K	234	248	210
one-half concentration	Ca	200	178	34
	Mg	48	37	24
* Fe-supplies are gener-	Fe	(1)*	4	3
ally problematic in	Mn	0.5	1.2	0.5
hydroponics, so that	Zn	0.05	0.01	0.2
Fe-additions should	Cu	0.02	0.01	0.15
be repeated, e.g.,	B	0.5	1	0.2
every three days	Mo	0.01	.	.

The small chlorine requirements of plants are covered by "impurities", and are therefore not added.

Hydroponics require either continuous *aeration* of the roots (blowing-through of air), or special containers or substrates ensuring adequate supplies of air without its separate introduction (root beds of granulated material, e.g., gravel, pumice stones, slags). The nutrient solution is renewed once a week or at longer intervals, depending on the hydroponic system. Constant renewal takes place in flow cultivation. The *pH* must be controlled. It must not become excessive, since iron and phosphates then precipitate (pH-values around 5.5 are optimal). Large scale hydroponics installations with automatic liquid renewal and comprehensive system controls have also been highly perfected.

4.5 Active-Agent Fertilizers and Soil Inoculants

Many fertilizers exert certain effects in the soil or on plants even at low concentrations. These effects are due to *growth regulation* by growth substances, promotion of the decomposition of organic fertilizers, *activation* of the "soil life", or some influences still difficult to explain.

Generally speaking, fertilization with these preparations serves less to improve the growth factors directly, than to control certain processes. These factors include:

- growth regulators;
- other active-agent fertilizers;
- processing aids for organic fertilizers;
- soil inoculants.

If these substances are subject to the fertilizer law, they are classified as *plant auxiliary substances* or *soil auxiliary substances*.

4.5.1 Growth Regulators

Growth regulators can be categorized as fertilizers since they promote plant growth. They were included in the fertilizer type list until 1969, but for legal reasons (licensing procedures) were later classified as plant protectives. They thus now subject to the plant-protection law, but are nevertheless fertilizers.

Growth regulators are organic active agents intervening in the biochemical metabolic processes of plants and thus causing chemical and morphological changes. They may be either active agents produced by the plant itself, in which case additional fertilization reinforces the effect, or synthetic agents with certain specific effects.

Better understanding of the biochemical processes in plants has greatly increased the possibilities of directed intervention, and thus the number of growth regulators. Plant growth substances (auxines) have been known for a long time, but the complex control mechanisms of plant metabolism by *phytohormones* are only now being recognized more clearly [87].

The most important fundamental compounds for growth regulation are

- *indoleacetic acid* and *-butyric acid;*
- *naphthylacetic acid* (synthetic growth substance);
- *Gibberilins* (promote growth in length);
- *kinins* (e.g., aging-retarding kinetin);
- *chlorocholine chloride* (stalk stabilizer).

With regard to application we can distinguish between growth substances or inhibitors on one hand, and agents promoting rooting, influencing the fruit, etc., on the other.

The following *application purposes* are possible:

- Promotion of germination (e.g., of seed potatoes);
- inhibition of germination (e.g., of potatoes in storage);
- promotion of rooting in cuttings;
- regulation of commencement of blossoming (blossoming control);
- fruit-dropping (e.g., for harvesting cherries);
- promotion of growth in length (reduction of stalk length in cereals, shorteners for grass and flowers);
- prevention of premature aging of plant parts (e.g., by kinetins);
- improving the displacement of products of photosynthesis from the leaves to reserve organs.

Stalk Stabilizers (Stalk-Shorteners)

Most growth regulators have their restricted field of application in horticulture. On the other hand, *stalk stabilizers* generally play an important part in agriculture. Lodging of cereals, due to storms and rain, causes large yield losses, so that increased stalk stability is of considerable importance [35].

Chlorocholine chloride (CCC) has proved effective in shortening and strengthening grain stalks. It slows down mitosis, thus retarding growth in length, but has no deforming side effects. It inhibits metabolism to a certain extent and thus acts as an *antimetabolite.* Chlorocholine chloride is marketed together with other substances as *Cycocel* (CCC preparation). The chemical formula of CCC is $Cl-CH_2-CH_2-N^+(CH_3)_3Cl^-$.

Application: for wheat during principal tilling (stages 2 to 4 = E to G); for rye and barley at the first and second stalk knot (stages 6 and 7 = I+J) (Chap. 7.1).

Quantities: 1 to 3 l/ha of Cycocel for relatively stable wheat strains (frequently best with "splitting");

2 to 3 l/ha of Cycocel for long-straw wheat, oats, rye.

The application of CCC alone as a "growth brake" slightly lowers the yield, but the latter is significantly increased if CCC is applied in conjunction with increased N-fertilization, which is possible (e.g., by about 10%), but would otherwise have to be omitted. Its principal significance is thus as a *yield-guaranteeing factor* under unfavorable weather conditions. In many regions this has made wheat growing more profitable. A side effect of CCC is the reduction of yield losses owing to stalk diseases. CCC is *decomposed* in the soil within a few weeks, so that less than 0.5 ppm is contained in the grain at the time of ripening; this concentration is far below the tolerance limit.

Substances used in horticulture for shortening the stalks of flowers will only be mentioned here, e.g., *Horticultural Cycocel* and *B-Nine*. The preparation *M-30* is used for keeping grass short, thus rendering frequent grass mowing unnecessary.

4.5.2 Soil Inoculants and Processing Aids

These fertilizers serve mainly to activate soil organisms, especially microbes.

Soil Inoculants

Natural fixation of nitrogen by certain soil microbes can sometimes be promoted by inoculation with suitable bacterial cultures [81]. The following should be considered for this:

- Nitrogen-fixing agents living freely, e.g., *azotobacter*;
- symbiotic bacteria, e.g., the *nodule bacteria* of legumes;
- bacteria for the decomposition of difficult to dissolve nutrients.

Free-living N-fixing microbes like azotobacter and others are common in all soils. Their activity depends mainly on environmental conditions, among which soil reaction is very important. Highly acidic soils are poor in *azotobacter*, and it may take considerable time after suitable improvements, until a normal population is re-established through new supplies. Inoculation with azotobacter spores considerably accelerates this process. However, the proper cost/benefit relationship should be ensured. The output of free-living N-fixing bacteria is about 10 kg N per ha, i.e., relatively little compared with plant requirements.

The *nodule bacteria* of legumes are much more important. They live as different strains of *rhizobium leguminosarum* in symbiosis with certain kinds of legumes. The higher plant supplies carbohydrates, etc., while the bacteria supply nitrogen to the plant. Spores for any legume are already present in the soil if this plant has been cultivated for a long time in a given field. Inoculation is then a natural process. However, the required spores are absent when the legumes considered are grown for the first time, e. g., when alfalfa grass or lupins are first introduced into a region where the only legume grown until then was clover. *Inoculation* of the seed with special legume spores is essential for N-fixation when legumes are newly introduced (Chap. 7.3.4).

The *output* of nodule bacteria in connection with legumes is about 200 kg N per hectare per year when the vegetation period is exploited fully (about one half with intercrops). Up to 400 kg/N per hectare per annum can be obtained in regions where there is growth throughout the year (e. g., New Zealand). Special soil inoculants for legumes are being marketed. They contain at least 1 million active bacterial cells per gramme, peat being used as a carrier substrate (Chap. 7.3.4).

The mobilization of phosphate and potassium by the use of nutrient-decomposing bacterial cultures is sometimes suggested. It is said that certain phosphate bacteria mobilize phosphate from, for example, apatite, which is dissolved with difficulty, and silicate bacteria are said to release potassium from silicates. Soil microbes undoubtedly do contribute to the decomposition of nutrients (Chap. 4.3.4), but intentional inoculation with special bacterial cultures has not yet given useful results in practice.

Processing Aids for Organic Fertilizers

Composts ripen through chemical decomposition in which soil microbes and soil fauna participate substantially. However, the wastes used for compost preparation are not always of optimal composition. Nitrogen, lime, earthworms, microbe-promoting agents, etc., may be absent for the optimal decomposition level; *ripening* or *conversion into soil* of raw compost can often be accelerated through additives, and the final product improved.

Additives of a special kind are the *processing aids* which only require to be identified but not licensed. According to the manufacturer, the *effects* of the material (e. g., *Nitralit, Eokomit, Terrasolin*) depend on the addition of certain beneficial bacterial spores and/or agents, etc. A beneficial effect on the ripening of composts is quite possible. On the other hand, bacterial spores are common and proliferate rapidly if conditions are favorable. *Inoculation* might therefore in many cases be less important than the provision of *optimal* basic conditions for conversion, i. e., proper aeration, avoidance of excessive acidification, etc. (Chap. 4.3.2).

4.5.3 Other Active-Agent Fertilizers

Directed control of the processes in soils and plants is only at the beginning of its development. Fertile soils are systems with high *self-regulating capacity,* but improved regulation would be necessary for less fertile soils (i.e., most soils). However, aimed intervention with only positive effects, is difficult in view of the complexity of the "soil" system. Some *biochemical processes* in plants could be rendered more effective, at least in theory.

Possible improvements in the combined *"soil"* and *"plant"* systems whose realization is just beginning, and will probably be further developed in the future, include the following:

A. Improvements in the *"soil"* system:

- Inhibition of nitrification (Chap. 2.1.5);
- inhibition of denitrification (to reduce fertilizer losses (Chap. 6.2.3));
- increased N-fixation by free-living microbes;
- increased mobilization of nutrients;
- reduced fixation of nutrients;
- promotion of the production of growth substances and antibiotics;
- reduced production of inhibitors.

B. Improvements in the *"plant"* system:

- Improvement of nutrient-absorption capacity;
- improvement of the displacement of immobile nutrients in the plant;
- improvement of the displacement of organic matter from the leaves to reserve organs;
- improvement of the utilization of light;
- increased resistance to harmful agents of all kinds.

On the Activation of Special Forces by Fertilizers
Plants consist of chemical substances, live in an environment of chemical substances, and can be influenced by chemical substances. However, plants also consist of electrical and magnetic "fields" or potentials, are surrounded by electromagnetic forces and "fields", and could possibly be influenced by such forces.

In principle it could be imagined that plant growth might be improved by "activation" of these forces. Several attempts have been made in this direction, either through direct radiation effects or by treating plants with certain "fertilizers" whose material components are insignificant in comparison with the "force effects" ascribed to them (e.g., spraying with silicic preparations). This problem will be discussed later in connection with the quality of food (Chap. 9.2.2).

The results of past attempts to promote plant growth through the activation of special "forces", if they exist at all, are not encouraging. "Research" in the admittedly difficult borderline area of present-day science sometimes violates elementary principles of experimentation, leading to absurd and easily refuted claims, or becomes lost in the mystic twilight of assumptions, beliefs, and desires, and thus cannot be checked at all.

Results obtained up until now can be expressed using an agricultural simile as: a grain of wheat that might possibly be present in a large amount of worthless straw.

5 Optimal Amounts of Fertilizer

One of the major problems of fertilization is the determination of the correct amount to be used. The object is to achieve optimal supplies while avoiding excesses. Excessive amounts of fertilizer are not only unnecessary but may even be detrimental.

Correct *dosage* of fertilizers is necessary in order to

- improve the fundamental *soil properties* controlling plant growth, especially the soil reaction (lime requirements and sometimes acid requirements, as well), but also the condition of the humus, the soil structure, etc.;
- improve the nutrient state of soil and plants.

5.1 Lime Requirement

This section should actually be entitled "Acid and Lime Requirement for Achieving the Optimum pH-Value". However, the pH of most soils is too low, so that emphasis is placed on *raising the pH*. Intentional lowering of the pH, acidification, by fertilization is only occasionally important (Chap. 4.1.3).

The correct *amount of lime* can be determined from the difference between the

- *pH* of the soil, i.e., *the existing pH*, and the
- *pH-target*, i.e., *the pH desired.*

5.1.1 Optimal Soil Reaction

Plants are tolerant to the reaction of the substrate (soil), as such, over a wide range. However, the soil reaction (Fig. 5–1) is important because it

- controls or indicates the availability of nutrients;
- indicates the presence of certain harmful substances;
- is a measure of structure-forming factors.

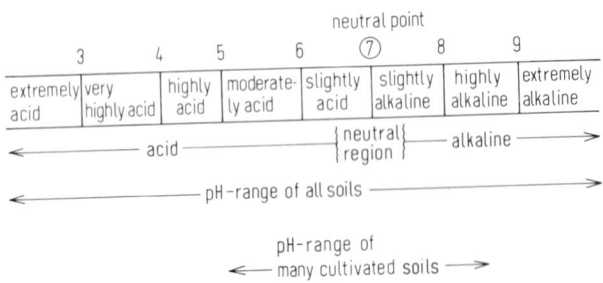

Fig. 5-1. pH-ranges in soils.

The importance of *optimal adjustment* of the soil reaction increases with the production level. The correct pH (Synopsis 5-1) is a precondition for effective utilization of the fertilizer and soil nutrients, which could partly or largely remain ineffective when the lime state is incorrect.

Achievement of the Optimal pH-Value

The optimal pH-value for plant growth is not the neutral point, as was assumed at the beginning of research in this area (50 years ago).

The *optimal soil reaction* primarily represents a compromise between

- the different optima of nutrient availability;
- the various requirements with regard to nutrient availability and structure.

In addition, the humus content plays a role in organic (bog) soils, mainly by its effect on the weight per unit volume. The optimal pH-values (or pH-ranges) are known *by experience* from many field trials and observations of growth. These values provide guidelines for large ranges, but may be modified for special production conditions. The optimal pH-values, on the whole, may be matched either to the requirements of individual plants, or to the similar requirements of a group of plants grown on certain soils (e. g., depending on the particle size). The latter principle is simpler and is therefore predominantly applied; it may be modified for special *plant requirements*.

Plants *preferring acid soils* or which are sensitive to lime, e. g., oats, rye, potatoes, and red and white clover, are

- sensitive to immobilization of heavy metal cations in the neutral and alkaline regions;
- able to mobilize phosphate and possibly molybdate in acid soils;
- tolerant to toxic substances in the more highly acid region (e. g., aluminum).

Calcicolous plants or plants sensitive to acids, e. g., barley, wheat, sugar beet, are

Synopsis 5–1. The pH-Value.

The *pH* is a measure of the soil reaction (liming state), i.e., it indicates whether the soil "reacts" acidly, neutrally, or basically. The term is derived from the Latin *potentia hydrogenii* (power of hydrogen). pH is the negative logarithm to the base 10 of the H^+-concentration or activity, which is measured in equiv. H^+ per liter or in (g H/l).

The *neutral point* (pH 7) represents the H^+-concentration of pure water, which has a H^+-concentration of 10^{-7} eq. (the negative logarithm is thus 7).

The narrow range around pH 7, approximately from pH 6.5 to pH 7.5, is sometimes called the neutral region. This is not quite chemically correct but is sometimes acceptable for agricultural purposes.

The *logarithmic* character of the pH-values should be stressed once more. Ranges of pH-values of 0.1 are thus 100 times greater around pH 5 than around pH 7.

Measurement of pH
Air-dried soil samples are treated with a dispersing agent in a 1 : 2.5 ratio. The pH is measured in this suspension; electrometric measurement yields exact values, whereas measurement with colored indicators can only provide approximations.

The following dispersing agents are used for pH-measurements:

- In Germany, dilute saline solution, i.e., 0.01 M $CaCl_2$, giving a *pH-value in calcium chloride* (formerly in 0.1 N KCl or 1 N KCl, with approximately the same values);
- outside Germany, most commonly pure water, yielding the *PH-value in water* (which is approximately 0.5 to 1 pH-points higher than the pH in saline solution).

Data on pH-values for German soils always refer to measurements in dilute saline solutions. They have the advantage of providing more stable values (on a time basis). The pH for soils should be indicated to only one decimal point, since more exact determination, although possible for soil samples, is impossible for the soil itself.

- sensitive to immobilization of phosphate in the acid region;
- able to mobilize heavy metal cations in the neutral region;
- sensitive to toxic substances in the more highly acid region (e.g., aluminum);
- sensitive to the effects of a poor soil structure on medium and heavy soils.

The pH-values now considered to be optimal are indicated in Table 5–1 according to particle size and humus content.

pH and Yield
There is only an indirect relationship between soil pH and yield, since the pH as such is not a plant-growth factor. However, a general relationship can be established.

Table 5-1. pH-values to be desired (pH-target) in agricultural land and grassland.
Mean values are indicated, from which deviations of up to ±0.2 pH-points are admissible.
Thus, the optimal range is between pH 5.8 and pH 6.2 when the pH-target is 6.0. Values
are slightly modified VDLUFA-guidelines.

Soil type	Clay content in %	pH-target Fields	pH-target Grassland
mineral soils (up to 4% humus)			
clay, loam, sandy loam rich in clay (loess)	> 17	7.0*	5.8
highly loamy sand and highly sandy loam	} 12–17	6.5	5.5
clayey and loamy silt			
slightly loamy sand	5–12	6.0	5.0
sand (acid sandy soils)	0– 5	5.5	5.0
	Humus %		
soils rich in humus (above 4% humus)			
sand rich in humus	4– 8	5.5	5.0
sand very rich in humus	8–15	5.0	5.0
half-bog sand	15–30	4.5	4.5
bog**	> 30	4.0	4.5

* pH 7 to 7.5: heavy soils should contain 0.2 to 1% finely dispersed carbonate
** lime-containing fens: pH 6 to 7

Fig. 5–2 clearly shows that maximum yields can only be obtained if the
pH is maintained within the optimal range. Any deviation from this optimum requires extensive special measures to eliminate the resulting problems, as indicated for plants "preferring acid soils" and "calcicolous plants".

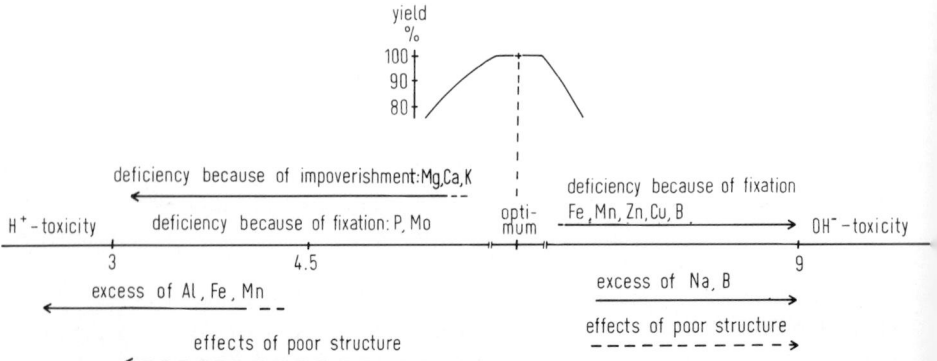

Fig. 5-2. Possible damage to plants owing to excessive or insufficient soil reaction.

Local and Time Variations of pH

The soil reaction is in several respects a variable property. *Local differences in pH* are smoothed out to form the averaged pH of a soil sample. However, a soil actually consists of a mosaic of regions having different pH-values. Thus, acid microregions also exist in calcareous soils, owing to the biotic activity of the root or the rhizospheric flora. Such differences are averaged out by soaking but reappear with increasing dryness.

Time differences in pH-value are caused by differences in the activities of crops and by increased biotic activity in spring and autumn, when the combination of temperature and humidity is optimal. A high activity implies a temporary reduction of the average pH. Such variations should be taken into account when the optimal pH-values are established, since the latter are averaged over the year.

It may be advisable to *increase the time variation* when production levels are raised. Such considerations can become acute because in medium and heavy soils which are limed to the upper limit of the optimal range, to improve the soil structure, this leads to immobilization of trace elements. It therefore seems advantageous to lower the optimal pH somewhat in the course of time starting from the higher values at the beginning of growth mainly for reasons of structure. The supply of important trace elements can thus take priority during the principal growth period, while the plant forms a certain structure for itself (Fig. 5–3).

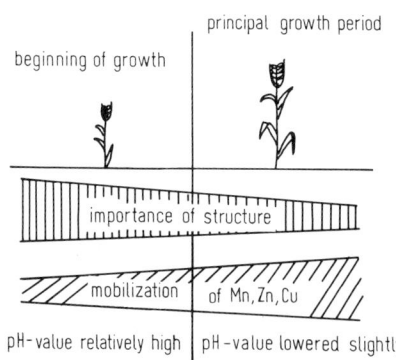

Fig. 5–3. Lowering of optimal soil reaction in medium and heavy soils during growth [8].

One simple way of temporarily lowering the pH is to use nitrogen fertilizers (Chap. 2.1.5) that have an acid reaction. Intentional deviations from the generally valid guidelines for pH may be indicated occasionally, i. e., when special soil properties such as a natural large lime content must be taken into account.

The optimal pH-values are not *natural constants* but empirical guidelines. They are valid for certain production forms and levels in soil with normal properties, and should therefore be under constant critical review.

5.1.2 Amount of Lime Required

The amount of lime required can either be determined precisely or estimated roughly. However, only in high-yielding agriculture should precise chemical determination be considered.

Rough estimation of the pH-value is possible with the aid of *indicator plants*, which grow in a certain pH-range, and whose presence and healthy growth thus indicate this pH-range [106]:

- Plants indicating "acid" are, e.g., corn spurrey, sheep's sorrel, and also wild radish;
- plants indicating "lime" are, e.g., wild mustard, wild oats, *Fumaria*.

A more precise determination is possible. A standardized application of lime is indicated if a soil appears to be "too acid". Estimation of the pH from weeds was important when pH-measurements were still cumbersome and expensive and weeds were still abundant on fields, but now the low cost of precise chemical determination of the pH renders this procedure largely obsolete.

Chemical Determination of Lime Requirements

Lime requirements unfortunately cannot be derived directly from the pH, since the latter indicates only part of the acidity. Lime requirements really depend on three factors:

- The *pH*, which indicates the actual acidity and thus, in a manner of speaking, the "active" acid content;
- The *total acid content*, expressed as potential acidity by the H-value and determined by special pH-measurements with 0.5-M Ca acetate (*Schachtschabel* method);
- the *pH-target*, i.e., the optimal pH-value aimed at.

The method of determining lime requirements is explained in Synopsis 5-2 (sampling is discussed in Chap. 5.3.2).

Lime Requirements

Lime requirements for some pH-targets can simply be read from Table 5-2 (p. 186), if the necessary initial data are available.

Synopsis 5–2. Methods for Determining Lime Requirements.

Step 1
The H-value is determined from the separately established acetate pH-value. Both values are closely related; empirical conversion values are given in Table 5–2. Thus a pH-acetate value of 6.5 corresponds to an H-value of 3.5 eq. H/100 g soil. (The H-value indicates the share of hydrogen (H) in the cation-exchange capacity.)

Step 2
Lime requirements for liming up to pH 7 are computed from the H-value, using the following formula:

1 meq. H/100 g soil = 8.4 dt/ha of CaO

Derivation: 1 meq. H/100 g = 1 meq. Ca/100 g
             = 28 mg CaO/100 g
             = 8.4 dt CaO for the 3 million kg in the soil layer
                between the surface and a depth of 20 cm

This already defines the lime requirements for soils with a pH 7 target.

Step 3
Lime requirements per hectare for soils with a pH-target below pH 7 are determined by means of the following conversion formula:

$$\text{dt CaO for pH-target} = \frac{\text{pH-target} - \text{pH-value}}{7 - \text{pH-value}} \cdot \text{dt CaO for pH 7}$$

Example: pH   = 5.3
         pH-target = 6
         acetate-pH = 6.5 which yields H-value of 3.5
         lime requirements for pH 7 = 3.5 × 8.4 = 29 dt CaO

$$\text{dt CaO for pH 6} = \frac{6 - 5.3}{7 - 5.3} \cdot 29 = 12 \text{ dt CaO}$$

Lime requirements are first indicated in "CaO". The correct lime form should be chosen according to the purpose involved (Chap. 4.1.2). Quantities refer to the volume of the A-horizon, of 0 to 20 cm depth. Corresponding additions are necessary if the topsoil goes deeper, e.g.:

• for an A-horizon between 0 and 25 cm depth: 25% addition;
• for an A-horizon between 0 and 30 cm depth: 50% addition.

The lime applied must obviously be mixed into the soil volume concerned by suitably working the soil. Separate lime-requirement determinations are indicated for melioration liming in greater depth or for special subsoil limings.

Table 5–2. Necessary lime quantities for A-horizon (0 to 20 cm depth) (according to [136]).

liming with dt/ha of CaO* for

potential acidity (acetate pH)	H-value (meq/100 g)	pH-target = 7	pH-target = 6.5 pH-value of soil					pH-target = 6.0 pH-value of soil					pH-target = 5.5 pH-value of soil					pH-target = 5.0 pH-value of soil				
			6.2	6.0	5.7– 5.5	5.4– 5.0	<5.0	5.7	5.5	5.3	4.9– 4.6	<4.6	5.2	5.0	4.6	4.4	<4.1	4.7	4.5	4.2– 4.1	3.7– 3.5	<3.5
6.9	0.7	6	2	4	4	4	5	1	2	2	3	4	1	2	2	3	3	1	1	2	2	3
6.8	1.4	12	5	6	8	9	9	3	4	5	7	8	2	3	5	5	6	2	2	4	5	6
6.7	2.0	17	6	9	11	12	13	4	6	7	10	11	3	4	6	7	9	2	3	5	7	8
6.6	2.7	23	9	12	15	17	18	5	8	9	13	14	4	6	9	10	12	3	5	7	9	11
6.5	3.5	29	11	15	19	21	23	7	10	12	16	18	5	7	11	12	15	4	6	9	12	14
6.4	4.3	36	14	18	23	26	28	8	12	15	20	23	6	9	14	15	18	5	7	11	15	17
6.3	5.4	45	17	23	29	32	35	10	15	19	25	28	8	11	17	19	23	6	9	13	19	21
6.2	6.5	55	21	28	35	40	43	13	18	22	31	35	9	14	21	23	28	7	11	16	23	26
6.1	8.3	70	26	35	45	51	55	16	23	29	40	44	12	18	26	30	35	9	14	21	29	33
6.0	11	90	34	45	58	65	70	21	30	37	51	57	15	23	34	38	45	12	18	27	37	43
5.9	14	120	45	60	77	87	94	27	39	49	67	74	20	30	44	50	59	15	24	35	49	56
5.8	23	190	71	95	122	137	149	44	63	78	107	120	32	48	71	81	95	25	38	57	78	90

* CaO is a measure of lime quantity: for choice of correct lime form, *see* Chap. 4.1.2
Fertilizer quantities are increased slightly in some newer tables.

Distribution of Large Lime Quantities

Uniform mixing is difficult if very large amounts of lime are required. The following maximum quantities should not be exceeded in single applications:

- For light soils: 20 dt calcium carbonate ($CaCO_3$);
- for medium and heavy soils: 30 to 50 dt quicklime (CaO) (at most 20 dt for grassland).

Liming may be repeated in the same year, if necessary. "Liming in advance" may be indicated for crop-rotation or other considerations, even if the pH of the soil is still in the optimum range.

Estimation of Lime Requirements from pH-Value

Since precise chemical determination of lime requirements is reliable and cheap, in principle it should be done; however, occasionally it might be necessary to use the pH-value alone for recommending the quantity of lime to be applied. Rough tentative values are given in Table 5–3 for the latter case.

Table 5–3. Estimation of lime requirements from pH-value and soil type (simplified according to [44]).

| Soil type | Use | Liming with dt/ha of CaO* at soil pH-value of | | | | |
		6.0	5.5	5.0	4.5	4.0
sand	field	—	5	5	5	10
	grassland	—	—	5	5	5
loamy sand ⎱	field	5	10	20	20	30
sandy loam ⎰	grassland	—	5	5	10	10
loam ⎱	field	13	20	33	45	55
clay ⎰	grassland	5	5	10	20	25

* CaO is a measure of lime quantity: for choice of correct lime form, *see* Chap. 4.1.2.

5.2 Optimal Nutrient Contents of Plants and Soils

High-yielding plants have yield potentials that are only incompletely exploited in practice. Optimization of all growth factors is an ideal state and can only be approximately achieved under field conditions.

Plant growth on most sites is limited less by climatic factors (light, carbon dioxide, etc.) than by water and mineral nutrients. Lack of water, even during droughts, is less harmful if sufficient mineral nutrients are available. The correct supply of *minerals* to plants and soils thus becomes decisive for plant production. The next important factor would then be protection against diseases and harmful organisms (this concerns plant protection).

As regards further attempts to increase yields, it should be mentioned that optimization of photosynthesis through correct mineral supplies does provide an essential precondition for maximum yields, but that rapid *removal of assimilation products* from the leaves into the reserve organs becomes more important precisely in conjunction with high photosynthesis performance. Proper mineral supplies also seem essential for this purpose; hormonal regulation (fertilization with growth substances) might provide further improvements in the future.

5.2.1 Fertilization and Yield Level

The *optimal nutrient content* is not a constant magnitude but depends to a considerable extent on the yield level. High-yielding plants require:

- Large contents of mobile nutrients in the soil;
- proper supplies of all necessary and beneficial nutrients;
- prevention of harmful overfertilization.

Large Contents of Mobile Nutrients

Wild plants obtain most nutrients from mineral reserves in the soil. This also applies to the major nutrients in the case of simple strains of cultivated plants. On the other hand, high-yielding plants require a "table abundantly set" with all nutrients. Their high genetic potential can be realized only when abundant supplies of mobile, easily available nutrients are continuously provided, i. e., when photosynthesis and other synthetic processes are optimized with a rapid absorption of minerals.

The aim should therefore be to "fill up" the plants with nutrients beyond the optimum nutrient level. However, under field conditions it is not sufficient for amounts to just exceed this level, since continuous nutrient supplies to the plant are not necessarily ensured during the principal growth period. Rather, nutrient supply should be so large that the plants can take in sufficient amounts even during transient stress situations (reduced nutrient supplies due to drought, water-logging, attacks by parasites). Alternatively, the plants should already possess nutrient reserves with which they can, at least to some extent, provide supplies to the tissues with maximum requirements, using internal displacements, as animals use their reserves of fat when they have to do without food.

It follows from this that in future fertilization must not only be abundant, but that the active components of fertilizers must become more important than total contents. A fertilizer must either directly increase the mobile nutrient content or be converted continuously into mobile forms through rapid reactions in the soil.

Proper Supplies of all Important Nutrients

Today, some 100 years after the beginning of mineral fertilization, the latter is still restricted mainly to the major nutrients nitrogen, phosphorus, and potassium (supplemented by liming) at an average yield level of 40 to 50 dt/ha for wheat. The other nutrients have for centuries been mostly supplied in adequate amounts from mineral reserves in the soil, or have been obtained in the organic-fertilization cycle.

Supplies from the reserve fractions flow slowly; it must therefore be expected that an increasing use of NPK-fertilizers at correspondingly higher yields will cause other elements to become minimum factors through dilution, especially if liming to the optimum further reduces the availability of some nutrients in the soil. Precisely because NPK-fertilizers have until now greatly raised yields, there is a risk of future relative overestimating this yield-increasing factor (the term "relative" here refers to the relationship to other nutrients). In future it might be necessary, when total fertilization is increased on some sites, to relatively increase fertilization with magnesium, manganese, copper, etc.

Doubling present-day use of NPK-fertilizers does not of itself double yields, or even raise them substantially. This is shown by the relatively few effects of large additional nitrogen supplies in fertilization trials. It is also demonstrated by the relatively greater increase in the use of NPK-fertilizers in the course of time, compared with yield increases (Chap. 1.3.4).

The minimum law, established at the beginning of scientific plant nutrition and illustrated clearly by the minimum cask (Chap. 1.1.2), has now again become highly topical in a different sense. It has been quite correctly pointed out that fertilization with N without simultaneous supplies of phosphate and potassium makes little sense. Today it should be stressed that intensive fertilization with NPK without ensuring adequate supplies of the other major and micronutrients will be increasingly ineffective. Only the correct ratio of primary to other nutrients ensures maximum yields.

Prevention of Overfertilization

It might be deduced from the above that in general application of fertilizers with major and micronutrients should steadily increase. However, large fertilizer doses involve the risk of damage, besides raising the problem of correctly distributing the total fertilizer amount over the various nutrients.

What should be avoided is not only the waste of fertilizer through possible nutrient losses, immobilization, or even fixation (accumulation of reserves difficult to mobilize), but undesirable oversupplies in mobile form in the soil, since they may at first cause luxury consumption (unnecessary intake) and finally, in extreme cases, toxicity (poisoning). This may be a result either of a single nutrient or of all mobile nutrients together as "salt damage".

Salt damage occurs more frequently in horticulture with intensive fertilization and is still quite rare in agriculture. However, "burning" due to high salt concentrations must be expected especially during dry early-summer periods when larger amounts of fertilizer are applied. It may be necessary to split fertilizer application or take other steps to prevent salt damage, especially when large fertilizer doses are used during sowing, since young plants are particularly sensitive. The salt load caused by fertilization should in any case also be taken into account in future.

Another risk inherent in large fertilizer doses might not as yet play a major role in agriculture. This is the mutual interference of fertilizer effects, e. g., a reduction of Mg-intake because of extremely large K-supplies, or mutual precipitation of nutrients in the soil (zinc deficiency caused by extremely heavy fertilization with phosphate, as is sometimes suspected in horticulture).

5.2.2 Ranges of Nutrient Supply

The nutrient supplies of plants and soils are divided into *supply ranges* for the purpose of quantitative classification, with the optimal supply range lying between deficiency and excess. The purpose of fertilization is to attain this optimum or maintain it without penetrating too far into the range of luxury supply or possibly cause damage.

The supply ranges for *plants* are divided into content classes or supply steps A to F (Synopsis 5-3). They are separated by the respective limit values, which can only be determined approximately, and should therefore in fact be called "limit ranges" (Fig. 5-4).

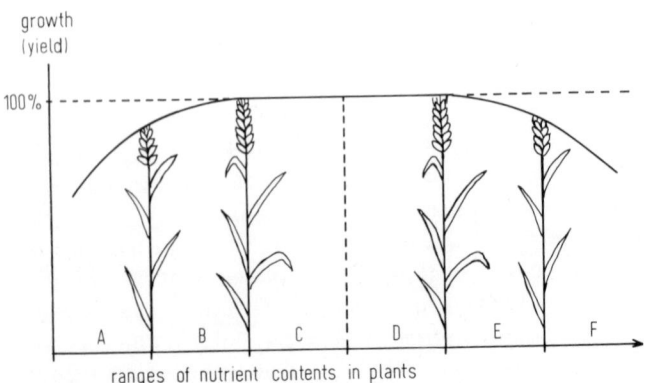

Fig 5-4. Plant growth as a function of its content of one nutrient.

Synopsis 5-3. Definitions of Supply Ranges and Limit Values [7].

Supply Ranges

Acute deficiency: clearly evident, typical deficiency symptoms, poor growth, considerable yield increase from fertilization (*pre-acute* deficiency indicates the appearance of first unspecific symptoms).

Latent deficiency: only hidden (latent) deficiency symptoms invisible to the naked eye, apparent proper growth, yield increase and usually also quality improvement after fertilization.

Optimal supply: no deficiency symptoms, optimal growth, generally highest quality, no yield increase from fertilization.

Luxury supply (luxury consumption): unnecessarily large ("luxury") intake of nutrients, proper growth, but quality sometimes reduced.

Latent toxicity: reduced growth owing to interfering slight excess.

Acute toxicity: damage symptoms due to large excess, poor growth and low quality, serious yield decreases after fertilization with nutrient considered. Countermeasures necessary.

Limit Values

Symptom limit value: nutrient concentration below which acute deficiency symptoms occur.

Yield limit value (= optimum nutrient level): nutrient concentration above which maximum yields are obtained.

Boundary region between optimal and luxury supply: no sharp differentiation is possible, since in theory the entire region could be included in the luxury consumption class. However, the great practical significance makes it desirable to separate these two ranges, even if only approximately.

Toxicity limit value: nutrient concentration above which yields are depressed because of excesses.

The supply ranges for *soils* largely correspond to those for plants (Synopsis 5-4). The small differences in the high ranges are due to the dominant role of phosphate and potassium in soil analyses. Complete harmonizing of the scales would be desirable in general. Moreover, contents of available nutrients vary in the course of time, both during the vegetation period and even, strictly speaking, during the course of the day. Nevertheless, these estimated values permit the classification of soils for assessing the state of nutrients.

5.3 Diagnosis of Nutrient Requirement

There are many possibilities for determining nutrient requirements, ranging from rough tentative values to comprehensive scientific investigations.

Synopsis 5–4. Nutrient-Supply Ranges and Fertilization.

	Plant		Soil	
	Supply ranges	Content class according to VDLUFA	supply ranges	recommended fertilization
Content class				
A	acute deficiency	A low	serious deficiency	greatly increased fertilization or special measures
B	latent deficiency	B medium	slight deficiency	increased fertilization
C	optimal supply	C high	optimal supply	maintenance fertilization as "normal fertilization"
D	luxury supply	D very high	luxury supply	reduced fertilization
E	latent toxicity	E extremely high	slight excess	no fertilization or
F	acute toxicity		large excess	countermeasures

Note: Classes D and E for soils correspond approximately to content class D for plants.

Synopsis 5–5. Possibilities of Plant and Soil Analysis.

1. Determination of Supply State by Soil Analysis
Soil analyses permit determination of the supply state of the soil, from which the required fertilizer amounts can be derived.

a) *Estimation of nutrient content from general soil properties* (initial material, extent of weathering, clay and humus contents, etc.). This procedure yields a rough approximation. Thus loamy soils frequently contain more nutrients than sandy soils.

b) *Estimation of nutrient content on the basis of indicator plants.* The presence and strong growth of certain weeds indicates the abundance, or scarcity, of certain nutrients. This procedure yields a rough approximation if various weeds grow on the field (pH-indicator plants, see Chap. 5.1.1).

c) *Chemical soil analyses for content of available nutrients.* Several procedures, that involve different degrees of effort and that yield correspondingly accurate results, are possible:
 – simple rapid soil tests for field testing;
 – standard soil analyses in the laboratory (Chap. 5.3.2);
 – special scientific analyses involving much effort.

2. Determination of Supply State by Plant Analysis

a) Visual *diagnosis* of *deficiency symptoms* (or excess symptoms). The degree of green coloration of the leaves as seen by the naked eye is an indicator of the state of nitrogen supplies or color charts may be used *(Munsell Color Chart),* provided that other nutrients are supplied as required. Certain symptoms point fairly unequivocally to certain deficiencies (Chap. 5.3.1). Diagnosis with the naked eye can be rendered more exact by using optical aids (magnifying glass, microscope).

b) *Rapid fertilization test* for identifying visible but not readily identifiable deficiencies. Thus, spraying yellow leaves with various nutrients makes it possible to determine the lacking nutrient from the ensuing increase in green coloration.

c) *Rapid chemical tissue tests* for immediate, rough estimation of the mobile component of nutrients in fresh plant material, e.g., a color reaction for nitrate in cut maize stalks.

d) *Chemical plant analysis* for determining total nutrient contents in the plant or plant part (Chap. 5.3.3).

The following diagnostic methods should be considered (Synopsis 5–5):

• analysing *soils* for available nutrients;
• examining *plants* for symptoms or nutrient contents.

Two other special methods, one simple, the other involved, will be discussed. Deducing fertilizer requirements from experience is particularly simple. There are local guidelines for the fertilization of certain plants (standard rules). They are based chiefly on nutrient removals, with additions or deductions according to local experience. The purpose of such fertilization is to compensate losses through removal and to maintain the proven nutrient level. Fertilization according to standard rules is a simple, rough method, but is not sufficient for the requirements of intensive production.

Field fertilization trials are particularly involved. This is the standard method for determining fertilizer requirements. The fertilizer dose that ensures maximum yield, or the economic optimum, is the optimal in case of differing fertilizer doses. The basic principle of field trials is very simple, but it may be very difficult to eliminate disturbing factors and to identify differences. Accurate performance of several parallel trials (repetitions), evaluation by using statistical methods, and critical interpretation (Chap. 6.6) are therefore essential. The results of field trials are in fact valid for the preceding year, but may also be extrapolated for future years.

Accurate field trials are often very involved and are therefore less indicated for determining the fertilizer requirements of individual farms. A simplified form of fertilizer trial is the *fertilizer pot experiment* under standard conditions. However, the results may only be applied to a limited extent under field conditions.

5.3.1 Diagnosis of Deficiency Symptoms

The main causes of symptoms, diseases or signs of damage in plants, are the following:

- Plant diseases and harmful organisms;
- Physiological damage, viz.,
  a) deficiency of general growth factors (light, heat, water, oxygen in the soil, etc.);
  b) *deficiency* or *excess* of mineral nutrients;
  c) special toxic influences (poisoning).

Disturbances of mineral nutrition are of special importance in connection with fertilization and manifest themselves in symptoms due to deficiency or excess. The latter are rarer but are also particularly difficult to diagnose. Frequently *diagnosis of a deficiency* is also only possible by approximation. A particularly useful method is to compare symptoms of damage with colored photographs of individual deficiency manifestations [103, 104, 105, 110, 140].

A simplified *key* for identifying deficiency damage is given in Synopsis 5-6.

The most important diagnostic characteristic is the site of the first occurrence of a deficiency, viz.:

Synopsis 5-6. Key for Identifying Deficiency Symptoms in Plants.

A. Cereals

I. Symptoms appearing first on older leaves

Lack of:

1. *Reddish discoloration* of leaves and stalks. Leaves at first dark green, later also brown — P

2. *Tip chlorosis* and necrosis: Leaves turn yellowish-brown from tip inwards, oldest leaves brown — N

3. *Marginal necrosis:* Leaves yellow or mostly brown at margin, also hang down limply (wilting attitude) — K

4. *Stripe chlorosis:* yellow lengthwise stripes between veins. Residues of chlorophyll aligned like string of pearls — Mg

5. *Spot chlorosis:* in *oats* (grey-speck disease): greyish-brown, stripeform spots on lower half of leaf — Mn

   in *barley:* dark brown, stripe-form spots predominantly on upper half of leaf — Mn

   in *rye and wheat:* whitish or grey, stripeform spots predominantly on upper half of leaf — Mn

II. Symptoms appearing first on younger leaves

1. Completely yellow-green leaves with bright yellow veins — S
2. Yellow to yellow-white leaves with green veins — Fe
3. *Withertip:* whitish leaf tips warped like filaments, especially in oats and barley — Cu
4. Yellowish leaves and dead *vegetation point*
   a) often combined with "acid damage" (brown spots, etc.) — Ca
   b) rarely in cereals — B

B. Beta Beets, Potatoes, Cabbage Types (also Rape), Legumes

I. Symptoms appearing first on older leaves

1. Leaves at first dark green, then often reddish — P
2. *Tip chlorosis:* Leaves turn yellowish-brown from tip inwards, oldest leaves brown, plant light green — N
3. *Marginal necrosis:* leaves with brown margins, hang down limply — K
4. Large *chlorotic spots*
   in beta beets, cabbage, legumes: leaves yellow between veins; finally brown — Mg
   in potatoes: leaf centre spotted yellow-brown, margin remains green for long time — Mg
5. Small *necrotic spots:* in *beta beets* and *cabbage:* respectively yellow and yellow-brown spots on entire leaf (marbling) — Mn

II. Symptoms appearing first on younger leaves

1. Yellow-green leaves with light-yellow veins — S
2. Yellow leaves with green veins — Fe

3. Small *necrotic spots*
   in *potatoes:* brown-black points especially on underside of leaf: Mn
   in *legumes:* brown or gray spots on light-green leaves   Mn
4. *Dead vegetation point,* etc.
   in *legumes:* yellowish-reddish leaves   B
   in *beta beets* (heart and dry rot): turning yellow, warping and turning   B
   black of youngest leaves, and on upper part of beet body
   in *rape:* yellowish leaves, fissured stalks   B
   in *Swede turnips:* (glassiness): watery appearance of beet tissue   B
5. *Clamped heart* in *cabbage:* deformed leaf and warped young leaves,   Mo
   in young plants spoon-shaped leaves
6. *White-leaf disease* in *legumes*   Cu

- first at the *older leaves:* this means that a mobile element in the plant is lacking, N, P, K, Mg, etc.;
- first at the *younger leaves:* this implies that an immobile element is lacking, Fe, Zn, Cu, B, etc.

This differentiation is possible only at the beginning of a deficiency; later all leaves are involved if the deficiency is relatively serious; the damage is also masked by secondary damage of parasitic origin.

An important aspect is also the differentiation between chloroses and necroses. *Chloroses* manifest themselves as yellow discolorations. Chlorophyll formation is disturbed, but the damage is *reversible*; that is, it can still be corrected by fertilization. Persistent chloroses are characteristic of N, Mg, S, and Fe deficiencies. They turn into necroses only in cases of severe deficiency.

Necroses are brownish discolorations indicating the presence of dead tissue. This damage is *irreversible*. Fertilization can only stimulate the formation of new leaves, provided that the plant is not yet completely dead. Necroses following a short transitional stage of chlorosis are characteristic of K, Mn and Cu deficiency.

The diagnosis of pronounced individual symptoms is still fairly easy but complex damage (syndromes = complexes of symptoms) is very difficult to identify. The "acid-damage" syndrome with brown and reddish discolorations on yellowish leaves is usually due to multiple deficiencies in combination with damage due to excess. Other symptoms due to excesses are "salt damage" in young plants or boron excess, especially in cereals, e.g., black spottiness in barley. Use of a magnifying glass is often indicated for visual diagnosis. The additional use of a microscope does permit a more detailed analysis of tissue changes, but requires specialized know-how [108]. Definite elucidation of damage that cannot be diagnosed visually is possible through plant or soil analysis.

5.3.2 Soil Analysis for Available Nutrients

Plants grow on nutrient substrates, most frequently on natural soils. The supply of mineral nutrients to plants depends on the content of available nutrients in the substrate, which can be determined by suitable methods. The required fertilizer quantities are then derived from this *available* content.

Available nutrients, i.e., nutrients absorbable by the plants, can roughly be divided into three fractions according to their binding, i.e., their mobility (Synopsis 5–7).

Synopsis 5–7. Binding Form and Availability of Mineral Nutrients.

Binding Form	Availability	Soil-Analysis Method Used to Determine
1. not bound, in soil solution	very easily	water-soluble nutrients
2. weakly bound, loosely absorbed to exchange complexes	easily (available)	exchangeable nutrients (mostly including water-soluble)
3. immobile but easily mobilizable	moderately	easily mobilizable nutrients
4. immobile and difficult to mobilize	practically unavailable	total nutrient reserves

In particular, the available fractions continuously interact by processes of nutrient dynamics (Fig. 1–2). Nutrients are usually more highly immobilized with increasing dryness and thus become less available, but become more readily mobilized with increasing humidity, thus becoming better available. For some nutrients these changes may be substantial, especially because the extent of availability of the three fractions differs considerably.

Determination of *available nutrients* is possible by various methods, some of which have become standardized in certain countries (Synopsis 5–8) [129, 131, 134, 114, 141].

Sampling of Soils

A precondition for reliable soil analysis is correct sampling.

a) The soil sample must be *representative* of the area to be analysed. A sample of 500 g should correspond to the mean of 3 million kg, formed by the weight of the topsoil (0 to 20 cm depth) in one hectare. This is only approximately possible even when many subsamples are mixed and represent the largest source of error in soil analyses. A random distribution of individual samples would be ideal,

Synopsis 5–8. Methods of Soil Analysis.

A. Biotic Methods
The available nutrients are extracted by micro organisms, e.g., fungi; the yield of mycelium is determined with and without fertilization respectively (the extent of dark coloration of the mycelium is sometimes used as a measure).

B. Biotic-Chemical Methods
The available nutrients are extracted through the roots and the nutrients absorbed by the plants are determined chemically. An example is the *germinating-plant method* due to *Neubauer*, which was introduced in Germany as a standard method around 1940, and is still used for scientific investigations.

C. Chemical Methods
The available nutrients are extracted chemically and the nutrients in the plant determined chemically. Various extraction solvents are in use:

1. *Water:* e.g., for phosphate in a 1:50 ratio at 20°C, water-soluble nitrogen (nitrate), boron soluble in hot water.

2. *Exchange solutions:* exchange, e.g., with:
   - N-ammonium acetate for cations;
   - 0.025 N $CaCl_2$ for magnesium (according to *Schachtschabel*);
   - sometimes also solid exchange complexes (according to *Tepe*).

3. *Inorganic acids:* e.g.,
   - 0.002 N H_2SO_4 with $(NH_4)_2SO_4$ at pH 3 according to *Truog* (USA) for phosphates;
   - 0.1 N HCl for potassium (Netherlands) or trace elements;
   - 0.43 N HNO_3 for copper in the FRG (according to *Westerhoff*).

4. *Organic acids:* e.g.,
   - 1% *citric acid* for phosphate;
   - 3% *acetic acid* with acetate according to *Morgan* (USA);
   - lactate solution (0.02 M Ca lactate and HCl, pH 3.6) according to *Egnér-Riehm*, used for more than 20 years as a standard method for phosphate and potassium in Germany, supplemented by
   - CAL-method (0.1 M Ca-lactate, 0.1 M Ca-acetate, and acetic acid, pH 4.1) according to *Schüller*, with higher buffering capacity for calcareous soils (in lieu of lactate method for soils with pH > 6.8).

5. *Alkalis:* e.g., 0.5 M $NaHCO_3$ for phosphate in calcareous soils according to *Olsen* (USA); also phosphate soluble in NaOH.

6. *Complex formers:* e.g., for available trace elements, 0.05 M EDTA (ethylene diamine tetraacetic acid) or 0.005 M DTPA (diethylene triamine pentaacetic acid).

7. *Reducing agents:* e.g., reducible manganese with 0.2% hydroquinone according to *Leeper* or $NaHSO_3$ and Na_2SO_3 according to *Schachtschabel*.

8. *Other methods:* determination of potentials, electro-ultrafiltration (EUF) according to *Nemeth*, isotope methods, e.g., A-value according to *Fried and Dean*, L-value according to *Larsen*.

but equally reliable information is obtainable with much less effort by reducing the sampling area when the fields are relatively uniform. Some of these methods are represented in Fig. 5–5.

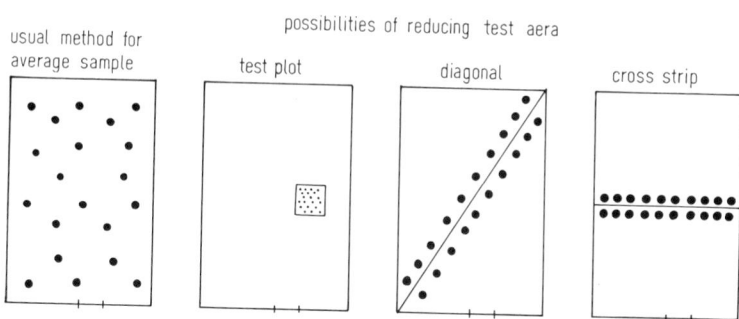

Fig. 5–5. Field Sampling procedure (20 individual samples for one mixed sample according to [7]).

b) The correct *sampling method* is to take at least 20 cylindrical soil samples per hectare, by soil auger and to combine them as a mixed sample. The normal depth of insertion is 0 to 20 cm for fields and 0 to 10 cm for grassland.

Samples of the subsoil are also taken for special investigations, e. g., from depths of up to 1 m for nitrate-nitrogen tests. Volume samples are indicated for bog soils. Results are then referred to 100 ml soil instead of 100 g.

c) The correct *time* is not specified rigorously. However, sampling directly after fertilization would be wrong. It is also important to sample at similar seasons during the year to compare analyses at time intervals. It is recommended that samples be taken after the harvest, before fertilizer application in autumn since the quantities applied are to be determined from the results. It is also possible to perform analyses during the principal growth period. However, the values obtained, e. g., for phosphate, are often higher than in autumn because of the increased soil activity, and this should be taken into account in the evaluation.

The Procedure of Soil Analysis (for available nutrients)

1. Sampling
On fields: 20 samples per hectare from a depth of 0 to 20 cm;
On grassland: 40 samples per hectare from a depth of 0 to 10 cm.

2. Processing
For most analyses the soil sample is air-dried, comminuted after removal of stones, and screened to 2 mm size. Only fine soil particles smaller than 2 mm are analyzed.

3. Chemical Analysis
a) *Extraction* of the available component or a representative fraction of the nutrient to be analysed is effected with a suitable solvent by shaking for one hour. In Germany, phosphate and potassium are extracted with *lactate solution* or with *CAL-solution* at high pH, and sometimes with water. The respective methods are indicated in the evaluation tables.

b) *Chemical analysis* of nutrients in the extracted solution can be performed rapidly and precisely photometrically and usually leads to fewest errors.

c) The value of the contents is *computed*, referred to air-dried soil, and for historical reasons is sometimes still given in the cumbersome oxide form, e. g., mg P_2O_5 per 100 g soil. However, contents of trace elements are already indicated in element form and in ppm.

4. Interpretation
a) The *supply state* or content class is determined with the aid of *evaluation tables* specifying the nutrient contents for the individual supply ranges. These tables are presented in the following chapters for the respective nutrients.

b) The *fertilizer amount* is determined from the content class, as shown for the various nutrients in the following chapters. The yield level must be taken into account, since it determines the removal level. There are different ways of carrying out this procedure in detail. This will be shown below by examples of evaluation procedures for various parts of the FRG. However, standardization would be advantageous.

Possibilities of Improved Soil Analysis
Soil analyses for available nutrients require compromise, since the methods should

- be both *cheap* and *simple* for routine investigations;
- provide *exact information* for precise fertilization.

The two aims are difficult to combine. Soil analysis for available nutrients, using simple field "testing kits", can be very cheap but are then, unfortunately, often inaccurate. They may however be carried out far more intensively than with the usual standard procedure (extraction of a fraction), i. e., by multiple-fraction analysis, with calibration for small soil units, etc. However, costs are then correspondingly high. The investigation may be worthwhile even in such cases, but a properly performed "standard" analysis could still be useful for many practical purposes in fertilization counselling.

The historical development from serious nutrient deficiency of soils to proper supplies through fertilization caused the emphasis in soil analysis to be placed almost exclusively on the establishment of *deficiencies*, i. e., in the lower part of the supply scale. Nowadays diagnosis in the upper part of the scale is gaining importance, especially in the interests of economy of fertilization, and prevention of overfertilization, caused by extremely large fertilizer doses or the accumulation of nutrients due to waste removal. In future it will also be necessary to determine the *upper limit* of proper supply and, thus, of useful fertilization. Abundantly supplied soils need either no fertilization at all or only little fertilization.

Soil analysis has developed historically. Many methods were temporarily introduced for certain regions, proved to be quite useful, and then became *standard methods* by virtue of long-time usage without checking to compare with competing methods. Consequently, some common methods of soil analysis lack the necessary reliability. Therefore methods should be calibrated to obtain a measure of their usefulness (Synopsis 5–9).

5.3.3 Plant Analysis for Nutrient Contents

Plant growth depends on the *concentrations* of nutrient elements in the plant. These concentrations are a measure of the supply state and may therefore serve to predict possible yield increases after fertilization. Analysis of leaves or other plant parts thus makes it possible to determine whether, at the time considered, supplies of a given nutrient are optimal or not. Fertilization is required if the analysed concentration is below the required concentration *(optimum nutrient level)*.

Plant analysis for the purpose of diagnosing fertilizer requirements has repeatedly proved to be reliable. If performed properly, it is more precise than soil analysis, but also more involved and generally more expensive. Its particular advantage is that limit values are largely universally valid for evaluation, which is not the case with soil analyses. However, the stipulated conditions must be precisely satisfied.

Plant analysis, except for special tests, e. g., of the plant sap [115], refer to the total content of a nutrient in the plant. The information obtained is thus less valuable with respect to highly immobile elements, especially calcium and iron. Luckily, precisely these two elements are less important in practical fertilization counselling. Moreover, data from plant analyses may be affected by extreme climatic conditions or extremely inharmonious plant nutrition. The analysis is

Synopsis 5-9. Calibration of Soil-Analysis Methods.

There are many methods for determining available nutrients, differing in their suitability for given soils and production conditions. In contrast to the uniform methods of plant analysis, any method of soil analysis must therefore first be tested for its usefulness. The quality of a method can be estimated only roughly and unreliably on the basis of theoretical considerations; empirical calibration, approximately corresponding to a product test, is therefore required. The suitability of a method can thus be determined mainly on the basis of three *calibration scales* (calibration standards):

a) Calibration on the basis of *plant yield,* i.e., based on the relative yield from field (possibly also pot) fertilizer trials, e.g., for phosphate:

$$\text{Relative yield} = \frac{\text{yield without P-fertilizer}}{\text{yield with P-fertilizer}} \cdot 100$$

Example:

$$\text{Relative yield} = \frac{40 \text{ dt/ha}}{50 \text{ dt/ha}} \cdot 100 = 80\%$$

The contents of available phosphate in the unfertilized plot, determined with the method being tested, are then plotted over the relative yields obtained in numerous trials. The extent of agreement is a measure of the quality of the method considered.

b) Calibration on the basis of plant *nutrient contents.* This indirect method makes use of the nutrient contents of plants at a suitable growth stage, instead of the yield. This procedure is much simpler, sometimes the only one possible, and quite useful, provided that certain basic conditions are ensured (see plant analysis).

c) Calibration based on the appearance of *deficiency symptoms.* This procedure is suitable only for distinguishing between acute and latent deficiency or proper supply. Calibration is also possible to some extent in this way, despite its shortcomings.

Lastly, the extent of *agreement* between extraction value and calibration standard must be determined. The established contents of available nutrients would coincide precisely with the calibration standard for an ideal method; expressed statistically, the correlation would have a B-value (the square of the correlation coefficient) of 100%. Proper methods should have B-values of at least 70%. (Indication of the *correlation coefficient* (r) is not recommended in view of its nonlinearity).

therefore only valid within the scope of the usually prevailing "normal conditions".

Evaluation may be performed in various ways, e.g., frequently on the basis of limit values (assuming an otherwise balanced nutrition), or by taking nutrient ratios into account [124, 132]. This discussion is based on the *evaluation of limit values.*

Optimal ranges of nutrient contents are not indicated uniformly in published works. Two concepts are of primary importance for the decisive value that should be attained as a minimum;

- The *yield limit value* (optimum nutrient level) (Chap. 5.2.2) should be estimated on the basis of pot and field trials of nutrient increases. It indicates the content necessary for maximum yield. This concept should be preferred for intensive production despite a certain experimental imprecision that, however, is usually insignificant for the purpose of counselling (Table 5-5).
- The *critical nutrient level* is frequently used in US literature. It indicates the content with which, e.g., 90% of the maximum yield can be achieved. It can be determined more precisely by experiment but is 10 to 20% (?) below the content required for maximum yield. This should be taken into account when comparisons are made with yield limit values [127].

The range of *normal contents*, usually indicated in tables for plant analyses, provides relatively little information for evaluation. These are values found in properly growing, apparently healthy plants. Such normal contents do not indicate whether low values already signify a latent deficiency, nor do they delimit luxury supplies.

The Procedure for Plant Analysis
1. Sampling at the proper time
Nutrient contents decrease in the course of time even with proper plant nutrition (dilution effect). Plant samples must therefore be taken precisely at the time indicated in the critical-values tables. Sometimes there is a choice of sampling times.

Table 5-4. Example of interpreting results of plant analysis of wheat (analysis at start of shoot growth).

Nutrient element	Fertilization during sowing	Determined content	Optimal range	Assessment
phosphorus	50 kg P/ha	5.5‰	3–5‰	luxury supply (range D) exists, no fertilization required at present, future fertilization may be reduced
magnesium	none	0.9‰	1.5–2.5‰	deficiency exists, immediate fertilization required, Mg-supplies should also be considerably increased in future
manganese	none	37 ppm	35–100 ppm	plants are optimally supplied but content is at lower limit, Mn-supplies should in future be increased for the sake of safety

Table 5-5. Optimal contents of nutrient elements in some cultivated plants.
(First figure indicates optimum nutrient level, i.e., minimum concentration in dry matter, r⃨ quired for high yields. Data of various authors were compiled) [7, 19].

Plant	Growth stage and plant part	‰ N
(Winter)	I. Start of shoot growth (1st to 2nd stalk knot, parts above soil surface)	30-5⃨
Wheat	II. Start of ear growth, parts above soil surface	20-3⃨
Oats	I. Start of shoot growth (1st to 2nd stalk knot, parts above soil surface)	40-5⃨
	II. Start of panicle growth, parts above soil surface	20-3⃨
(Winter) *Rape*	Shortly before budding stage, upper fully developed leaves	40-5
Sugar Beet	Fully developed plant, intermediate leaf lamina	35-4
Potato	Beginning of blossoming, older leaves	(45-6
Alfalfa	Beginning of blossoming, parts above soil surface	30-6⃨
Grass: for yield	Beginning of blossoming (10 to 15% raw proteins), parts above soil surface	25-4
Grass: desired value for cows		.

Table 5-6. Adequate contents of nutrient elements in some fruit trees and flowers in Californ⃨ USA. (First figure indicates *critical nutrient content* in dry matter, which is slightly less than ⃨ optimum nutrient level according to [10], partly supplemented).

Plant	Growth stage and plant part		‰ N	P
Orange (Citrus sin.)	5-7 months old end leaves (mostly September to October)	nonfruit-bearing branches	24-26	1.2-
Apple (Malus sylv.)	fully developed leaves		20-24	(1.3-
Olive (Olea europ.)	older and medium-age leaves		15-20	(1.2)-
Poinsettia (Euphorbia pulcher.)	young fully developed leaves		30-	2-
Rose (Rosa hybrid.)			30-	2-

Table 5-7. Optimal contents of nutrient elements in maize (contents of dry matter, data for ⃨ according to [117]).

Growth stage and plant part	‰ N	P	S
1. 3rd to 4th leaf stage, parts above soil surface	35-50	4-8	2-3
2. during blossoming, leaf at cob	27-35	2-4	1-3
3. during blossoming, stalk above cob leaf	.	1-2	.
4. when ripe, grain	10-25	2-6	.

able 5-5. (Continued)

S	K	Ca	Mg	Fe	Mn	Zn	Cu	B	Mo
	‰					ppm			
3-5	2-4	25-40	5-15	1.5-2.5	50-200	35-100	20-50	5-10	3-10	0.4-3
2-4	1.5-2	20-35	3-10	1-2	40-200	30-100	15-40	3-10	3-10	0.3-3
5-6	2-4	30-45	4-10	1.5-2.5	50-200	40-100	20-50	5-10	5-20	0.4-3
3-5	1.5-2	25-30	2-5	1-2	40-200	30-100	15-40	3-10	5-20	0.3-3
)-	6-	30-50	(3)-	1.2-	(50)-	30-	20-30	3-5	30-60	0.3-
3-6	1-10	25-60	5-15	2.5-10	50-200	40-200	15-50	5-10	30-100	0.3-3
3-6	·	25-40	·	2.5-10	60-	40-	30-90	5-10	30-50	0.3-
3-6	(2)-	20-40	10-30	3-10	40-200	40-200	30-90	8-15	30-100	0.5-5
5-7	2-4	20-40	·	1.5-4	50-200	40-200	20-50	5-10	5-20	0.3-2
4-	·	20-30	6-	·	60-	60-	40-	8-	·	0.2-

ble 5-6. (Continued)

K	Ca	Mg	Fe	Mn	Zn	Cu	B	Mo
	‰				ppm			
7-11	(30-55)	2.6-6	60-120	25-200	25-100	5-16	31-100	0.1-3
12-(20)	10-	2.5-(4)	(50)-	(25-200)	14-(50)	(5-12)	25-70	(0.1)-
8	10-	1-	(50)-	(25)-	(12)-	(5)-	20-150	·
10-	5-	2-	50-	30-	15-	1-	20-	·
18-	10-	2.5-	50-	30-	15-	5-	30-	·

le 5-7. (Continued)

Ca	Mg	Fe	Mn	Zn	Cu	B
‰				ppm		
9-16	3-8	50-300	50-160	·	7-20	7-25
4-10	2-4	50-200	20-250	(20)-	3-15	4-15
1-3	1-3	50-75	20-70	·	3-15	4-12
0.1-0.2	1-2	30-50	5-15	·	1-5	1-10

2. Sampling of the proper plant part

The critical-values tables specify whether leaves, older, medium-age, or younger, stalk parts, or, in the youth stage, even all plant parts above the surface of the soil are to be analysed. The root is hardly ever analysed. The required fresh substance is 300 to 500 g. In general the analysis should be performed only on plant parts not yet completely necrotic.

3. Representative sampling

The sample must be characteristic of the population to be investigated (plot or field). Sampling of fields is discussed in the instructions for soil analysis (Chap. 5.3.2).

4. Immediate and careful processing

The plant material should be processed, if possible in the laboratory, on the day of sampling. The procedure is as follows:

a) *Washing:* Plant samples should be clean when taken, but must in addition be thoroughly washed with tap water and distilled water.

b) *Drying:* Predrying at 60 to 70 °C, followed by after-drying at 100 to 105 °C until the weight remains constant.

c) *Grinding:* comminution without losses or additions in the grinding process.

5. Chemical Analysis

Common chemical procedures first require dry or wet ashing. *Wet ashing* in concentrated acid is somewhat more accurate but cumbersome and dangerous. *Incineration* at 500 °C or additionally at 700 °C is therefore usually preferred.

Actual *chemical determination* of the mineral constituents can be performed very accurately photometrically and is generally the smallest source of error.

6. Computation of Concentrations (contents)

The contents, or more precisely, concentrations, are commonly calculated with respect to dry matter. For major nutrients, they are best expressed in ‰ (parts per thousand), and for trace elements, in ppm (parts per million).

7. Interpretation of Results Obtained

The contents found are classified into suitable *supply ranges*, using limit-value tables. This shows whether the plant is supplied insufficiently, optimally, or excessively with the nutrient involved. This simple evaluation is obviously only possible if reliable data on optimal contents are available. The evaluation procedure is illustrated in Table 5-4 (p. 203). Examples of yield limit values are given in Tables 5-5, 5-6, and 5-7 (p. 204).

A clear interpretation procedure is illustrated in Fig. 5-6.

nutrient supplies in % of
optimum nutrient level

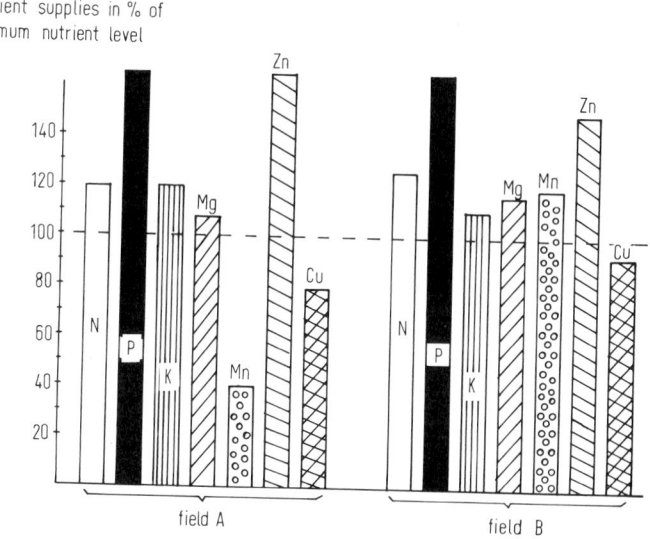

Fig. 5-6. **Nutrient supply of young wheat plants in two fields near Kiel** (1976, field A = medium yield, field B = high yield).

Determination of Required Fertilizer Amounts

Supplies should be improved if the analysis indicates that they are insufficient. This usually implies the need for fertilization. The fertilizer amounts to be applied immediately are indicated in the guidelines for the deficiency range (Chap. 5.4).

Supplementary fertilization may be needed when there are large current requirements, even if supplies are optimal at the time of analysis. This may be the case especially with nitrogen. Countermeasures are to be considered if the analysis indicates a harmful excess. Some toxicity limits necessary for assessment are given in Table 5-8. Other data are contained in [7].

Table 5-8. Toxicity limit values for oats, alfalfa, and rice (approximate values for parts above soil surface during principal growth stage, according to [7, 229]).

Plant	Contents in ppm (dry matter)			
	Mn	Zn	Cu	B
oats	1 000	400	20	(30)
alfalfa	500	(200)	(30)	200
rice	1 500	.	30	100

Conclusions on *fertilization during the coming year* can also be drawn from plant analysis during the current year. However, it should then be remembered that plant analysis also takes the influence of preceding fertilization into account. The following conclusions concerning the next crop would be possible:

- Greatly increased fertilization (higher than amounts removed) is necessary when there is a deficiency;
- maintenance fertilization is indicated for the future when there are normal supplies and fertilization (see fertilizer quantities for range C of soil analysis);
- no fertilization is necessary ordinarily in the next two years where normal supplies are achieved without any fertilization (e. g., with trace elements);
- greatly reduced fertilization with major nutrients and no fertilization with micronutrients is indicated for the future when luxury supply occurs.

5.4 Amounts of Nutrients Required

The correct fertilizer amounts should provide sufficient nutrients for high yields, without causing unnecessary accumulation or losses.

The problems involving fertilizer requirements differ for individual nutrients, so that separate discussion seems appropriate. For a long time, the main subject in determining fertilizer requirements was research on phosphorus and potassium, with occasional additional work on magnesium and some of the trace elements. Today there are clear signs that fertilization with N is also placed on a diagnostic basis, and this appears to be urgently necessary at least for high-yield production, in view of its economic importance.

This chapter deals mainly with the determination of fertilizer requirements on the basis of soil analysis, but the results of plant analysis may also be used. Fertilizer requirements are deduced in a similar way after the supply state has been established by either plant or soil analysis.

The evaluation of soil analysis has in recent years been put on a new and more comprehensive basis in the FRG, the main purpose being to ensure uniformity. Until now, this objective has only been achieved in part, because allowing for local differences in soils etc. would require a considerably larger evaluation scheme. However, a uniform evaluation system would be desirable and feasible in the final analysis. The fertilizer amounts recommended in the following chapters apply principally to normal conditions, i. e.,

- proper penetrability of the soil by roots, about 20 to 25 cm deep topsoil;
- adequate precipitation and medium water-storage capacity;
- no extremely dry periods ;
- no extreme nutrient fixation.

These conditions are at least approximately satisfied on most sites in the FRG. The relatively rare exceptions require special determination of fertilizer requirements.

5.4.1 Nitrogen Requirement

Nitrogen is the principal stimulant of plant production. Natural nitrogen supplies from the decomposition of humus are generally only sufficient for low yields and must therefore be supplemented by fertilization. This can be done with either organic fertilizers, most often mainly produced on the farm itself, or mineral fertilizers.

Proper dosage of N-fertilizers in amount and distribution over time, as a function of soil, climate and plant properties, is a "wide field". Many factors are difficult to quantify or predict. It is therefore not surprising that fertilization with N is still largely determined by rule of thumb, i. e., values derived from experience, and that are often only valid locally. However, it is precisely in the *range of large N-doses* that *precise fertilizer application* would be indicated, both to increase production and to avoid unnecessary environmental effects. Amounts and distribution of N-fertilizers obviously depend largely on the plant to be fertilized. Fertilization of the different plants is discussed in Chap. 7 and 8. The *tentative values* in Table 5-9 provide general information.

Table 5-9. Tentative values for fertilization with N (Mineral and organic fertilization: additions and deductions according to circumstances) [111, 102].

Crop	for Northern Germany		for Southern Germany	
	Yield target (dt/ha)	N-Fertilization dt/ha	Yield target (dt/ha)	Fertilization dt/ha
Winter wheat	50–65	120–160	50	80–180
Summer wheat	45–60	100–140	50	80–180
Winter barley	50–65	100–140	50	80–150
Summer barley	40–55	80–100	45	20–80
Wintes rye	40–55	80–120	40	60–120
Oats	45–55	80–100	40	60–120
Maize	silage maize	180–240	60	120–180
Winter rape	25–35	160–200	30	120–200
Fodder beet	600–900	180–200	900	140–200
Sugar beet	400–500	160–200	550	140–200
Potatoes	300–400	120–160	300	100–140
Fodder grass	•	240–360	•	•

The tentative values given for N-fertilizers in part represent relatively wide ranges. The principal reason for this is the *unknown N-reserve* at the start of growth. The greater the reserves, the less N has to be supplied. The development of reliable diagnostic methods would therefore be welcomed, since they would eleminate this factor of uncertainty. However, this investigation is more involved than "ordinary" soil analysis, since samples must be taken from depths of up to 1 m. The amount of mineral nitrogen (N_{min}) available at the beginning of the vegetation period can be deducted from the required ferti!izer quantity, as shown for example in Fig. 5–7.

requirement

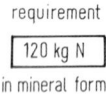

in mineral form

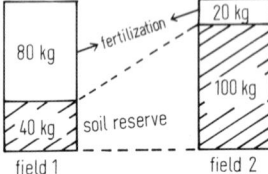

field 1   field 2

Fig. 5–7. Determination of N-fertilizer requirement for winter wheat (example according to [133]).

Synopsis 5–10. N-Supplies from the Soil and Diagnosis of N-Requirement.

General Estimation of N-Supply

- The N-content of many soils is 0,3–3‰ of which more than 95% is contained in organic matter. This represents a reserve of 900 to 9000 kg/ha in the topsoil (from the surface to a depth of 20 cm);
- The mineralization rate is 1 to 3% decomposition of humus per year depending on the biotic activity which is mainly controlled by temperature and humidity;
- This results in about 10 to 300 kg/ha of N being supplied per year, of which about one third, i.e., 3 to 100 kg, can be used by plants.

Estimation of *natural N-supplies* in a field requires a knowledge of two values:

a) *Easily available reserves* (in mineral form, mostly as nitrate, sometimes also as ammonium) at the time of sowing or in spring at the beginning of the vegetation period. Nitrate reserves in the entire root region down to a depth of 60 cm or 1 m are frequently in the range of 20 to 100 kg/ha of N ("N_{min}" content);

b) *Additional quantities of N*, supplied during the vegetation period, i.e., the content of easily mobilized N, which can be determined in about a fortnight by *incubation methods* with intensive decomposition of humus. This procedure is involved and is therefore only used in special cases.

Mobile N-reserves represent only one possible measured magnitude, as seen from Synopsis 5-10 (p. 210). The amounts needed for supplementary fertilization may be determined from the N-quantities derived from soil reserves or from the N-content of plants. This can be done primarily on the basis of the N-content but also from tissue tests for nitrate and leaf color (Chap. 5.3.3).

5.4.2 Phosphate Requirement

The content of available phosphate in the soil in the FRG is determined by the *lactate method*, or in the case of higher pH-values, by the *CAL-method* (Chap. 5.3.2). Sometimes the *water-soluble phosphate* method is used, which originated in the Netherlands [120]. Evaluation for determining fertilizer requirements will be explained for the *lactate method*, which has been in use for 30 years.

New guidelines were established in 1974/75 by various agencies for determining the amounts of phosphate and potassium fertilizers needed, the aim being improved and more uniform evaluation. Clear schemes for determining fertilizer quantities now exist, at least for larger regions:

- as proposed by the Association of Chambers of *Agriculture* for large parts of Northern Germany, North-Rhine-Westphalia and Hessen [111], as well as for the Weser-Ems region as proposed by LUFA Oldenburg [139];
- for Bavaria as proposed by the *Bavarian State Institute* (Bayerische Landesanstalt).

Table 5-10. Grouping of phosphate contents by content class (tentative values of Chambers of Agriculture (ACA) for Weser-Ems region and Bavaria; provisional values, lactate and CAL-methods respectively).

| | Mineral soil mg P_2O_5/100 g soil | | | Bog soil mg P_2O_5/100 ml soil |
| | Fields and grassland | | Fields | Fields and grassland |
Content class	Tentative values of Chambers of Agriculture; for grassland in Bavaria	Tentative values for Weser-Ems region	Tentative values for Bavaria	Tentative values of Chambers of Agriculture
A	1 – 10	4 – 6	1 – 10	1 – 5
B	11 – 20	7 – 15	11 – 18	6 – 10
C	21 – 30	16 – 30	19 – 35	11 – 15
D	31 –(40)	31 –(50)	36 –(60)	16 –(20)
E	(40)–	(51)–	(61)–	(21)–

(Values in parentheses are relatively unreliable)

Classification of soil-analysis results according to the corresponding content classes is shown in Table 5-10 (p. 211). The resulting tentative values for the fertilization of fields are given in Table 5-11, and for grassland in Table 5-12.

All values, in contrast to the original publication, have been designated as *provisional*, since they were initially established as first approximations and therefore require checks and further development. The extensive evaluation material can only be reproduced here as extracts of essential parts. The reader is referred to the original publication for details.

5.4.3 Potassium Requirement

The content of available potassium in the soil is determined in the FRG by the *lactate method*, as for phosphate. Classification by contents differs greatly according to the soil type. Adequate supplies on light soils require smaller contents of available potassium than heavy soils rich in clay, since a certain content of exchange potassium is needed at the exchange complexes (about 5% in sandy soils, 2% in clay soils, with corresponding intermediate values). However, 5% potassium is quantitatively less at the low exchange capacity of sandy soils than 2% for clay soils. Most lactate potassium (70 to 90%) is exchange potassium.

As for phosphate, the Tables contain two different proposals for Northern and Southern Germany respectively. We only mention the detailed guidelines by LUFA Oldenburg [139]. Grouping of soil-analysis data by content classes is demonstrated in Table 5-13 (p. 214). The resulting tentative values for fertilization are given in Table 5-14 (p. 215) for fields and in Table 5-15 (p. 216), for grassland. All values are designated as *provisional* for the same reason as in the case of phosphate. For details the reader is again referred to the original literature.

5.4.4 Magnesium and Other Nutrient requirements

Diagnosis of magnesium requirement, among major nutrients, is also important, but that of calcium and sulfur requirements is less so.

A useful measure of available *calcium* could be the exchange-Ca, in analogy to magnesium. Available *sulfur* could be determined similarly to nitrogen, i.e., as sulfate (easily available component) and resuppliable sulfur (from decomposed humus).

Magnesium Requirement

Magnesium requirement is simpler to determine than that of primary nutrients, since abundant fertilization for storage is possible. No damage will result even from large doses in ordinary fertilization; moreover, the price is less important.

Table 5-11. Provisional standard values for fertilization of fields (mineral soil) with P (kg/ha of P₂O₅ as mineral and organic fertilizer).

Tentative values of Chambers of Agriculture (ACA)

Content class	Requirement group 1 Root crops		Requirement group 2 Rape, legumes, etc.		Requirement group 3 Cereals		Tentative values for Bavaria
	Fertilization for normal yield	Addition per 50 dt yield increase	Fertilization for normal yield	Addition, e.g., for 10 dt rape yield increase	Fertilization for normal yield	Addition per 10 dt grain yield increase	Fertilization
A	200		175		150		Removal + 90
B	150	10	125	15	100	15	Removal + 40
C	**100**		**75**		**50**		**Removal + 10**
D	50		40		25		only for root crops (removal)
E	0		0		0		0

Normal yields (dt/ha) on loamy soils (according to ACA)

Fodder beets = 900 (beets)   Rape = 30 (grains)   Cereals = 50 (grains)
Sugar beet = 500 (beets)   Legumes ·   (for sandy
Potatoes   = 350 (tubers)   Field fodder ·   soils = 40)
Silage Maize = 600

Addition for phophate-fixing soils (e.g., acid podsols in amounts of 10 to 20 kg/ha P₂O₅

Tentative values for cereals (at 45 dt/ha) in Weser-Ems region class:
C = 80 kg
D = 40 kg

Removal values for Bavaria

Crop	Yield dt/ha	Removal in kg/ha		
		grain Beets Tubers	Straw Leaves	Total
Wheat	50	40	20	60
Brewer's barley	45	40	15	55
Rye/Oats	40	35	15	50
Kernel maize	60	35	20	55
Potatoes	300	50	10	60
Sugar beets	550	55	45	100
Fodder beets	900	75	25	100
Rape	30	55	25	80
Peas and beans	30	35	20	55
Clover, grass (dry matter)	100	·	·	80

Table 5–12. Provisional standard values for fertilization of grassland with P (kg/ha of P_2O_5 as mineral and organic fertilizer).

Content class	Tentative values of Chambers of Agriculture					Tentative values for Bavaria			
	Pastures		Mowed	Meadows (mowed areas)		Meadows (mowed areas)			Pastures
			pastures			2 Cuttings	3 Cuttings		
	extensive	intensive	pastures	2 cuttings	3 cuttings	50 dt dry matter	70 dt dry matter	90 dt dry matter	
A	60	120	140	160	200	70	100	120	·
B	30	80	100	120	160	50	80	100	·
C	0	40	60	80	110	**30**	**60**	**80**	(0–20) = net removal
D	0	0	20	40	60	·	·	·	·
E	0	0	0	0	0	·	·	·	·

removal fertilization

Table 5–13. Grouping of potassium contents by content class (provisional values, lactate method).

Content class	Tentative values of Chambers of Agriculture						Tentative values for Bavaria			
	Mineral soil mg K_2O per 100 g soil Fields					bog soil (mg K_2O per 100 ml soil)	Mineral Soil mg K_2O per 100 g soil Fields			Grass-land
	Sand	sand rich in humus	sand very rich in humus and loamy sand	half-bog soil and sandy loam	Fields (loam to clay) grassland (total)		light	medium	heavy	
A	1 – 5	1 – 6	1 – 7	1 – 8	1 – 10	1 – 5	1 – 7	1 – 11	1 – 14	1 – 9
B	6 – 10	7 – 12	8 – 14	9 – 16	11 – 20	6 – 10	8 – 15	12 – 24	15 – 30	10 – 20
C	**11 – 15**	**13 – 18**	**15 – 21**	**17 – 24**	**21 – 30**	**11 – 15**	**16 – 25**	**25 – 40**	**31 – 50**	**21 – 30**
D	16 –(20)	19 –(24)	22 –(28)	25 –(32)	31 –(40)	16 –(20)	26 –(50)	41 –(60)	51 –(70)	31 –(40)
E	(21)–	(25)–	(29)–	(33)–	(41)–	(21)–	(51)–	(61)–	(70)–	(40)–

... K (kg/ha K₂O as mineral and organic fertilizer).

Tentative values of Chambers of Agriculture (ACA)

Content class	Requirement group 1 Root crops			Requirement group 2 rape, legumes, etc.		Requirement group 3 cereals		tentative values for Bavaria
	fertilization for normal yield (light and bog soils / heavy mineral soils)	addition per 50 dt yield increase	for fodder beet	fertilization for normal yield (light and bog soils / heavy mineral soils)	addition per 10 dt rape yield increase	fertilization for normal yield (light and bog soils / heavy mineral soils)	addition per 10 dt grain yield increase	fertilization
A	Ø400 500	} 25	} 100	Ø330 400	} 20(−40)	200 300	} 20	removal + 100
B	Ø330 400			250 300		150 200		removal + 50
C	**250 300**			**150 200**		**100 125**		**removal**
D	150 200			100 100		50 75		removal for root crops
E	0 100			0 50		0 0		0

normal yields (dt/ha) on loamy soils (according to ACA)

fodder beet = 900 (beets)	rape = 30 (grain)	cereals = 50 (grain)
sugar beet = 500 (beets)	silage maize = 600	(for sandy soils = 40)
potatoes = 350 (tubers)	legumes, field fod-	
cabbage .	der, field vegetables	

removal values for Bavaria

crop	yield dt/ha	removal in kg/ha — grains, beets, tubers	straw, leaves	total
wheat	50	25	85	110
brewer's barley	45	20	80	100
rye/oats	40	20	90	110
kernel maize	60	40	130	170
potatoes	300	160	40	200
sugar beet	550	125	200	325
fodder beet	900	270	90	360
rape	30	30	120	150
peas and beans	30	65	65	130
clover grass (dry matter)	100	.	.	300

additions for potassium-fixing soils
deduction for intensively potassium-resupplying soils

Table 5-15. Provisional standard values for fertilization of grassland with K (kg/ha of K_2O as mineral and organic fertilizer).

Content class	Tentative values of Chambers of Agriculture				Tentative values for Bavaria			
	Pastures		Mowed Meadows (mowed areas)		Meadows (mowed areas)			Pastures
					2 Cuttings	3 Cuttings		
	extensive	intensive	2 cuttings	3 cuttings	50 dt dry matter	70 dt dry matter	90 dt dry matter	
A	60	120	240	400	200	300	390	
B	30	80	180	320	170	270	360	
C	0	40	120	220	**140**	**240**	**330**	
D	0	0	60	120				
E	0	0	0	0				

removal fertilization

(0-30) = net removal

Table 5-16. Grouping of magnesium contents by content class.

Content class	Tentative values of Chambers of Agriculture				bog soil (mg Mg/100 ml) fields and grassland	Tentative values for Bavaria			
	Mineral soil (mg Mg per 100 g soil)					Mineral soil (mg Mg per 100 g soil)			
	Fields			Grass-land		Fields			Grass-land
	sandy soils	sandy loam and half-bog soils	clayey loam and clay			light soils	medium soils	heavy soils	
A	1-3	1-4	1-6	1-5	1-3	1-2	1-5	1-7	1-9
B	4-5	5-7	7-10	6-10	4-6	3-6	6-12	8-14	10-21
C	**6-7**	**8-9**	**11-14**	**11-15**	7-9	**7-**	**13-**	**15-**	**21-**
D	8-	10-	15-	16-	10-	.	.	.	.

The content of *available* magnesium is determined in the FRG by extraction with 0.025 N $CaCl_2$ according to *Schachtschabel*. This corresponds approximately to the exchange magnesium. Two different proposals, valid for Northern and Southern Germany respectively, have been made for magnesium as for phosphate [111, 102]. The detailed guidelines by LUFA Oldenburg can only be mentioned here [139]. Grouping of soil-analysis data by content class is demonstrated in Table 5-16 (p. 216). The resulting standard values for fertilization are given in Table 5-17. Mg-removal during one harvest is 10 to 40 kg/ha.

Table 5-17. Tentative values for fertilization with magnesium (kg/ha Mg)*.

Content class	Tentative values of Chambers of Agriculture field soils requirement group		grassland soils mineral soils potassium-content class			bog soils	Tentative values for Bavaria	
	1	2	A–C	D	E	A–D	fields	grassland
A	100	50	120	120	120	120	25–60	40–50
B	50	25	60	90	120	90	20–25	20–25
C	25	10	30	90	120	30	only if potassium
D	0	0	0	60	90	0	abundant

requirement group: 1. beet, potatoes, maize, field fodder
2. grain, rape
* for choice of correct Mg-form, see Chap. 2.4

Sodium and Silicon Requirements

Sodium requirement is deduced from the content of available sodium in the soil, determined by extraction with $CaCl_2$, as for magnesium. Classification of soil-analysis data and tentative values for fertilization are given in Table 5-18.

Table 5-18. Tentative values for sodium contents and fertilization of grassland with sodium (extraction with $CaCl_2$, in analogy to magnesium).

Content class	Tentative values of Chambers of Agriculture Content of available Na in soil ppm	Fertilization with Na kg/ha Na sandy soils	loamy soils	Tentative values for Bavaria Content of available Na in soil ppm	Fertilization with Na kg/ha Na
A	1– 15	60	40	1– 50	45–60
B	16– 45	45	30	50–100	25–30
C	46– 95	30	20	100–	0
D	96–135	15	10	.	.

Silicon requirement can be determined analogously to phosphate requirement. Most soils contain absorbable silicon in the form of ortho-silicic acid (H_4SiO_4). *Imaizumi* and *Yoshida* found that extraction of the soil with Na-acetate buffered at pH 4 gives a good correlation with silicon fertilizer requirements [64].

5.4.5 Requirement of Trace Elements

It is of primary importance to estimate the *supply state* of trace elements in plants and soils by a suitable diagnostic method. *Fertilizer amounts* depend less on the extent of any existing *deficiency* and hence on requirements but can largely be handled with standard methods.

Information on standard fertilizer amounts was given in Chap. 3.1 when fertilization with trace elements was discussed. We shall therefore now consider soil analyses for available trace elements.

Soil analysis for available trace elements, used as a basis for determining fertilizer requirements, has gained much importance during the last 20 years. Some methods are internationally standardized (e. g., determination of available boron), but various methods exist for some elements, such as manganese. Some methods yield at best *approximations*, and it is frequently necessary to check soil-analysis data, especially for heavy metals.

Iron requirements are less easy to determine by plant analysis, but even the determination of *available iron* in the soil is problematic because of the large total content and the different mobilization conditions. Soil analysis for available iron hardly plays any role at all in the FRG. Extraction of the available component with the chelating agent *DTPA* (diethylene triamine pentaacetic acid) has proved successful for soils in arid regions (Chap. 6.3.3), where Fe-deficiency is common [109].

The similarity of the behavior of iron and manganese in the soil is the reason similar methods are used for these elements.

Manganese requirement can be reliably determined by plant analysis, and approximately so by soil analysis. Nearly all the methods explained in Chap. 5.3.2 have been tested (more than 50 different methods for manganese). The strong dependence of Mn-mobilization on the redox potential has led to the large-scale use of slightly reducing extraction solvents. The *Leeper* method with 0.2% hydroquinone in N ammonium acetate is historically important and widely used.

Available (active) manganese is usually determined in the FRG with the *Schachtschabel* method. This involves extraction with two reducing agents (2 g sodium sulfite ($Na_2SO_3 \cdot 7H_2O$) and 1 g sodium hydrogen sulfite ($NaHSO_3$) per liter) in N $MgSO_4$ as an exchange solution. This permits the determination of *available* and *easily reducible* manganese [136]. Provisional limit values for proper supplies are, at

pH 6.5–7.5: 50 to 70 ppm Mn
pH 6.0–6.3: 25 to 40 ppm Mn
pH   5.8: 15 to 20 ppm Mn

There is practically no longer any deficiency in most soils at pH-values below 5.7. It should be stressed that limit values for certain regions may differ from the above.

Zinc requirement can also be detesmined reliably by plant analysis (Chap. 5.3.3). Soil analyses play a minor role in the FRG, since zinc deficiency is still rare. Chelating agents are generally used as extraction solvents in countries where zinc deficiency is more common, e.g., DTPA (as for iron) or EDTA (Chap. 3.1.1). Diluted acids are also useful for acid soils e.g., as for copper. Concentrations of 8 to 10 ppm of available zinc represent proper supplies when HNO_3 extraction is used (see copper).

Copper requirement can be calculated from plant contents (Chap. 5.3.3), but soil analyses are also frequently performed. The standard method in the FRG is extraction with 0.43 N nitric acid (HNO_3) according to *Westerhoff*. However, it should be applied only to soils containing no lime, since the solution is not buffered. Chelating agents are especially suitable at high pH-values (as for zinc), e.g., *EDTA* (Chap. 3.1.1). The classification of copper values according to *Westerhoff* is given in Table 5–19 [139, 102].

Table 5–19. Standard values for copper contents and fertilization of fields and grassland with copper (*Westerhoff* method).

Content class	Content of available copper (ppm)					Tentative values for stock fertilization (kg/ha of Cu)
	Tentative values of LUFA Oldenburg				Tentative values for Bavaria[*]	
	0–4% humus		8–15% humus			
	loamy sand	sandy loam and loam	loamy sand	sandy loam and loam		
A	0–2	0–2.5	0–3	0–3.5	0–2	5–10
B	2.1–4	2.5–5	3.1–6	3.6–7	2–4	2–5
C	**4.1–**	**5.1–**	**6.1–**	**7.1–**	> 4	–

* values for soils in regions of Cu-deficiency

Boron requirement can be determined relatively reliably by soil analysis. Available boron is determined as boron soluble in hot water, according to *Berger* and *Truog* (essentially standardized, internationally). Tentative values for content classes and fertilization with boron are given in Table 5–20.

Table 5–20. Standard values for boron contents and fertilization of mineral-soil fields with boron (unified values for FRG in conformity with international values; boron soluble in hot water, according to *Berger* and *Truog*).

| Content class | Content of easily available B in soil ppm | | Tentative values for fertilization kg B/ha | | | |
| | light soils | medium and heavy soils | light soils | | medium and heavy soils | |
			beets, alfalfa	maize, rape, cabbage, potatoes	beets, alfalfa	maize, rape, cabbage, potatoes
A	0.1 –0.3	0.1 –0.4	1.5	1.0	2.5	1.5
B	0.31–0.6	0.41–0.7	1.0	0.5	1.5	0.7
C	**0.61**–1.1	**0.71**–2.0	0–0.5	0	0–1	0
D–E	1.2 –	2.1 –	0	0	0	0

Molybdenum requirements can be estimated from the results of soil analyses. Extraction with *oxalate solution* (*Grigg* method) has proved suitable for the determination of easily mobilizable molybdenum. However, other methods can also be used, e.g., extraction with water. The strong dependence of Mo-availability on the pH led *Müller* [104] to propose that combined allowance be made for available Mo and pH-value; this is expressed as Mo-soil number:

$$\text{Mo-Soil number} = \text{pH} + (\text{ppm Mo} \times 10)$$

e.g., at a soil pH of 6.5 and a Mo-concentration of 0.3 ppm, the Mo-soil number is $6.5 + 0.3 \times 10 = 9.5$. Mo-soil numbers between 6 and 8 represent medium supplies, values above 8.2 indicate good supplies.

Cobalt requirement can be reliably deduced from plant contents, like copper requirement. Pasture forage for cows should contain at least 0.1 ppm Co in dry matter. It is desirable to have soil concentrations above 0.2 to 0.3 ppm (available cobalt) according to the *Bönig-Heigener* method (extraction with 0.1 N HCl).

5.5 Fertilizer Recovery and Nutrient Removal

Optimal fertilizer amounts depend on two important factors. The *recovery rate* influences fertilizer requirements insofar as the less a fertilizer is utilized, the more of it is required. Only fertilizer amounts equal to requirements would have to be applied if recovery were complete (this can hardly ever be achieved), whereas twice as much fertilizer would be required at a utilization rate of only 50%.

Removal forms the basis for determining fertilizer requirements. Fertilizers may be applied at the level of removal if the supply state of the soil is good, allowing for losses and utilization rate.

5.5.1 Recovery of Fertilizer Nutrients

Fertilizer nutrients introduced into the soil are disposed of in four ways:

- they occur in available form and are absorbed by the fertilized plants (this is the utilized fraction);
- they remain in available form but are not absorbed;
- they are fixed and thus removed from circulation for longer periods;
- they are lost from the root space.

Thus, only part of the total amount of nutrients supplied is recovered by the (fertilized) plants.

The recovery rate of a fertilizer is the nutrient fraction which is absorbed by the plants (expressed in % of the nutrient amount supplied).

The utilization rate is a factor determining the fertilizer amounts supplied. Therefore at least its order of magnitude must be established correctly. However, available data are contradictory, largely because of different concepts and methods of determining the recovery. The utilization rate of fertilizer nutrients may be referred to different time intervals:

- in the narrower sense, to the growth period of the fertilized plant (i.e., one year in the case of annual harvests);
- in a *wider* sense, including residual effects during several years (as far as they can be established);
- in the *widest* sense, for a long period of several decades (this is decisive for long-term considerations).

The recovery rate also depends on the extent to which the soil is supplied with nutrients, i.e., whether the soil is *deficient* or *well supplied*. Moreover, *true* recovery must be distinguished from *apparent* recovery.

Recovery rates in the ordinary sense, cited in published works, usually refer to *true* utilization during *one* year on *deficient soil*. On the other hand, the reference basis for fertilization of well supplied high-yielding soils should be the apparent utilization over longer periods on well supplied soils.

The Recovery Rate in the Ordinary Sense

The fraction of a fertilizer nutrient (actually) absorbed by plants during one vegetation period can be determined by different methods [107, 122]. Most data are based on the older commonly used *difference method* (Synopsis 5–11).

This method consists in comparing the nutrient intake of plants on fertilized and unfertilized plots, on the assumption that fertilization does not affect the intake of nutrients from the soil, although this is only roughly correct. The fraction of nutrients absorbed from the soil in many cases might be reduced by fertilization, so that the difference method indicates exaggerated values. This is confirmed by comparison with the *isotope method* in which *labelled* nutrients are added [128]. However, the isotope method also has its weak points (non-verifiable isotope exchange!), so that the utilization rates always represent only rough approximations.

However, the differences between the results obtained with the two methods can be compared as to the order of magnitude, so that the utilization rate can be estimated as an approximation for practical purposes.

Some average values have been compiled in Table 5–21. It is quite possible to achieve higher values, but a recovery rate of 80% for N, which is sometimes reported, is probably exaggerated. The utilization rate (in the first year) can be increased by certain methods of fertilizer application (Chap. 2.2.3), but this is only important for inadequately supplied soils [101].

Table 5–21. Recovery of fertilizer nutrients (true utilization in first year after application as soil dressing) (modified according to [52]).

Fertilizer	Utilization rate (%)
nitrogen, mineral	50–60
nitrogen, organic	20–30
phosphate, mineral	15
phosphate, organic	20–30
potassium (mineral or organic)	50–60
manganese, copper, zinc (mineral)	0.5–5

Other Aspects of Fertilizer Recovery

The utilization rate determined over longer periods is obviously higher than that for one year. A fertilizer acts not only during the first year but has residual ef-

Synopsis 5–11. Determination of Recovery Rate.

1. Difference Method

a) The *difference between nutrient intakes* of fertilized and unfertilized plants (obtained from a fertilization experiment) is determined and related to the fertilizer quantities. The utilization rate is then given by the following formula:

$$\text{Utilization rate (in \%)} = \frac{\text{total removal} - \text{removal from soil reserves}}{\text{nutrient amount of fertilizer}} \cdot 100$$

Example (quantities per hectare):
Total removal from fertilizer and soil reserves $= 100 \text{ kg N}$
Removal from soil reserves (without fertilization) $= 40 \text{ kg N}$
Fertilizer quantity $= 120 \text{ kg N}$

$$\text{Utilization rate} = \frac{100-40}{120} \cdot 100 = 50\%$$

b) A variant of the difference method is to *include available nutrients in the soil,* which must have the same availability as the nutrients in the fertilizer, e.g., nitrate fertilizer and nitrate-N in the soil).

Utilization rate (modified)
$$\text{(in \%):} = \frac{\text{total removal} - \text{removal from soil reserves}}{\text{nutrients in fertilizer} + \text{nutrients in soil}} \cdot 100$$

This variant yields values that are lower, the more nutrients of equal availability are present in the soil.

2. Isotope Methods

These also require a fertilization experiment, but the recovery is determined only on one plot by labelling the fertilizer nutrient with isotopes in order to distinguish it from nutrients in the soil.
The utilization rate is derived in the following manner (e.g., for phosphate per ha):

$$\% \text{ fertilizer-P in plants} = \frac{\text{specific P-activity in plants}}{\text{specific P-activity in fertilizer}} \cdot 100$$

(The specific activity is the ratio of ^{32}P- and ^{31}P-isotopes).

The quantity of P in the plant, originating in the fertilizer, is obtained by the following formula:

$$\text{kg fertilizer P in plants} = \frac{\text{kg P in plants} \times \% \text{ fertilizer-P in plants}}{100}$$

This yields the utilization rate in the following way:

$$\text{Utilization rate (in \%)} = \frac{\text{kg fertilizer-P in plants}}{\text{kg fertilizer-P}} \cdot 100$$

fects of varying duration and intensity. A general tentative value is a residual effect during the second year equal to one tenth of the utilization rate in the first year (for N, P, K).

The mean utilization rate of *fertilizer phosphate*, for example, is 15% during the first year (with differences for the various P-forms and application techniques). Residual effects during the second year are 1 to 2%, and about 1% during the following years. We thus obtain for longer periods:

- about 10% for 10 years;
- about 45% for 30 years;
- approximately 100% over very long periods.

Computation of the recovery rate over very long periods is only meaningful with respect to the apparent utilization, which will be discussed below. Fertilizer utilization on well supplied soils is significantly lower than on deficient soils at least during the first year, since the soil already contains sufficient nutrients for the plants and fertilization serves primarily to replenish reserves. This, however, applies to true utilization, which is primarily of theoretical interest. The situation is quite different with regard to the apparent recovery.

The Apparent Recovery Rate

The *apparent (effective)* recovery rate of a fertilizer is important for nutrient balances under special production conditions. The content of available nutrients in soils well supplied with a certain nutrient attains an equilibrium value in the same way as production tends to an equilibrium value, provided that supplies of other nutrients are sufficient.

The apparent recovery rate is obtained from data for fertilized plots according to an abbreviated formula for the utilization rate (see difference methods:

$$\text{Apparent recovery rate (\%)} = \frac{\text{nutrients removed}}{\text{nutrients in fertilizer}} \cdot 100$$

In contrast to the customary formula, the above equation does not take removal from soil reserves into account, since these are assumed to be constant. This means that it is not the true recovery (referred to one year) that is measured, but only an apparent recovery; or expressed another way the true recovery over a very long period is determined. The apparent recovery rate should approach 100% if there were no fertilizer nutrient losses. Values of up to 85% (for one year) have been found for phosphate [118].

The concept of apparent recovery has the advantage of being highly relevant to fertilization conditions in well supplied soils. Thus, assuming a true utilization rate of 15% for phosphate during the first year shows that seven times as much fertilizer would be needed to cover requirements. On the other hand, assuming an apparent utilization rate of 100% indicates that fertilization at the level of removal is sufficient, i.e., replenishment of all losses is indicated in principle.

This concept is valid, however, only if the mobility of available reserves in the soil is maintained with respect to plant absorption. This is not obvious in the case of phosphate [125]. However, approximate constancy may be assumed for fertile soils, if their transformation capacity is constantly activated by suitable measures (supply of humus, etc.).

5.5.2 Removal of Nutrients

Plants remove nutrients from the soil. These nutrients leave the soil together with the harvested products; a knowledge of nutrient removal is therefore very important for establishing a nutrient balance (in which supplies and losses should be approximately balanced over long periods). The first comprehensive compilations of removal values were done by *Wagner* and *Remy* (quoted in [119]).

Average values are determined for the various crops from data obtained by analyzing harvested products. This procedure provides rough tentative values, if a relatively uniform yield level can be assumed. However, more precise removal values are better referred to one *basic unit* of a harvest, e.g., 10 dt cereal grain. The overall removal is then obtained by multiplication by the total yield. Until now, removal values for any crop have usually been summarized as averages for different sites. This concept appears to be of practical use if nutrients in the soil are scarce rather than abundant. However, there is a tendency towards increasing removal values by *luxury consumption* of plants, when supplies in the soil increase, or even earlier, when well-supplied sites are included.

However, because of the dilution effect there is also an opposite tendency, since mineral contents in the harvested products must of necessity decrease [119] with increasing yield level, if there is no compensation by suitably raising supplies. However, the latter tendency is largely masked by the tendency to luxury consumption. Another change to be mentioned is a shift within the yield components. Thus, removals 70 years ago were referred to a grain/straw ratio of 1 : 2. This ratio was about 1 : 1.5 some 40 years ago and now approaches 1 : 1.

Removal values play an important role nowadays as a basis for determining fertilizer requirements. *Critical consideration* is therefore indicated.

Customary data on average removals are not satisfactory because of their variability. A range of varying extent has to be indicated to allow for different sites and fertilizer amounts. This is not enough for a basic value of this importance, especially since there is a tendency to adopt the upper rather than the lower limit of an indicated range. This seems even less desirable if removal values have a tendency to increase accompanying the trend to luxury consumption.

In theory there are at least two possibilities for defining removal values (Fig. 5-8), i.e.:

- necessary *minimum* removal based on optimum nutrient levels;
- *luxury* removal that may be several times the minimum.

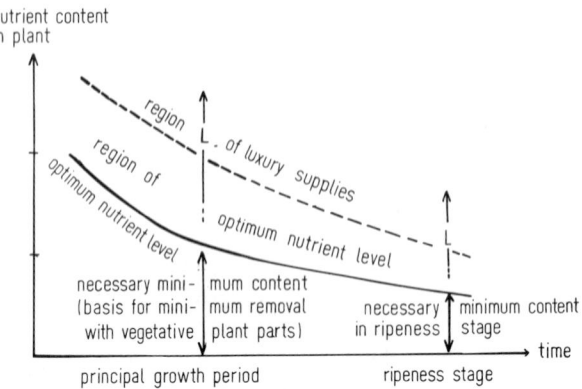

nutrient content in plant

Fig. 5-8. Derivation of removal from nutrient contents (L = excessive contents causing luxury removals).

Minimum removal is much more suitable as the basis of fertilizer-requirement calculations. The demand to disregard luxury consumption is frequently made but has not been sufficiently considered, with a few exceptions [191]. The various removal possibilities are explained in Synopsis 5-12.

Determination of removal requires a knowledge not only of nutrient content but also of the yields of principal and minor harvested products. The latter are not generally determined but may be deduced from known relationships, e.g., the grain/straw or beet/leaf ratios.

Fertilizer Amounts According to Removal and Recovery

The following deductions apply to the use of removal values for determining fertilizer requirements: Fertilizer balances should be based on *minimum removals. Average removals* may be used if minimum removals are not yet known (this would not cause excessive differences in the case of plant reproduction organs).

Synopsis 5-12. Various Kinds of Nutrient Removal.

1. Necessary Minimum Removal
This removal can be defined precisely in theory as the nutrient quantity that must be absorbed to cover necessary requirements. This means that supplies to the plants must at least attain the optimum nutrient level (Chap. 5.2.2 an Synopsis 5-4). We must distinguish between

- the theoretically determined *necessary removal* at the principal growth stage in the case of harvesting in the vegetative stage (from optimum nutrient level and dry matter);
- a *reduced necessary removal* in the case of harvesting in the ripeness stage. This reduction is due to the dilution effect etc., caused by decreasing nutrient contents prior to ripening (this magnitude might also be defined as "optimum nutrient level at ripening time". No accurately determined values of this magnitude are yet available. However, the (slightly higher) average removals may be used as approximations.

2. Average Removal
General tentative values are usually compiled from the results of analyses by numerous authors. They obviously vary within more or less wide ranges. The greater the general validity aimed at, the wider the range, but the less useful the data for exact balances in definite cases. A problem with average removals is to determine the supply range of the plants to which they should be assigned. They might frequently be above the optimum nutrient level in range C (optimal supply) or D (luxury supply), for vegetative plant parts and minor harvested products when ripe (e. g., straw), and largely in range C for reproductive plant parts (Synopsis 5-4).

3. Luxury Removal
Removal in the range of luxury consumption are characteristic of intensively fertilized agriculture. This implies an unnecessarily high removal rate of nutrients from the farm and is at best wasteful. However, removals in the upper part of the optimal supply range (C) must also be considered to be luxury removals, since they sometimes by far exceed the necessary minimum removals. Luxury removals signify unnecessary losses.
   Luxury removal of plants in supply range D is in any case undesirable, whereas a certain removal above the optimum nutrient level might be required to ensure a certain yield.

   Use of removal values from the upper range of normal supplies and especially from the range of luxury consumption leads to continuous excessive fertilization that is not advisable for well supplied soils. Average removal values must therefore be thoroughly *revised*. This has been done as a first approach in discussing fertilization of the various plants (Chap. 7 and 8).
   The removal value forms the basis of fertilization especially for phosphate and potassium (sometimes also magnesium). Replacement fertilization in supply

range C (optimal supply) should compensate for losses due to removal and leaching.

In practice leaching is only important for potassium, on light soils. Removal is thus the decisive loss factor. Fertilization should in every case replace the necessary minimum removal in the optimal supply range. Replacements may be below minimum removals if supplies in the soil are larger, and vice versa (Chap. 5.2.1).

The problem is whether the minimum supply proposed for range C should be corrected by introducing a utilization coefficient, i. e., whether they should be increased. The low utilization rate of trace elements would indicate this, but it would be wrong to take the normally given utilization rate into account (Chap. 5.5.1). *Apparent* utilization seems to be more suitable as a basis for well supplied soils (this should again be stressed), since otherwise there would be a continuous accumulation of nutrients in the soil. This would either lead to undesirable luxury consumption or would have to be considered as lost through increasing immobilization.

In conclusion, we maintain that fertilizer balances, especially for potassium and phosphate, should normally be based on *necessary minimum removal* (i. e., without additions) for the yield target; assuming complete utilization, provided that the supply state can be maintained by removal-compensating fertilization.

5.6 Profitability of Fertilization

Fertilization has the dual aim of increasing plant production and income. The diminishing yield increases accompanying increased fertilization mean that the economic optimum is reached before the possible maximum yield is attained.

The economic limit is less important in times or areas of food scarcity. The costs of food production become increasingly less important when famine has to be relieved. However, the *economic optimum* represents a suitable upper limit of fertilization when there is a *food surplus*. The profitability of fertilization [121] is in general given by the ratio of profit to costs:

$$\text{profitability in \%} = \frac{\text{profit achieved}}{\text{capital invested}}$$

Profitability calculations are based on laws of yield formation, which will therefore be discussed first.

5.6.1 Laws of Yield Formation

Plants grow in accordance with certain rules which have been formulated mathematically as *laws of yield formation*. They characterize the relationship between increased fertilization and yields. The most important approaches for computing the yield as a function of fertilization are:

- *Liebig's* minimum law;
- *Mitscherlich's* law of diminishing yield increases.

The minimum law states that the yield increases in proportion to the minimum factor until the latter, alone, limits the yield (Synopsis 5–13). This rule largely reflected conditions in practice when fertilization first began to be used (from 1880). Yields often increased in proportion to the relatively small fertilizer doses; even abundant supplies of other nutrients caused no further increases as long as the minimum factor imposed a limit. The minimum law is figuratively illustrated by the "minimum cask". The water level in the cask cannot rise beyond the height of the shortest stave (Fig. 1–1). This simile still applies in principle today, despite the fact that the yield does not continue to increase in proportion to the minimum (Fig. 5–9).

The law of *diminishing yield increases* states that the yield does not increase linearly with the fertilizer quantities applied, but that the increase gradually becomes smaller. The curve tends to a yield maximum and beyond that drops (Synopsis 5–13). The yield variation is given by a curved, logarithmic line resembling the saturation curve for chemical reactions (Fig. 5–9) [123].

This law describes the actual variation of yield formation as a function of the increased supply of one nutrient far more accurately than the minimum law.

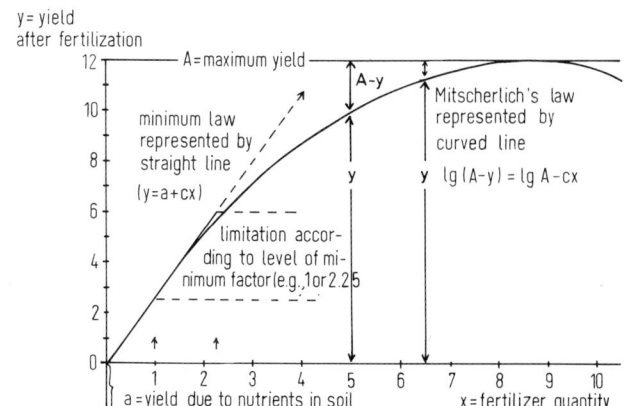

Fig. 5-9. Minimum law and *Mitscherlich's* law in graphic representation.

Synopsis 5–13. Laws of Yield Formation.

1. The Minimum Law (according to *Liebig,* etc.)
The yield increases with the minimum factor until a deficient growth factor limits the growth (yield), even if all other factors are available in sufficient amounts.
Formula for total nutrient supply:

$$y = c \cdot x$$

Formula for the separate consideration of nutrients in fertilizer and soil respectively:

$$y = a + c \cdot x$$

where:

x = fertilizer quantity (nutrient quantity)
y = yield due to fertilization
c = constant coefficient
a = yield due to nutrients in soil without fertilization

Increased fertilization thus causes the yield values to lie on a straight line; the effects of given fertilizer quantities can thus be calculated.

The minimum law may be applied for approximate yield calculations only in the region of serious deficiency. However, the significance of the minimum factor in limiting yields is given fairly correctly by this law even in the region of intensive fertilization (and is still valid now).

A slight correction should be introduced for a proper interpretation of the "cask analogue". Nutrients interact to some extent, so that the staves should not be considered as rigid, but as slightly elastic: an increase of a certain "optimal factor" may sometimes slightly increase the minimum factor.

2. The Law of Diminishing Yield Increases (according to *Mitscherlich,* 1906)
Increasing a minimum factor raises the plant yield (in accordance with a logarithmic law) in proportion to the amount lacking from the maximum yield, i. e., yield increases diminish when the rate of fertilization increase is constant.
Formula in general form:

$$\text{yield increase per fertilizer unit} = c_1(A - y)$$

Mathematical transformation then yields the usual formula:

$$\log(A - y) = \log A - cx$$

where:

A = maximum (theoretical) yield
y = yield after fertilization
x = fertilizer quantity (nutrient quantity)
c_1 = factor of proportionality
c = efficiency factor

The *efficiency factors* c are constant magnitudes for the various nutrients, e. g., 0.2 for N, 0.4 for K, 0.6 for P, 100 for Cu.

This law too has been extended to make allowance for nutrients in the soil. The variable x is then replaced by the sum $(b + x)$, where b is the quantity of nutrient in the soil.

Other extensions provide for fertilizer utilization and damage caused by overfertilization.

The *problem* with this law is that its decisive precondition, i. e., constancy of the efficiency factors, is not satisfied in a general form. It is nevertheless possible to obtain fairly reliable information on yield increases, within certain limits of validity.

However, the "*Mitscherlich*-curve" is almost straight in the lower section for cases of relatively severe deficiency; this region therefore corresponds to the minimum law (Synopsis 5-13).

Besides the logarithmic *Mitscherlich*-curve, there are other mathematical functions suitable for describing yield increases [126], ranging to general *polynomials of higher degree*.

However, for all functions, including the *Mitscherlich*-curve, it is true that they are based on natural law, but they describe it less in a physically exact manner than as an *approximation*. However, this approximate validity is sufficient. The curves can be evaluated from various aspects, especially for profitability calculations.

5.6.2 Economic Optimum of Fertilization

Fertilization is only profitable within certain limits. This range is the wider, the less fertilizers cost in relation to the products obtained. The problems concerned with the profitability of fertilization are explained by means of a model fertilization trial with arbitrary data that approximate reality. Table 5-22 gives the basic data for a trial in the first columns. 30 dt grain were obtained (without fertilization) because the soil was well supplied with N, whereas the maximum yield was achieved with 210 kg N. All magnitudes relevant to the profitability are then derived from these basic data.

The same trial is represented graphically in Fig. 5-10. The following concepts are defined for this purpose:

- The *lower limit of profitable fertilization* is due to the fact that application of very small fertilizer quantities, relative to requirements, is not worth the effort; this case is only important in practice under special conditions.
- *Maximum interest* on the fertilizer investment is usually obtained in the low region of moderate fertilization: the capital invested bears the highest interest in this case (Table 5-22).
- The *maximum net profit* per unit area is the most important profitability factor: this is the economic optimum for which the ratio of value to costs is most favorable. It can be represented graphically by drawing a tangent to the yield curve, parallel to the cost curve.
- The *upper limit of profitable fertilization* is located at the point where the cost of a further fertilization step is equal to the value thus gained (Table 5-22); additional fertilization then implies a financial loss despite the further yield increase.
- The *maximum yield* beyond profitable fertilization is mentioned for the sake of completeness.

Table 5-22. Fertilization profitability, represented by data from examples.

fertilization with N kg/ha	grain yield dt/ha	DM/ha for overall fertilization		per step of 30 g N		interest on fertilizer investment (%)		value/cost ratio (VCR) (per fertilization step)	1 kg N produces kg grain per fertilization step
		fertilizer costs	yield value	fertilizer costs	yield value	for overall fertilization	per fertilization step		
—	30	—	—	—	—	—	—	—	—
30	39	50	360	50	360	620	620	7.2	30
60	47	80	680	30	320	750	967	10.7	27
90	54	110	960	30	280	773[a]	833	9.3	23
120	60	140	1200	30	240	757	700	8	20
150	63	170	1320	30	120	676	300	4	10
180	63.75	200	1350	30	30	575	0[b]	1[c]	3
210	64	230	1360	30	10	491	negative	negative	1
240	60	260	1200	30	-160	362	negative	negative	negative

assumptions: 1 kg N   = 1 DM
1 dt grain = 40 DM
fixed costs of fertilization = 20 DM/ha

a = maximum interest
b = upper limit of profitable fertilization
c = profitability limit according to VCR (at 2)

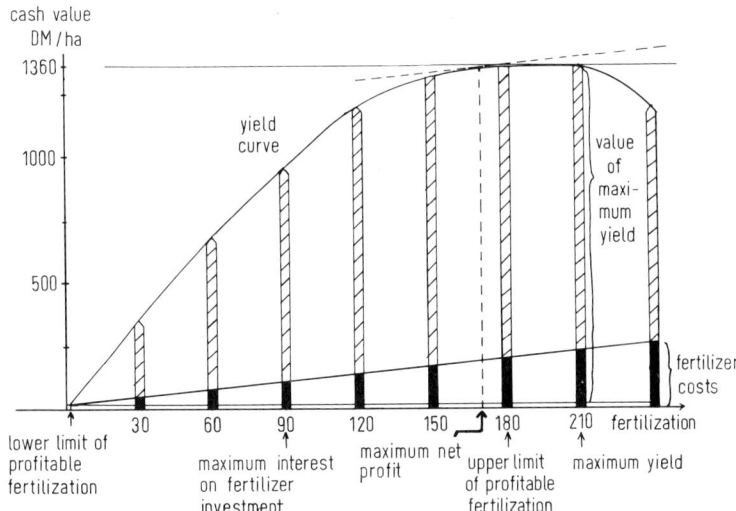

Fig. 5-10. Fertilization profitability as a function of expenditure and yield (data from Table 5-22).

The influence on the yield of increasing one factor, as discussed until now, permits derivation of an economic fertilization optimum from the yield functions (for the mathematical derivation see [121]). The maximum yield will be closer to the economic optimum, the lower the price of fertilizer in relation to the prices of agricultural products. The *practical rule* valid in the case of high grain prices and cheap fertilizers is that *any fertilization providing measurable yield increases is still worthwhile.*

The FAO recommends fertilization only if the *value/cost ratio* (VCR) is at least equal to 2, i.e., if fertilization yields a profit of 100%, especially in less developed countries, where fertilizer prices and growers' risks are often high [112]. In the example of the "fertilizer trial" (Table 5-22) this would be the case at a fertilizer input of approximately 160 kg N.

However, in intensive agriculture it is quite possible to consider profitability directly when determining the fertilizer quantities, i.e., when establishing the *economic fertilization optimum* [139]. However, this additional calculation effort yields no decisive advantage under conditions in Germany.

The economic fertilization optimum is in practice much more difficult to determine than would appear from a consideration of just one increase factor. Only *multifactorial* trials with increased supplies of several nutrients, and other yield-limiting factors, permit correct estimation of the proper level of further fer-

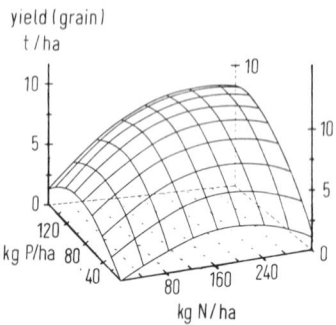

yield (grain)
t/ha

Fig. 5–11. Maize yield as a function of fertilization with N and P [126].

tilizer application (Fig. 5–11). However, the costs of experiments and computations required may be higher than the fertilizer input possibly saved.

In conclusion, we shall discuss the often posed problem of the *average* production value of fertilizer nutrients. It should in principle be stressed that the reply depends on experimental conditions and that the values vary within a wide range (as seen from Table 5–22).

The average value for nitrogen obtained from many trials with grain, is from *10 to more than 20* at a medium level of fertilization (i.e., one kg N, "produces" 10 to 25 kg grain). However, this value is much lower when fertilization is intensive, and may even become negative (Table 5–22). Overall calculations with such variable production values should therefore be regarded critically.

Fertilizer inputs, expressed in financial terms are some 10% (sometimes up to 20%) of total inputs in agriculture, and about 5% (2 to 8%) in horticulture. In concluding the discussion of fertilization profitability, it should be stated that fertilization is a highly profitable means of production. Fertilizer capital, if invested properly, on average shows a two- to five-fold return per year.

5.6.3 Computer-assisted Fertilization

Determination of the correct fertilization including cost calculations requires a considerable mental effort, based on a good knowledge of fertilization problems. The fertilization plan for a farm must be established in all details on the basis of reliable diagnosis, usually to be entrusted to experts by the farm manager:

• the required *amounts of fertilizer* per hectare, for the various production branches, and lastly for the entire farm (allowing for the farmyard manure produced on the farm itself);

- the fertilizer form, allowing for the plants to be grown, the respective site conditions, and fertilizer prices at the time calculations are made;
- instructions as regards fertilizer *distribution in time*;
- instructions for *application*.

In addition special features of the farm should be taken into consideration. These are, e.g., general profitability and organizational, etc., capabilities of the farm manager. Thus high-yielding production requires complex detailed fertilization plans, and these must be carried out expertly and effectively. However, elaboration of complex fertilization instructions may be unnecessary or even harmful, if these lead to errors that might be avoided with simpler plans. In principle, there are two possibilities for working out a *fertilization plan*:

- individual step by step calculation;
- with a computer, using a fertilization programme (electronic data processing).

Individual calculation may require several workdays on a farm, but this mental investment may be highly profitable. It not only provides a better understanding of the chemical aspects of plant production, nutrient balances, and *nutrient flows* on the farm; more general and production factors can be taken into account than with rigid general programs, if suitable know-how of fertilizers and fertilization is available.

A correctly established *individual fertilization plan* should be far superior to any automatically produced plan. On the other hand, a possible shortcoming should be stressed. The less expert one is in fertilization, the more will erroneous, subjective influences reduce the value of an individual fertilization plan.

The trend to simplification and reduction of effort nowadays favors the establishment of fertilization plans by electronic data processing (computer programmes). The result of a computer-produced fertilization plan is:

- as good or as bad as the basic (input) data, especially the diagnosis of requirements;
- as good or as bad as the adequate and correct allowance made for the many necessary auxiliary data.

The computer cannot improve the quality of fertilization plans. It can only compute more quickly and with fewer errors, according to the expert data fed into it. This can be done by the farm manager himself (with or without a calculator).

The value of a fertilization plan prepared by a computer depends primarily on the data stored in it. Fertilization plans produced by computers therefore require strict quality checks to prevent farmers relying on computers guaranteed to operate faultlessly, and from fertilizing according to faulty plans produced by computers.

The more factors are properly taken into account, the more useful will be the information in each case. A simple basic scheme of the possibilities of using electronic data processing is presented below. Detailed programming is referred to in [116, 138, 44].

The computer will gain in importance as an aid for easing the work involved in fertilization, and also because rapid price fluctuations will have to be taken into account.

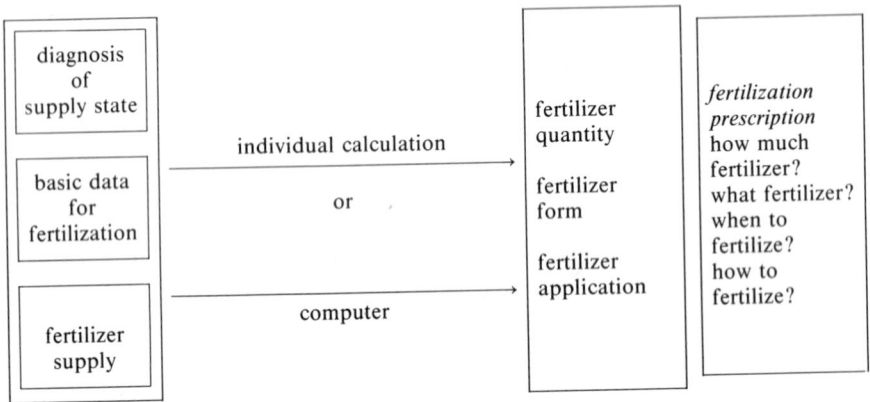

5.6.4 Fertilizer Prices and Fertilizer Consumption

The fertilization level should depend on the economic optimum (Chap. 5.6.2). However, the cost-yield curve is ordinarily not precisely known to the farmer, and even the achievable maximum yield, which might be used as an index in its stead, can often only be defined as an inexact range. Exact calculation of the optimal fertilization level is therefore only rarely possible. However, estimation is facilitated by the fact that supplies of many nutrients are kept above minimum requirements by a safety margin.

A continuously managing farmer knows from experience the fertilization level that allows a certain yield. These values may or may not be scientifically optimal, but they nevertheless serve as useful guidelines. A farmer often assumes that the fertilizer amounts applied by him slightly exceed the optimum.

It may be assumed that the price of the fertilizer concerned is an important factor determining the fertilization level. From the farmer's point of view, higher nutrient prices should cause fertilization to be reduced. This does happen with certain nutrients; more expensive fertilizers at least are applied more sparingly [137]. A historical survey of fertilizer prices is given in Table 5-23. Present prices are indicated in price lists [113].

Table 5-23. Fertilizer prices in FRG (according to [1]).

pure nutrient	prices in DM per 100 kg pure nutrient					
	1935/38*	1950/51	1959/60	1969/70**	1973/74	1975/76
N	53	93	107	103	119	140
P_2O_5	27	38	56	71	100	135
K_2O	13	23	28	34	42	49
CaO	2	4	5	7	9	13

* prewar price in RM
** prices after 1969 include tax

On the other hand, fertilizer prices determine the fertilization level far less than might be expected. However, the economic reason for this is quite different than might be expected from the agrochemical viewpoint. Investigation of the *economic determinants* of fertilizer consumption shows that farmers apply (and thus purchase) fertilizers in amounts depending far less on their prices or marginal costs, than on the liquidity situation [135]. The income from the sale of agricultural products, i. e., the overall liquidity situation, thus becomes an important determinant of fertilizer demand. This influence is mutual, since fertilization affects the income from sales, which in its turn influences the fertilization level for the next harvest. Every year a farmer must buy different means of production in varying amounts; fertilizers thus compete with capital goods. Fertilization is sometimes reduced or even omitted when some other large expenditure has to be incurred. Thus, in the USA the level of fertilization with lime shows a clear correlation with annual farm income. Economies are made in liming when cash is short.

The very common procedure of matching the fertilization level to available funds is only advisable, however, when it involves "postponing" stock fertilization at times of strained liquidity. On the other hand, it would be a gross error from both the agrochemical and economic viewpoints to postpone necessary fertilization because of liquidity problems. The high return on (correctly applied) investments in fertilizers makes it reasonable to borrow, when necessary.

Substitution of fertilizers should also be mentioned in this connection. Substitution (in the economic sense) of one fertilizer by another is sometimes used in profitability calculations, but is justified only to a very limited extent. The same yield increase may be achieved on a given field by fertilizing a plant with e. g., 50 kg nitrogen or 40 kg phosphate. The cheaper fertilizer will then be preferred for reasons of cost.

However, disregarding the fact that nitrogen and phosphate can in no way replace each other from the agrochemical aspect, fertilization with P may sometimes improve N-supplies, and vice versa, but considerations of economic substitution are frequently highly inadequate. A cumulative effect is often caused by the interaction of nutrients (Chap. 5.6.2), rendering all considerations of individual effects meaningless. The best decision on fertilization is not a choice between alternatives, but the proper combination of fertilizers. All *substitution* proposals should therefore be critically considered with regard to their real justification.

To summarize, the *relationship between fertilizer price, economic situation of the farm, and fertilizer input* is such that when there is temporary non-liquidity:

- Economies in the necessary fertilizer amounts are a gross error that has to be paid for dearly;
- reducing "above-optimal" stock, fertilization is quite justified;
- temporary suspension of fertilization, at least with some nutrients, with which the soil is abundantly supplied, may definitely be considered.

The aim of determining the optimal fertilizer input by economic methods depends essentially on correct initial data. Inadequate diagnosis of fertilizer requirements, neglecting the various factors influencing fertilizer effects, does not provide useful fertilization recommendations even with the best and fastest economic computation methods.

6 Special Fertilization Problems

6.1 Technology of Fertilizer Application

Fertilization problems extend beyond proper selection of the fertilizer form and determination of the correct fertilizer quantities. The fertilizer must, in addition, be easy to apply and be applied properly. The *physical* properties of a fertilizer are especially important for its optimal application, since fertilizers must be transported, stored if necessary, and finally applied so as to ensure optimal effects.

The variety of solid, liquid and gaseous fertilizers, poses many problems in application. Transportation and storage should be as economical as possible without causing chemical or physical changes in the fertilizer, so that its dispersibility and effects are maintained. Application should be such that the nutrients penetrate into the root zone and become optimally effective there.

6.1.1 Transportation, Storage, and Mixing of Solid Fertilizers

Fertilizers are conditioned during production and given a form suitable for easy application, which should not be lost in transportation and storage. Fertilizers are mixed with additives so that they remain free-flowing and dispersible.

Besides *granulation, powdering* is most important with non-hygroscopic substances like lime, diatomaceous earth, etc. These powdery substances prevent fertilizer particles from sticking together when humidity is high, and prevent the collapse of granulates under pressure, or liquefaction of the fertilizer as a whole.

Hygroscopic fertilizers absorb moisture from the atmosphere, and this may alter the pre-determined grain size of granulated or powdered fertilizers. Some fertilizers are extremely sensitive to moisture, harden easily, or even become liquefied, e. g., calcium nitrate. The hygroscopicity of fertilizers is indicated in Table 6–1 by two characteristic values that provide the same information. Mixtures of single fertilizers are usually relatively more hygroscopic than the most highly

Table 6-1. Water absorption of fertilizers from the atmosphere (according to [41, 44]).

Fertilizer	Hygroscopic coefficient at 20°C	30°C	Critical relative air humidity at 20°C	30°C
calcium nitrate (4 H₂O)	45	53	·	47
ammonium nitrate	33	41	63	61
sodium nitrate	23	28	·	72
urea	20	28	79	74
ammonium sulfate	19	21	81	81
potassium chloride	14	16	·	84

Hygroscopic coefficient = 100 – relative humidity of air above saturated solution (high values indicate intense absorption of water)
Critical relative atmospheric humidity: fertilizer absorbs water from air when humidity is higher (high values indicate little absorption of water)

hygroscopic single component. Only few fertilizers, e.g., ammonium sulfate, keep their good flow properties at increased air humidity. Such fertilizers are therefore particularly suitable for regions where they have to be transported over large distances or stored in humid climates, e.g., in the tropics.

The hygroscopicity of ammonium sulfate increases only slightly with temperature, which is an additional advantage for its use in the tropics. The (undesirable) hardening of fertilizers after the absorption of water is due to crystal bridges being formed between the fertilizer particles after wetting and later drying, especially when they are compressed during storage. Large, well-rounded particles tend less to hardening. Hardening may cause the formation of voids in fertilizer silos, so that "load-bearing bridges" obstruct flow during discharge. Temperature variations, as well as the absorption of moisture affect the stability of granulates and lead to disintegration.

Storage of fertilizer in bags obviously poses fewer problems than storage in bulk. Storage of fertilizer in bulk over long periods may require measures to prevent the absorption of moisture from humid air. The small unavoidable amounts of moisture absorbed are more undesirable the higher the compression in storage and the longer the duration of storage.

The fertilizer *weight* is important in transportation, storage, and application. The bulk weight (weight of loosely filled fertilizer) of most solid fertilizers is about 1 kg per liter. However, urea is considerably lighter, with a bulk density of 0.7 kg. Thomas phosphate is considerably heavier, with a bulk density of 2 kg.

Care must be taken during transportation and storage not only to avoid detrimental effects to the fertilizers but also to avoid *danger*. Some fertilizers become heated when they absorb moisture (fire hazard), others are explosive, many are corrosive, etc., and may release noxious gases. Fertilizers are also conditioned against such undesirable effects, but such conditioning is only possible to a certain extent. The authorities have therefore issued regulations for the storage and transportation of various fertilizers, especially in large quantities.

Solid fertilizers are usually transported in *bags*, which is advantageous with respect to labor economy, especially for small quantities. On the other hand, bags are expensive, about DM 1.— for one 50 kg plastic bag.

Transportation by bulk fertilizer system becomes important when labor is scarce. The fertilizer is transported mechanically in bulk, thus eliminating the physical handling of bags. However, suitable equipment is required where the fertilizer is stored in order to move it to the distributor and expenditure on storage is higher because protection against moisture is necessary [175].

Mixing of Solid Fertilizers

Many fertilizers can be mixed without detriment and can then be distributed. Ready-for-use mixes are often available as multiple-nutrient fertilizers. However, there are three reasons for not mixing fertilizers indiscriminately [168]:

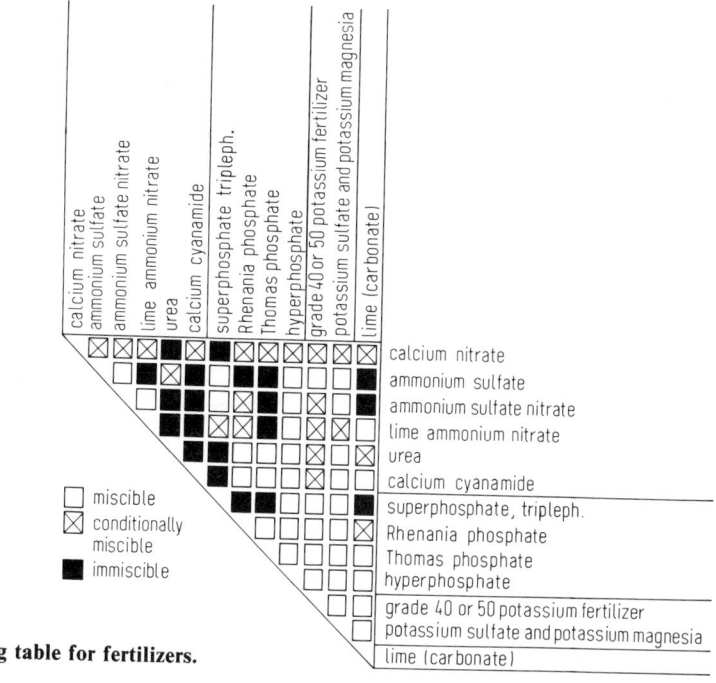

Fig. 6–1. Mixing table for fertilizers.

- losses of nitrogen by chemical reactions (i. e., immiscible);
- immobilization of water-soluble phosphate (i. e., immiscible);
- deterioration of distribution properties owing to the absorption of moisture (conditionally miscible).

Miscibilities, allowing for these factors, are indicated in the mixing table (Fig. 6–1) (p. 241).

Losses of nitrogen by moistening are due to reactions of ammonium fertilizers with alkaline acting substances (lime), with ammonia escaping in gaseous form. A noticeable exception is the mixture of lime ammonium nitrate. Other N-losses are due to the formation of nitric oxides, e. g., the reaction of urea with nitrate or with acid-containing fertilizers, etc.

Water-soluble phosphates should not be mixed with lime-containing fertilizers (in the broadest sense), since secondary and tertiary phosphates, which are not water-soluble, are then formed. The relatively expensive P-form thus loses its effectiveness.

Highly hygroscopic fertilizers are *conditionally* miscible. As far as possible, they should not be mixed with other fertilizers, or, if necessary, should be mixed only in dry weather shortly before distribution. The mixing rules apply 'mutatis mutandis' also to contact in the soil. Thus, ammonium sulfate or superphosphate should not be applied to freshly limed soils.

6.1.2 Distribution of Solid Fertilizers

Fertilizers containing plant nutrients should be applied in such a way that the nutrients penetrate to the main root zone of the plants or so that they can reach it. Depending on the properties of the fertilizer, this can either be done by simply scattering it on to the soil surface, if the nutrients can be displaced to the root zone with percolating water, or if necessary, by working the fertilizer into the root zone. Some fertilizers should be as finely dispersed as possible in the root zone, while others should be concentrated at given points (placed). The possibilities of placement (row, strip, layer, etc. fertilization) have been discussed in connection with P-fertilizers (Chap. 2.2.3).

Another essential requirement for proper fertilizer application is a uniform distribution over the entire surface to be fertilized. This applies in particular to N-fertilizers, but to others too, especially if applied in small doses to insufficiently supplied soils. Uniform application depends mainly on the quality of the fertilizer distributor and its correct operation.

Granulation of Fertilizers
Granulation increases the dispersibility of fertilizers (originally of water-soluble fertilizers). The grains may or may not contain additives. Because of the consid-

erable advantages of granulated fertilizers in application, non-water-soluble fertilizers are also granulated. However, these granulates are fine particles loosely bonded to form grains, which, after moistening in the soil, rapidly disintegrate again into their fine constituents.

The grains may be either as round as possible or of irregular shape. Those spherical in shape have the highest stability under pressure, the highest density, and the least contact area, which is desirable for preventing hardening. One disadvantage is the high tendency to rolling, which makes the grains accumulate in depressions of uneven surfaces with the result that there may be a very nonuniform distribution. On the other hand, the goal of an ideal distribution is approached more closely if the grains have irregular shapes.

The optimal *size* of fertilizer grains is 1 to 3 mm diameter. However, up- or downward deviations from this are possible.

One gram of a granulated fertilizer contains 100 (lime ammonium nitrate) to 50 (NPK-fertilizers) granules (grains). A dose of 100 kg N in the form of lime ammonium nitrate corresponds to 40 million grains per hectare or 40 grains per dm^2. The same dose of N in the form of NPK-fertilizer represents about 35 grains per dm^2. Assuming 1 dm^2 to equal 1 liter of soil, this corresponds to one fertilizer grain per 30 ml soil.

The uniform *distribution* of the fertilizer in the root zone is ensured almost automatically in the case of fertilizers which dissolve easily in the soil moisture (at least to the extent required). Special working-in is therefore unnecessary. However, uniform working-in to the A-horizon of poorly supplied soils is especially important if the fertilizer must first be mobilized in the soil. When there are good supplies, it suffices if the fertilizer grains are mixed into the soil in the course of routine working of the soil.

The requirement of *uniform* distribution of fertilizers, whether granulated or in powder form, is not easily satisfied [162]. Thus, granulated fertilizer spread on the surface can be introduced into the soil with

- the *harrow*, in which case, however, most of the fertilizer remains in the uppermost 2 to 3 cm;
- with the *cultivator*, in which case the penetration depth is approximately doubled;
- with the *plough*, in which case the fertilizer is predominantly introduced into the lower root zone.

Application of fertilizers for soil improvement is effected into either the A-horizon or the subsoil, depending on the purpose, or as a protective layer, on the soil (e.g., organic fertilizer as mulch).

Equipment for Fertilizer Distribution

Solid fertilizer can be applied manually without any equipment, and trained personnel achieve an approximately correct dosage and uniform distribution. The first step in mechanization is represented by portable distributors operated by muscle power, which in principle function like centrifugal distributors. The distributors now in use for mineral fertilizers can be divided into [147]:

- wide-sweep distributors for application over large areas;
- row distributors for the precise fertilization of plant rows.

Wide-sweep distribution of fertilizers predominates in agriculture. Dosage for fertilization with plant nutrients mostly varies between 20 and 1 500 kg/ha (up to 50 dt/ha when lime is used as the fertilizer).

A good-quality mineral-fertilizer *distributor* should:

- permit correct dosage, uniform distribution over the area, irrespective of surface slope and travel speed, with deviations not exceeding $\pm 10\%$;
- be easy to handle and to maintain;
- resist corrosion and wear.

Mineral-fertilizer distributors are of the box, centrifugal, or pneumatic type.

Box-type distributors consist of an oblong box for the fertilizer, which is drawn transversely to the direction of travel. A distributing device at the bottom of the box ensures uniform discharge during the travel. The working width depends on the length of the fertilizer box. The distributing devices work on differing operating principles. The *fertilizer quantities* are regulated, e. g., in

- slot distributors, by an adjustable slot;
- endless-chain distributors, by the slot opening and the speed of the chains moving along the bottom of the box;
- plate distributors, by the slot opening and the speed of the plates rotating at the bottom of the box;
- grid distributors, by adjusting the advance of the (sliding) grid.

The main advantage of box-type distributors is their universal suitability for flour-like and granulated fertilizers. Their principal drawback is the small working width (up to 5 m).

Centrifugal distributors at present play the principal role in fertilization. They usually include a distributor disk. The fertilizer drops from a conical container onto a high-speed rotary disk provided with throwing bars. A baffle plate ensures that the fertilizer is spread in a semicircle only to the rear. The main advantage of centrifugal distributors is the large working width which is determined by the quantities to be applied and the grain size of the fertilizer. Spreading widths of 12 to 14 m are possible with coarse-grained fertilizers. This represents an effective working width of approximately 10 m. One of the shortcomings of this type of distributor is that in practice only granulated fertilizers can be used. There are many designs for centrifugal distributors but it is the device for mounting on the tractor that is most important.

Large-area distributors with containers of 3 to 4 m³ capacity serve efficiently for application on large areas or for large fertilizer amounts. They also operate as disk or pneumatic distributors.

Pneumatic fertilizer distributors apply the fertilizer to the planted rows in a strong air current via a dosing device over baffle dishes (up to 15 m working width) or through hoses.

Fertilizer application from airplanes has lately gained importance in certain countries [147]. The advantages are the high performance (fertilization of large areas in a short time), the suitability for inaccessible areas (paddy fields), and lastly the avoidance of "damage by machinery", caused by the usual methods of fertilizing growing crops.

The proper dosage can be adjusted approximately in many devices, but it is recommended that it be checked. Typical *mistakes* in fertilizer application are faulty joining of individual fertilizer strips and undesirable *overlapping*, causing striations, particularly during fertilization with nitrogen.

Solid *organic fertilizers* like farmyard manure and compost must also be applied, in addition to mineral fertilizers. Fully mechanized *farmyard manure distributors* are trailers of 20 to 40 dt-carrying capacity (3 to 5 m³). The farmyard manure is gradually pushed towards the distributing device, arranged at the back or at the sides, by a *feed system* while the trailer is in motion. The distributing rollers have prongs for comminuting the material, and they distribute it over a width of 2 to 4 m.

The *costs* of fertilizer distribution amount to some 10 to 20% of the total fertilization costs. Costs are divided as follows: 3 to 7% for labour, for machinery 11 to 16% for distributing fertilizers in bags, 5 to 12% for distributing fertilizers in bulk [152]. Bulk fertilizer system save not only distribution costs but also packing materials.

6.1.3 Distribution of Liquid and Gaseous Fertilizers

Liquid fertilizers are more easily handled, from the labor aspect, whereas the application of gaseous fertilizers is more expensive. Liquid fertilizers are stored in tanks and transferred by pumps. Gaseous fertilizers should be considered the same as liquid fertilizers with regard to storage and transportation. Their application requires special methods of dosing and introduction into the soil. A scheme for their classification is given in Synopsis 6-1.

Synopsis 6–1. Classification of Liquid and Gaseous Fertilizers According to Distribution.

1. Distribution of non-pressurized liquids *on* the soil (or sometimes on plants):
   a) Solutions: N-solution, NP-solution, NPK-solution
   b) Suspensions: mineral NPK-fertilizers, organic fertilizers (liquid and semi-liquid manure)

2. Introduction *into* the soil:
   a) Solution: Ammonia water (non-pressurized or at low pressure)
   b) Liquefied gas: ammonia (transported as liquid under pressure, introduced into the soil as gas).

Application of gaseous carbon dioxide (CO_2) to increase the CO_2-concentration in the air of greenhouses requires continuous checking of the optimal concentration (Chap. 3.4.3).

Distribution of Liquids on the Surface
The procedure will be explained for N-solution (Chap. 2.1.3). Other solutions are applied correspondingly. Liquid fertilizers may be transported in tanks of fiberglass-reinforced polyester resin (2 to 4 m^3), and they may be stored in large containers made of the same material, or in plastic-coated or foilclad concrete silos. Safety regulations published by the authorities should be adhered to. The liquids are transferred by corrosion-resistant pumps. Distribution is effected with sprayers as used in plant protection, but special requirements are imposed with regard to corrosion resistance.

Pressures of 1 to 3 bar are used for *spraying*, with variously designed nozzles matching varying doses of N. The discharge can be varied to between 0.5 and 4 liters per nozzle. Different travel speeds thus permit the application of between 10 and more than 300 kg N per hectare (at a nozzle spacing of 50 cm).

The great advantage of spraying solutions lies in the speed (up to 10 ha/hr) and high *application precision*. Marking aids are required, at least at working widths exceeding 10 m), to avoid overlapping. Distribution by means of small wheel paths is the best guarantee for properly joining the individual fertilizer strips for growing crops.

Suspensions of mineral fertilizers are *distributed* similarly to solutions. However, the solid particles necessitate special spraying devices. Precise distribution of liquid or semi-liquid manure is less important, but the method used must be suitable for large quantities.

Fertilization by sprinkling is a combination of supplying water and mineral nutrients [165]. With irrigation by ditches this is usually referred to as *fertilization by irrigation.* In this case, fertilizer is added to the water and thus reaches the plants. The concentration of a sprinkler solution should generally not exceed 3 to 5‰, to prevent plant corrosion. Higher concentrations are permissible for fertilization by irrigation.

Introduction of Solutions and Gases into the Soil

Water-free ammonia is transported under pressure in liquid form and leaves the distributing device as gas after the pressure is released. It must then be introduced into the soil to prevent serious losses. The problem of applying it thus consists in correctly dosing the liquefied gas at a pressure of about 10 bar from the field tank with the aid of pumps, despite its conversion into a gas, allowing for the speed of travel, temperature, etc.

After dosing, the gaseous ammonia must be introduced into the soil at a sufficient depth, to avoid serious losses. This is done with devices that have *special injection prongs* to which the gas is supplied by hoses. The prongs should disturb the soil as little as possible, so that no gaps open to the surface are formed. A precondition for this is a correct moisture content of the soil, i. e., the soil must be neither too wet nor too dry.

Pressurized ammonia is subject to special compressed-gas regulations concerning the strength of containers and pipelines, corrosion damage, etc., allowing for the toxicity of the gas. Ammonia water, being a nonpressurized solution, is subject to less strict transportation and storage regulations. However, it must also be introduced into the soil because of the volatility of the ammonia [150].

The introduction of fertilizers into the soil under trees poses special problems, particularly if grass grows beneath them. *Lance fertilization* may be indicated in such cases. Metal pipes, lances with fixed tips and lateral nozzles, are used to introduce a pressurized fertilizer solution at the desired depth in the main root zone [160].

Application is in quantities of 1 liter per injection with 10 to 50 injections, depending on tree size. The concentration of the solution may be higher (20%) during growth interruption than during growth (5%). More modern devices operate at pressures of 20 to 40 bar. The injection depth is important. It must at least be distinctly below most grass roots.

6.1.4 Leaf Fertilization

Leaf fertilization will be considered separately in view of its importance. Plants absorb mineral nutrients chiefly through the roots, but through their *micropores* the leaves too can absorb water and substances dissolved in it. In theory, plants could be completely nourished via the leaves, but the major importance of foliar nutrient application is the additional supply of nitrogen, magnesium, and trace elements. Leaf fertilization consists in spraying leaves with diluted nutrient solutions or suspensions.

The advantage of leaf fertilization is the high recovery rate; its shortcoming is the limited amount of nutrients that can be supplied. Fertilizer salts and organic fertilizers corrode when applied in concentrations above a given level, although different plants are sensitive to different degrees [115]. The aim of leaf fertilization is to distribute small amounts of nutrient uniformly over the leaves. The commonly employed procedures are as follows:

- spraying about 400 l/ha of a solution in fine droplets of 0.1 to 0.2 mm [diameter];
- fine spraying 100 to 200 l/ha of a solution (blowing at the leaves) in sometimes even smaller droplets.

The procedures may involve either maximum adhesion of the solution to the leaves, or dripping-off of the excess solution (excess method). The latter method particularly has a role to play in fruit growing. Concentrations are then correspondingly low and solution quantities are large.

The *main problem* involved in the spraying of leaves is the sensitivity of the latter to excessive solution concentrations. Fertilizers can *corrode* leaves by osmotic action (removal of water) and other effects, i. e., attack the leaf tissue and damage it, in some places causing necrosis. Urea, being a "nonsalt", has a relatively slight osmotic action and is therefore well tolerated by plants. This permits its application at concentrations of up to 15%. However, N-solution (Chap. 2.1.3) appears to be tolerated even better, probably because of the low ammonia stress on the metabolism.

The risk of corrosion decreases with the growth rate and corresponding nutrient requirements. Rapid drying of the solution on the leaves in dry weather reduces possible corrosion damage, but it also reduces the fertilizer effect. Application of highly diluted solutions or after-washing with water is indicated if corrosion damage is to be avoided altogether (e. g., in the case of ornamental plants). A new development in leaf fertilization is depot foliar nutrients. They are only slowly converted into soluble forms, so that the corrosion hazard and sometimes also the effect is smaller.

The following points should be observed in *practical leaf fertilization*;

- the proper concentration of the solution (tentative values are given in Table 6-2);
- critical growth periods, during which the application of foliar nutrient is contra-indicated.

Table 6-2. Instructions for leaf fertilization (at 400 l/ha of solution for solid fertilizers) (according to [161 and others]).

Fertilizer	Formula	Concentration of sprayed solution (%)	Amounts in kg/ha Fertilizer	Nutrient element	Application
urea	$CO(NH_2)_2$	8–16	32–65	15–30 N	cereals, rape
urea	$CO(NH_2)_2$	0.5–1	2–4	1–1.8 N	fruit, vegetables, vines, etc.
N-solution	$NH_4NO_3 + CO(NH_2)_2$	.	50–100	14–28 N	cereals, rape
N-solution	$NH_4NO_3 + CO(NH_2)_2$	.	25–35	7–10 N	potatoes, beets
triple phosphate	$Ca(H_2PO_4)_2$	2	8	1.6 P	P and K as foliar nutrients
potassium sulfate	K_2SO_4	1	4	1.7 K	relatively seldom
calcium nitrate	$Ca(NO_3)_2 \cdot 4H_2O$	0.5–1	2–4	0.3–0.6 Ca	e.g. for fruit
magnesium sulfate	$MgSO_4 \cdot 7H_2O$	2	8	0.8 Mg	fruit, cereals
iron chelate	e.g. Fe-EDTA (5% Fe)	0.1–0.2	0.4–0.8	⌀0.03 Fe	fruit, vegetables, vines
manganous sulfate	$MnSO_4 \cdot 4H_2O$	1–2	4–8	1–2 Mn	cereals
manganous sulfate	$MnSO_4 \cdot 4H_2O$	0.5	2	0.5 Mn	in horticulture
zinc sulfate	$ZnSO_4 \cdot 7H_2O$	0.5	2	0.5 Zn	cereals
zinc sulfate	$ZnSO_4 \cdot 7H_2O$	0.2	0.8	0.2 Zn	in horticulture
cupric sulfate	$CuSO_4 \cdot 5H_2O$	0.5	2	0.5 Cu	cereals
cupric sulfate	$CuSO_4 \cdot 5H_2O$	0.2	0.8	0.2 Cu	in horticulture
copper calx	$Cu(OH)Cl$ etc.	0.1–0.3	0.4–1.2	0.2–0.6 Cu	in horticulture
borax	$Na_2B_4O_7 \cdot 10H_2O$	1	4	0.4 B	beets, etc. (large B-requirements)
borax	$Na_2B_4O_7 \cdot 10H_2O$	0.5	2	0.2 B	fruit, vegetables
molybdate	NH_4- or Na-salt	0.1	0.4	0.03 Mo	cabbage, half amount for trees
Fetrilon-Combi foliar "complete fertilizer"	.	0.5–1	1–4	.	cereals
	(Chap. 3.3.3)	1–2	.	.	.

N-solution should not be applied as foliar nutrient in the following cases [173]:

- *cereals:* from sowing to the third leaf, and after the appearance of ears or panicles;
- *rape:* at the very beginning of development;
- *beets:* up to the fifth-leaf stage;
- *grassland:* one day after mowing.

Application of N-solution to maize as a foliar nutrient is generally problematic, so that liquid fertilization between rows in the form of soil dressing is to be preferred (use of movable hoses).

All plants can nearly always be sprayed with micronutrients if dosing is carried out with proper care. Leaf fertilization can often be combined with plant protection spraying, which is obviously desirable in terms of labor economy.

6.2 Influence of Fertilization on the Environment

Fertilization increases agricultural production. This of itself represents an influence on the environment: corrective ecology.

However, the effects of fertilization on the three large environmental factors, *soil, water,* and *air,* merit detailed consideration (Synopsis 6–2). They are, in varying degrees, positively or negatively affected by fertilization activities in agriculture, just as any intervention in "nature" has two-sided effects. Assess-

---

Synopsis 6–2. Important Environmental Effects of Fertilizers.

1. Effects on the soil:
   a) effects on soil reaction;
   b) effects on soil structure;
   c) effects on soil life;
   d) possible accumulation of toxic substances.

2. Effects on the water:
   a) eutrophication of surface water;
   b) accumulation of fertilizers in groundwater.

3. Effects on the air:
   a) improvement of air quality;
   b) possible accumulation of noxious substances;
   c) possible disturbance of the Earth's ozone shield.

ment therefore requires weighing advantages and disadvantages against one another; it is especially important to establish the influence of fertilization properly in relation to other factors [154]. Fertilization is not to blame for all relevant *damage to the environment,* nor are its effects always positive.

6.2.1 Fertilization Influences on the Soil

Fertilizers affect the soil in various ways. Application of soil and plant fertilizers is expressly intended to improve the soil for the sake of plant production. These effects are considered in detail in the relevant chapters. However, some particular problems will be considered here in conclusion and to complement the information already given.

Many fertilizers contribute to *soil acidification* and this is generally considered to be a detrimental effect on the soil.

The extent of acidification is indicated for particular fertilizers (Chaps. 3 and 4). Reducing the soil reaction may represent a serious problem in slightly buffered soils, if no suitable countermeasures are undertaken before this becomes detrimental to nutrient dynamics and soil organisms. Slight temporary soil acidification through fertilization may even be highly advantageous (Chap. 2.1.5). *Acid damage* caused by fertilization can easily be avoided and should therefore not be considered as a detrimental influence.

The same holds true for the *soil structure.* Many fertilizers improve the structure, but some also influence it negatively, especially fertilizers with a large sodium content and some potassium fertilizers. Acidifying fertilizers also harm the soil structure if they are used for long periods. However, damage to the structure can easily be avoided through countermeasures, and therefore need not necessarily be considered a detrimental effect.

Accumulation of *toxic substances* in the soil may be caused by uncontrolled large-scale application of certain fertilizers. This applies primarily to heavy metals. They are supplied only in small amounts, compared to those already present in the soil, by usual fertilization, but special intensive fertilization measures could lead to undesirable accumulations. The most important example is the increased supply of heavy metals in fertilizers produced from domestic waste (Chap. 4.3.3), even disregarding the temporary salt load due to such composts. The decisive point is the *tolerable load* exerted on soils by noxious substances or excess plant nutrients, especially if application serves less for fertilization than for waste disposal.

A rough tentative value is provided by the total amount already present in the soil. Maximum contents should be exceeded only in justifiable cases (Table 6-3). However, in future more precise scales will have to allow for the *availability* of each particular substance.

Table 6-3. Normal and tolerable total contents of some elements in cultivated soils [163].

Element		Normal content ppm	Tolerable content ppm
arsenic	As	2–20	20
beryllium	Be	1–5	10
lead	Pb	0.1–20	100
boron	B	5–30	25
bromine	Br	1–10	10
cadmium	Cd	0.1–1	5
chromium	Cr	10–50	100
fluorine	F	50–200	200
cobalt	Co	1–10	50
copper	Cu	5–20	100
molybdenum	Mo	1–5	5
nickel	Ni	10–50	50
mercury	Hg	0.1–1	5
selenium	Se	0.1–5	10
vanadium	V	10–100	50
zinc	Zn	10–50	300
tin	Sn	1–20	50

The effects of mineral fertilizers on soil organisms are manifold [172]. Sometimes particularly negative effects are ascribed to mineral fertilizers. This is also adduced as proof of the "unbiological" effect of mineral fertilizers (Chap. 6.2.1).

There is no doubt that *soil life* can also be affected by the already mentioned acidification and damage to the soil structure. This is also possible through incorrect fertilizer application, e. g., salt damage. Occasionally there are fewer earthworms and microbes in intensively fertilized fields than in moderately fertilized ones. However, this proves very little, in view of the many causal factors. Moreover, many examples prove the opposite, that abundant mineral fertilization really activates soil life.

Organic and mineral fertilizers considerably affect the quantity and composition of soil organisms whose activity is so important for soil fertility. Mineral fertilizers in particular, if correctly applied to eliminate or prevent deficiencies, almost always promote soil life.

Certain *shifts* undoubtedly do take place. Organisms preferring a higher soil reaction will increase in number in limed soils; the activity of N-fixing bacteria decreases with increasing amounts of applied N-fertilizers, etc. However, total activity is generally increased by fertilization, i. e., the activity increases with soil fertility, which is the purpose of fertilization.

6.2.2 Fertilization Influences on the Water

Water in its many forms is an important environmental factor. Possible detrimental effects of fertilization on drinking or surface water, brooks, rivers, lakes, are to be considered most critically.

The purpose of fertilization is to enrich with nutrients (eutrophicate) the natural nutrient substrate of plants, i.e., the soil, but not the water discharged from the soil. This would hardly be possible in humid zones where there is a constant water discharge through the soil, if most soils did not have a substantial filtering capacity.

All soils, whether fertilized or not, nevertheless release some nutrients, usually in very small amounts, into the percolating water. Increasing losses of nutrients thus seem to be an unavoidable consequence of increased fertilization. This may be so, but it is not generally the case. *Leaching of nutrients* depends not only on the amounts of fertilizers applied, but to a far greater extent on many other factors determining the degree of mobility, etc. An intensively fertilized field may release much smaller amounts of nutrients than unfertilized wasteland or forest. There are also specific differences between the nutrients.

Negative effects of fertilizers on water can be expected from

- *nitrates* in groundwater, affecting the quality of drinking-water;
- *phosphates* in surface water (in some regions also nitrates), causing eutrophication.

Drinking water should not contain more than 20 ppm nitrate-N. This limit might be exceeded when there is intensive fertilization near wells (deep wells too) and large losses of N from soils with little filtering capacity. This is why restrictions are imposed on fertilization in groundwater-protection areas.

We shall therefore limit our discussion to the influence of N-fertilizers on areas utilized in the ordinary way. Additional leaching due to fertilization varies within wide limits, as could be expected, e.g., between 5 and 50 kg/ha of N. These values may be even higher in special cases. Leaching of N in the Netherlands is about 50 kg/ha for fields and 13 kg for grassland [164], the average being about 30 kg/ha. A lower average value might be assumed for the FRG, but the values are similar in certain regions. In any case, these are rough estimates from which considerable deviations are possible in special cases.

It should be stressed that leaching of N is not automatically linked to the amount of fertilizers supplied. The greatest losses are due to incorrect application and can be reduced. The following *measures* are recommended for this reason:

- Reduction of nitrate residues in autumn, since leaching occurs mainly in winter and spring;
- precise application of N after diagnosis, in order to avoid unnecessary fertilization (which is hardly used because of other minimum factors);
- application of fertilizers less readily soluble or at least less mobile in the soil;
- prevention of fertilizer losses through washing-off from frozen soil on slopes;
- increasing the storage capacity for nitrogen by promoting soil life (temporary storage of nitrate);
- increasing the water-storage capacity to avoid unnecessary losses of water.

Eutrophication (accumulation of nutrients) in surface water is largely due to increased phosphate supplies since this is the most important growth-restricting minimum factor. Increased P-supplies promote the growth of algae. Decomposition of dead algae masses consumes oxygen, excessively, and fish die because of lack of oxygen.

There is now no doubt that most P is supplied from domestic waste. The share of fertilization in the increased P-load is highly exaggerated at one third, as given by earlier data. Values below 10% around 5%, would be more realistic. These figures apply primarily to mineral fertilization with P. However, far higher values may occur when there is more intensive fertilization with semi-liquid manure, of soils close to groundwater. The P-component due to fertilizers should not be equated to the P-component due to agriculture, which may be considerably larger in some areas.

Phosphate enters bodies of water through *leaching* and washing-off. Leaching is insignificant, at least in most soils far from groundwater. It amounts to less than 1 kg/ha of P, often only 0.1 to 0.2 kg/ha. These values apply equally to soils fertilized more or less intensively with phosphate. It is therefore completely unjustified to connect increased use of P-fertilizers in agriculture with the increasingly detrimental effects on bodies of water.

Fertilizer phosphate, even in easily soluble form, is largely immobilized in the soil. This is precisely the problem with P-fertilizers. Washing-off of the latter by water erosion may cause larger local supplies to bodies of water but is most often due to incorrect fertilizer application.

In *concluding* this discussion of the detrimental effects on water, it should be stated that eutrophication of the soil, necessary for high yields need not necessarily lead to correspondingly higher losses of nutrients from the soil. Precise fertilization, which also means intentional eutrophication of the soil, can in fact maintain losses within the range of naturally occurring losses of nutrients. Damage to the environment is chiefly caused by incorrect fertilization.

6.2.3 Fertilization Influences on the Air

Fertilization increases not only the yields of organic matter, but also of oxygen (O_2) through photosynthesis. The yield of O_2 is of the same order of magnitude as the total yield. A field producing 30 dt grain and 30 dt straw also produces about 60 dt O_2 during the vegetation period (all figures per hectare). However, about 120 dt/ha of O_2 are produced when the yield is doubled by fertilization. Fields therefore produce much more O_2 than forests or wasteland. A person requires about 1.3 kg O_2 per day, i. e., 5 dt O_2 per year. Thus a hectare of well fertilized field produces oxygen for 10 more persons than a poorly fertilized field.

Detoxification of the air, due to the absorption of carbon dioxide (CO_2) by plants, in addition to the production of O_2, should be mentioned. Plants also remove many toxic substances from the air. This function is also enhanced by fertilization, so that the latter definitely has a positive overall influence on the composition of the air.

The intensive air-improving effect of fertilization is only accompanied by a slight release of fertilizer constituents that have detrimental effects, such as the occasional release of ammonia and nitric oxides. This, however, is more than balanced by the *air purification* effected by plants.

Denitrification and Cosmic Ultraviolet Radiation

Denitrification in the soil (Chap. 2.1.5) causes the formation of nitric oxides (NO, NO_2, N_2O) in addition to molecular nitrogen (N_2). NO and NO_2 return to the soil with rain and water in the form of dilute acids. They thus cause a slight additional acidification that is however, insignificant.

The relatively slow-reacting nitrous oxide (N_2O), however, sometimes rises to the ozone layer in the stratosphere, which forms an effective filter against harmful cosmic ultraviolet radiation. N_2O promotes the decomposition of ozone, where equilibrium between decomposition and formation had existed for millions of years. A predominance of decomposition would increase the ultraviolet radiation and might lead to an increased incidence of dangerous skin diseases [142].

The connection between this problem and fertilization consists in the fact that increased application of N-fertilizers by necessity leads to the formation of more N_2O. However, any realistic consideration must be based on quantification. About 200 million tons of N are required each year (by the vegetation of the earth). Commercial N-fertilizers at present account for about 20% of this amount (i. e., 40 million tons). Losses from fertilizers, caused by denitrification, vary between 5% and more than 30%, a rough average being 10%.

The share of N_2O in the products of denitrification is about 5% (1 to 10%). Denitrification of fertilizers thus produces about 0.2 million tons of N in the form of N_2O per year. This figure should be compared with the 7 million tons of N from total denitrification on dry land on earth. Fertilizers thus contribute only a few percent to increases in the "natural" value.

However, this calculation refers only to the denitrification of fertilizers not absorbed by plants. All fixed nitrogen is eventually again denitrified (most of it in the oceans). This "fertilizer share" in the total production of N_2O on earth nevertheless appears to be insignificant at present if referred to immediate losses.

6.3 Fertilization as a Function of Soil Type

The many soil types provide quite different preconditions for the application of fertilizers. The better one knows the natural fertility of a soil type, the more precisely one can fertilize.

Only very few soils are ideal plant substrates "by nature". On the other hand, many soils are quite the reverse. In suitable climates they may even support an abundant vegetation, but the latter is matched to existing deficiencies. However, crops impose specific, sometimes very severe, requirements on the fertility of the soil. The latter must therefore frequently be considered to be deficient for crops. Most soils are thus *problem soils* requiring correction by fertilization.

Soil types are grouped to permit a logical discussion of their fertility and the necessary corrective measures. To begin with the following rough classification is adopted:

A. Soils with good natural fertility; there are only a few and they require minimal fertilization measures.

B. Soils with inadequate natural fertility, requiring major, sometimes even meliorative fertilization measures. They comprise most of the dry-land surface of the earth, and will be divided into the following groups:

- Humid soils in cool, temperate zones;
- humid soils in tropical zones;
- arid soils in subtropical zones (also when irrigated);
- saline soils;
- organic soils (moor soils).

On the other hand, we shall not discuss soils with very low fertility, which should be considered as only marginally useful for agriculture, i. e., unworked soils of all kinds (shallow rocky soils, dune sands, etc.).

Fertilization of Soils Having High Natural Fertility
This category includes (in terms of area) a small minority of soils, especially the following soil types (however, only their best subtypes are mentioned):

- Black earth of steppes (prairies);
- brown earth in temperate zones;
- red earth in Mediterranean and tropical zones;
- lime soils (deep-soil forms rich in humus);
- alluvial soils (riverine soils and marshy soils on coasts);
- volcanic soils (with their nutrient-rich forms).

The best variants of these soil types have medium particle sizes (i. e., the range from loamy sand to sandy loam) and nearly ideal fertility properties, viz:

- optimal soil reaction in the slightly acid to neutral range;
- sufficient humus content with good humus forms and abundant and active soil life;
- good structure down to great depth;
- high water and air capacities;
- abundant nutrient reserves and readily available nutrients;
- good storage and regulating capacities for mineral nutrients (resupply when required, buffering of surpluses).

Fertilization of these soils may be limited to maintaining the favorable natural properties and replacing some losses due to removal, and therefore poses no special problems.

6.3.1 Fertilization of Soils in Humid Temperate Climates

Soil types of these climatic zones often have a grey or grey-brown color as an external characteristic. They are mostly *podsols* and transitional forms to other types, with iron and clay displacement [94]. They occupy large parts of Europe, Asia, and North America, being predominantly former forest soils. Their average fertility covers a wide range from medium to low (Synopsis 6-3). Their natural yield potential is often low, but precise improvement of minimum factors offers great possibilities for yield increases.

The possibilities of fertilizer application are large, corresponding to the many deficiencies in soil fertility. Basic melioration may be required first in ex-

Synopsis 6–3. Fertility Problems of Podsols, Podsolic Soils, etc.

1. Low natural soil reaction, often causing acid-damage syndrome in cultivated plants.

2. Sometimes small workable profile depth because of hardened strata (ortstein in case of podsols), or compaction in subsoils rich in clay.

3. Humus forms of low value (especially poor in nitrogen).

4. Low water capacity in sandy forms.

5. Insufficient aeration of insufficiently drained soils.

6. Small nutrient reserves in sandy forms.

7. Small contents of available nutrients (due to impoverishment of original material or leaching).

8. Low storage capacity for mineral nutrients in sandy forms, or due to unfavorable humus forms.

9. Partial fixation of nutrients (e.g., Mo, P, Cu).

10. Low transformation and regulating capacity.

treme cases, e.g., removal of interfering strata, drainage, etc. The first decisive fertilization measure is proper *liming*. This puts into motion a process of multiple fertility improvements, ranging from activation of soil life to improvement of nutrient and humus forms to a better regulating capacity.

However, sandy soils in particular are very sensitive to *overliming*, so that the optimal soil reaction is below pH 6 (Chap. 5.1.1). Higher pH-values often cause secondary damage, of which manganese deficiency is most frequent.

The general *lack of nutrients*, especially in sandy soil forms, requires relatively intensive fertilization with major nutrients (NPK). Considering the *low storage capacity*, division of the nitrogen fertilizer dose is indicated. Stock fertilization is therefore unsuitable for nitrogen and potassium, especially since losses through leaching may be appreciable. Owing to the low optimal pH-values, easily mobilized phosphate forms are generally more suitable than water-soluble ones. Magnesium deficiency is common. Magnesium lime is therefore the material of choice for liming.

Deficiencies of Mn, B, and Cu among trace elements are important. Mn- and B-deficiencies are frequently caused by immobilization due to excessive pH-values. Cu-deficiency is characteristic of halfbog soils with certain Cu-fixing humus forms. Molybdenum deficiency is rare in properly limed podsols.

Soils react favorably to organic fertilizers applied to improve their physical properties, as well as to other soil conditioners.

Climatic *stress* factors to be taken into account, especially during fertilization, are the cold, which primarily damages inadequately nourished young plants, and dry periods in spring.

Full utilization of the production potential thus requires a comprehensive fertilization programme with relatively large inputs. However, this also offers large possibilities for yield increases.

6.3.2 Fertilization of Soils in Humid Tropics

Soil types of these climatic zones often have a red or red-yellow-brown color, as an external characteristic. They are mostly *red loams* and *red earths* (*latosols, ferralsols,* etc.) [156, 157, 171].

These soils cover large areas on both sides of the equator (tropical rain forests, moist savannahs). Their average fertility has a wide range from low to medium, except that some red earths belong to the already-mentioned soils that have an especially high fertility (Synopsis 6–4).

Crop yields are frequently low, despite a sometimes abundant natural vegetation. However, these soils too offer great possibilities for improvement of the natural yield potential [158, 176].

Synopsis 6–4. Fertility Problems of Humid Tropical Soils.

1. Low natural soil reaction, often causing acid-damage syndrome in crops.

2. Sometimes small utilizable profile depth because of hardened strata (iron-oxide concretions, stoney layers).

3. Small humus content and rapid decomposition of humus during agricultural use.

4. Small content of available nitrogen and small N-mobilization.

5. Small content of nutrient reserves.

6. Small content of available potassium, calcium, magnesium, and sulfur.

7. Partly strong fixation of phosphate and molybdate by mobile iron and aluminum oxides.

8. Low storage capacity for mineral nutrients, due to clay minerals with low sorption capacity.

9. Rapid disintegration of the otherwise quite stable structure in agricultural use.

10. Danger of soil erosion by water on slopes.

Various deficiencies can be eliminated by fertilization in humid tropical soils, too [153, 159]. A basic measure is proper *liming*. However, the many experiments that have good lime effects may be contrasted with numerous trials giving no significant yield increases. The discrepancy between a theoretically expected but, in practice, absent lime action can be explained by the many effects of liming and the limits of its action. It is only when the new minimum factors that become decisive after liming are eliminated by additional fertilization that liming can finally become effective (Fig. 6-2).

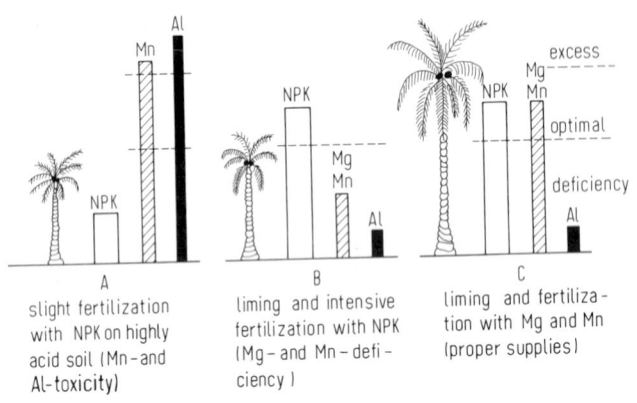

A
slight fertilization
with NPK on highly
acid soil (Mn - and
Al-toxicity)

B
liming and intensive
fertilization with NPK
(Mg - and Mn - defi -
ciency)

C
liming and fertiliza -
tion with Mg and Mn
(proper supplies)

Fig. 6-2. Liming of tropical soils and growth, and mineral contents of leaves (full effects of liming only after additional fertilization with Mg and Mn, schematic) [159].

The optimal *soil reaction* depends on the soil type, as with other soils. However, the structure of these soils is usually relatively good, because of the bonding by iron oxides, so that the pH-target may be lowered. A moderately acid pH-range between 5 and 6 is frequently optimal. Problems are involved in improving extremely acid soils, if there are no lime supplies in a large area, as is the case in tropical forest regions. Limes are usually not worth the expense of transportation over large distances. Substitution by plant ash as a fertilizer is possible to some extent.

Organic fertilization has a favorable effect on humid tropical soils, partly as a result of the reduced danger of erosion, especially when it is used as mulch. Nutrient deficiencies are frequent, so that relatively intensive fertilization with N, P, and K is required for high yields.

However, the possibilities of stock fertilization are limited because of leaching losses. The tendency to immobilization is the reason easily mobilizable phosphate forms should be preferred to watersoluble ones. Magnesium deficiency is common, depending on the original rock. Sulfur deficiency occurs mainly at large distances from the sea.

Deficiencies of trace elements are relatively rare, except in extremely poor substrates, because of the usually low soil reaction. They are only of practical importance with intensive cultivation. An exception to this rule might be molybdenum deficiency, but it is only at higher yields that is likely to become acute.

Climatic *stress* factors to be taken into account, in view of fertilization, are intensive light irradiation and dry periods.

The natural fertility of humid tropical soils is usually sufficient for low yields. More intensive fertilization and a comprehensive fertilization concept are required only when intensive production starts. However, the favorable climatic conditions then provide the potential for substantial yield increases.

6.3.3 Fertilization of Soils in Arid Tropics (Subtropics)

Soil types of these climatic zones have a light brownish color as their external characteristic. The *arid soils* belong to quite a variety of soil types, depending on the original material and the climate. However, they are all similar with regard to their fertility.

These soils cover large parts of the arid regions on earth (savannah, prairie, and steppe belt). Generally they support only a sparse vegetation, or none at all, owing to the lack of water. However, production potential is often considerable, if sandy and stony forms are disregarded, but more than water is needed to make "the desert flourish like a rose".

They are characterized by a neutral to slightly alkaline soil reaction because of the lime content. They also usually have a good soil structure, are well supplied with potassium, magnesium, sulfur, boron, and molybdenum, and are not subject to leaching [155, 158]. Special problems are indicated in Synopsis 6-5.

The fertilization of arid soils poses special problems and will therefore be considered in some detail.

The following consequences concern fertilization with *major nutrients*: fertilization with N is almost always necessary when there is intensive cultivation; all common N-fertilizers act relatively quickly because of the high transformation capacity of the soil. On lighter soils fertilization with P and K is often necessary to obtain high yields, but it is less so on medium and heavy soils. Superphosphate has proved effective in covering P-requirement for immediate fertilizer action; it also has a slight soil-acidifying action, and in addition increases the supply of sulfur. Highly concentrated forms of K-fertilizers should be preferred to reduce the salt load. Fertilization with lime, and, usually the supply of magnesium and sulfur can also be omitted. Forms with a large N-content should be preferred when NPK-fertilizers are applied.

Synopsis 6–5. Fertility Problems of Arid Tropical Soils.

1. Nearly always lack of water, so that irrigation is usually required.
2. Sometimes little utilizable profile depth because of hardened lime crusts in subsoil.
3. Small to very small humus content.
4. Very small content of available and mobilizable nitrogen.
5. Phosphate deficiency in sandy forms.
6. Frequent iron and zinc deficiencies because of fixation.
7. Low storage capacity for mineral nutrients in sandy forms (important only in cases of intensive irrigation).
8. Low regulating capacity because of small clay content and only slightly developed soil life.
9. Occasionally excess of salts, sodium, and boron.
10. High susceptibility to wind erosion.

Supplies of *trace elements* depend to a great extent on soil reaction. Arid soils, which usually contain lime, therefore pose many problems.

Acute *iron deficiency*, in particular, occurs frequently in the form of *lime chlorosis*, and latent deficiency may be common. Lime chlorosis may be due not only to the immobilization of iron in the soil, but also to its immobilization in the root or in plant parts above the surface of the soil. The large bicarbonate content of the soil plays an additional detrimental role in this case. The capacity for mobilizing iron from the soil differs for individual varieties of the same crop. It is sometimes possible to eliminate an iron deficiency by changing the plant strain.

Acute *manganese deficiency* is rare despite the reduced availability. This may be due to the readier reducibility of Mn-compounds, in comparison with iron, the higher mobility of manganese carbonate, or a better capacity of most cultivated plants to transfer the Mn within the plant.

The availability of *zinc* and *copper* is less dependent on pH. However, zinc requirements are large, due to the high light intensity. Zinc deficiency in acute form is therefore frequent and probably plays an even more important role in the latent form. However, there are hardly any reports of copper deficiency on arid sites, probably because of the absence of large amounts of Cu-immobilizing organic matter. Nor does *boron deficiency* play any significant role, despite its reduced availability when there is high soil reaction, since the boron content of many arid soils is above average.

Temporary, limited *lowering* of the high soil reaction is possible through the application of nitrogen fertilizers that have a physiologically acid action. Ammonium sulfate and urea are therefore indicated in arid zones. However, yield-reducing losses of gaseous ammonia occur at pH-values above 7.5 or thereabouts when ammonium fertilizer is applied; these losses vary with the form of application.

A special soil type in subtropical zones is a *dark clay soil,* called *grumusols* or, now, more often *vertisols.* They cover about 3% of the surface of Africa and often have low productivity, because of a lack of water. However, their high potential productivity places them among the best soil reserves still available.

Vertisols are very deep and have a soil reaction that depends on their lime content; they are well supplied with most mineral nutrients, and have a high storage capacity. Their large clay content (frequently above 50%) is a disadvantage. It not only makes working the soil difficult but also reduces the permeability to water and particularly interferes with aeration. However, the clay is montmorillonite, so that these shortcomings are less serious than might be expected; cracking as the soil dries often ensures considerable additional aeration.

Nutrient Supplies and Fertilization of Irrigated Arid Soils

Mineral supplies also undergo typical variations in the irrigation rhythm. This will be considered below, for the wet and dry phases respectively.

Wet irrigation phase causes considerable dilution of the soil solution by the water supplied, despite certain supplies of nutrients. This has the advantage of lowering the osmotic pressure, but the disadvantage of a lower supply intensity of nutrients whose concentration reduction cannot be compensated rapidly. Fortunately, in arid soils there is no deficiency of calcium and boron, which are continuously required in every case.

There is a relative increase in the concentration of monovalent cations in the soil solution caused by exchange processes. This increases the potassium supply but temporarily reduces the supply of magnesium. The Ca-concentration also decreases, but this has no detrimental effects in view of the large total supply.

Waterlogging of the soil causes intensive *mobilization* and resupply from nutrient reserves at high temperatures. Nutrients mobilized are in particular those converted into mobile forms because of the reducing conditions, i.e., *iron* and *manganese.* Supplies of these two elements, only slightly mobile because of the high soil reaction, occur under irrigation in pronounced pulses. Manganese compounds are more easily reduced when the redox potential is lowered; Mn-supply may thus even become excessive. Moreover, manganese is sufficiently mobile in most plants so that this pulsating Mn-supply is usually adequate for a continuous supply to the active tissues, and deficiencies are rare.

The slow drying-out of the soil in the *dry phase* between irrigation periods increases concentration of the soil solution by evapo-transpiration and thus intensifies nutrient supplies.

The concentration of monovalent cations decreases relatively in the soil solution, and absorption of the antagonists potassium and calcium is shifted at the expense of potassium.

More intensive drying-out finally causes *immobilization* of mobile nutrients, i.e., conversion from the mobile phase to the reserve fraction. Phosphates precipitate, iron and manganese are converted to oxides, and thus immobile forms, potassium is more or less immobilized, depending on the content of clay minerals in the soil. However, plants can bridge temporary slight deficiencies at the end of the dry phase with reserve nutrients, which are mobile in the plants. Latent and even acute deficiency occurs fairly often only in the case of iron, because of a low intake and additional immobilization in the plant.

These features of irrigated soils must be taken into account in *fertilization*. The relatively high production level in any case requires more intensive fertilization. However, it must be primarily aimed at the minimum factors. Thus, intensive use of NPK-fertilizers will not be fully effective if, e.g., a trace element periodically becomes the minimum factor in the irrigation rhythm.

Fertilization by irrigation, representing a combination of the two measures, will only be mentioned here. It is less effective than leaf spraying with regard to the supply of trace elements.

Special climatic and biotic stress factors must be taken into account in fertilization of arid soils. These are, e.g., heat, dryness, relatively cool periods for thermophilic plants, intensive attacks by insects, etc. (Chap. 6.5). Large production reserves still exist in arid zones; they can be fully utilized only by combining fertilization and irrigation.

6.3.4 Fertilization of Saline Soils

Saline soils occur in some places in arid zones. They are either of natural origin, or sometimes there is a secondary salinization caused by irrigation. In extreme cases, they support only a sparse salt-plant vegetation. Cultivated plants at best can grow with only low yields if at all, if no desalinization is carried out. However, the fertility may otherwise be considerable, if toxic salt effects are disregarded.

The *salt* must be removed from the soil through *melioration*, or the salt effect must be reduced by suitable fertilization; and at the same time proper nutrition of the plants must be ensured despite it. The latter case of *management with a higher than normal salt content* of the soil will be considered below. This refers both to saline soils in the narrower sense (with increased concentrations of soluble salts) and to soils with a larger content of exchange sodium (Na-soils) [155].

Plants in saline soils suffer not only from *lack of water*, but also from *nutritional disturbances*. There is both a considerable excess of certain ions (sodium, chloride, sulfate, sometimes magnesium or borate) and a real or induced deficiency of some nutrients, e.g., potassium, calcium, and/or some trace elements. This discrepancy between large supplies of partly necessary, partly undesirable ions on one hand, and deficiencies of necessary ions on the other, causes *nutritional stress* manifesting itself in yield reductions to a varying extent.

Correct *fertilization* can reduce this stress. Increased N-supplies, and thus protein formation, can inactivate part of the chloride in the tissues. However, it might be more important to correct deficiencies and reduce excess salt intake at the root. Thus, increased application of potassium fertilizers, beyond the usual amounts, not only considerably reduces the sodium intake through antagonistic effects but may also raise the often insufficient potassium supply to the optimal region. This fertilization measure thus permits simultaneous correction of two serious shortcomings.

Calcium deficiency may occur particularly in limecontaining saline soils rich in magnesium, when certain plants are grown. Increasing the supply of soluble calcium by fertilization or via the leaves may be indicated.

The high soil reaction of saline soils often causes deficiencies of some *trace elements*. This should also be corrected by proper fertilization. On the other hand, it is difficult to reduce the effects of *boron excess* by fertilization. Fortunately, this rarely occurs.

The possible effects of improved nutrition of plants on saline soils by fertilization are shown in Fig. 6–3, where the concentrations in plants during the principal growth stage are plotted in relation to optimal supplies, with or without precise fertilization. The sodium content without fertilization is above the

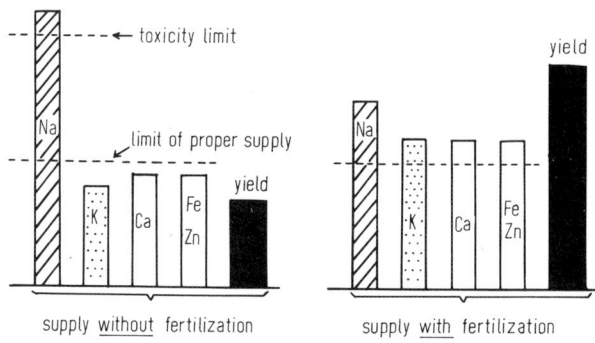

Fig. 6–3. Fertilization of saline soils to correct imbalance of mineral contents of plants (schematic).

toxicity limit in this example; on the other hand, supplies of potassium, calcium, iron, and zinc are below the normal level. The resulting yield is therefore low. Additional fertilization with K thus not only lowers the sodium concentration in plants to the harmless normal level but also raises the K-content to the optimal region. Additional calcium supplies may be required. Considerably higher yields may be expected if there is supplementary fertilization with iron and zinc. This model serves to show that precise fertilization can ensure satisfactory yields despite a certain soil salinity.

Unnecessary *salt loads* should in any case be avoided when applying the necessary fertilizers. Fertilizers with large amounts of water-soluble ballast ions are undesirable. Potassium sulfate thus appears to be preferable to chloride (potassium nitrate would be even better). Highly concentrated K-salts are to be preferred to low-concentration ones. However, the minimum factors cannot be readily guessed for each saline site, so that correct *diagnosis* is even more important in these cases than otherwise, since unnecessary fertilization would only increase the salt load. It should be remembered that the usual optimum nutrient level for proper nutrition may sometimes be considerably increased under salt stress (e. g., for potassium).

The dissolution of Na-carbonate (soda) by fertilizers with strong acid action may be indicated especially in alkaline (soda) soils (Chap. 4.1.3).

6.3.5 Fertilization of Organic Soils

Soils with large contents of organic matter occur in many climatic zones when waterlogging is intense, but it is the bogs in temperate zones that are best known. *Bog soils* were originally formed on wet sites, however drained for agricultural purposes. Volumetrically they consist largely of organic matter; true bog soils contain more than 30% humus by weight, and half-bog soils 15 to 30% humus.

The *water capacity* of bogs is considerable. Much of it remains unavailable, but water supply in general is adequate. Deficient aeration may limit growth if drainage is insufficient. Fertilization problems should be considered from the volumetric aspect, since the specific weight of peat is only 0.1 to 0.2.

All data concerning *nutrient supplies* are referred to in volumes, e. g., nutrient contents per 100 ml soil. Bog soils are best divided into *raised-bog* and *fen soils*, as far as fertilization is concerned. Organic fertilizers are generally not required on bog soils. *Raised-bog soils*, which are highly acid and poor in minerals, often require some liming as the basis for further fertilization, but the soil reaction should only be raised to a limited extent (Chap. 5.1.1).

Supplies of N, P, and K are small because of the absence of reserves, so that suitable fertilization is necessary. *Potassium* deficiency, especially, should be taken into account. Relatively small amounts of trace elements are present, but their availability is quite good, except for molybdenum.

Copper deficiency often plays a role in the transition to half-bog soils (heath-bog disease). Raised-bog soils utilize fertilizers well, since immobilization is insignificant. The deficiency of mineral matter can often be corrected by sand-mixing amelioration. Formerly, nutrient reserves were also mobilized by *burning* bog layers.

Fen soils are often rich in lime and mineral nutrient reserves. Therefore, they do not require fertilization with lime; acidifying fertilizers should rather be applied instead. An *abundance of nitrogen* is characteristic of many fens, and this provides abundant supplies from the soil. The lime content causes *immobilization* of phosphate and some trace elements. Thus, manganese deficiency is frequent. Copper deficiency is also frequent in the transition to half-bog soils.

Bog soils in temperate zones can be transformed into highly productive soils by appropriate fertilization. Such soils react to increases of minimum factors with high yields. An important climatic stress factor is the *cold* in the early growth periods of plants. This increases the importance of proper supplies of nutrients to improve the resistance (Chap. 6.5.1).

Tropical bog soils originating from forest bogs, papyrus marshes, etc. are usually more fertile after drainage than corresponding bogs in temperate zones, as the intensive decomposition of humus causes a relatively larger accumulation of minerals in the drained layer. The principal problem in using such soils after drainage is the often intense *acidity* caused by the formation of sulfuric acid from sulfide oxidation. pH-Values as low as 2 have been measured. Such extreme acidity prevents any plant growth either directly or by mobilizing toxic quantities of metal ions. Liming is essential for any practical use, even in less extreme cases.

6.4 Fertilization as a Function of the Cropping System

Fertilization is not an isolated measure in agriculture but jointly with others serves to obtain the maximum possible *quantity* and quality of a plant product, and thus also a *profit* in the economic sense. Fertilization must therefore be considered not only from agrochemical but also from economic aspects within the framework of a comprehensive enterprise.

This leads to quite different priorities. No additional fertilization is needed for agrochemical or economic reasons if a small population utilizes a large area of fertile soil. On the other hand, any sensible fertilizer input is justified in cases of famine and the accompanying high prices; considerations of cost

Synopsis 6-6. Four Stages of Fertilization as a Function of Cropping System.

role and utilization of nutrients in soil	cropping system	soil fertility	additional fertilization	yields
1. a) exploitation	exhaustion cropping (exploitation cropping) with regeneration	degeneration	none	decrease
1. b) utilization	utilization cropping	maintenance by natural capacity	none	constancy in "steady state" equilibrium
2. replacement	permanent stable crop sequences with soil-improving plants	maintenance through replacement	replacement fertilization to compensate for losses	constancy in "steady state" equilibrium
3. enrichment	permanent stable demanding crop sequences	increase through improvements	differentiated fertilization according to supply state with aim of optimal supplies of all nutrients	increases to high yield level
4. substitution	arbitrary	optimization or substitution	artificial combination of optimal nutrient substrate	high yield level

are then unnecessary. Food is absolutely vital and lies outside the accepted monetary scale when there is a serious shortage.

However, fertilizer input depends on *agrochemical and economic* factors within the scope of the cropping system involved, if the two above-mentioned, rare, limit situations are disregarded. These systems have evolved historically and may be associated with different stages of fertilization.

This is nevertheless not a discussion of historical interest only, since all three stages already existed in antiquity. All cropping systems and fertilization stages still exist today simultaneously in different regions on earth, sometimes even on one and the same farm. This is one of the principal reasons for the differences in fertilizer input.

In 1869, *Mayer* was the first to consider fertilization from the point of view of farm economy. He distinguished between three fertilization stages according to the cropping system [170]. These stages correspond to certain degrees of soil fertility [169].

Synopsis 6-6 shows the relationships between cropping system, fertilization, soil fertility, and yield, using, modifying, and supplementing existing definitions and concepts (p. 268).

6.4.1 Exploitation and Utilization of Soil Nutrients

Agriculture based on exploitation (exploitation cropping) and utilization of nutrients stored in the soil is the oldest production form. Even today it still plays an important role in food production in many regions on earth. Common to these exploitation and utilization systems is that hardly any additional fertilization is undertaken beyond the supply of harvest residues.

Exploitation of Soil Nutrients

This *exhaustion cropping*, justly also known as *exploitation cropping*, makes agricultural use of the natural nutrient capital of the soil. This causes soil impoverishment, and yields decrease from year to year. The many causes of the rapidly diminishing yields have been compiled in Synopsis 6-7.

The soil is left to regenerate through natural forest and grass vegetation when cultivation is no longer worthwhile after 2 to 5 years. There is practically no fertilization if the unavoidable supply of harvest residues is disregarded, since it was not known in the early periods of agriculture, is not possible at present, nor is it profitable. Exploitation cropping permits high net profits to be achieved temporarily because of the small input (e. g., after clearing or on leaseholds).

The technical procedure for utilizing accumulated nutrients is based on the following principles:

Synopsis 6-7: Causes of Large Yield Decreases Through Agricultural Use Within the Framework of Tropic Shifting Cultivation [158].

1. **Changes in Nutrient Supplies**
   a) *Mobilization* of nutrients
      - through burning of vegetation at time of clearing;
      - through humus decomposition (mineralization) in use;
      - through weathering of mineral reserves (of little significance, since cultivation phase is short).

   b) *Losses* of nutrients
      - through removal during harvesting;
      - through leaching;
      - through erosion;
      - indirectly through reduced accessibility due to deterioration of the structure.

2. **Effects of Humus Decomposition**
   a) *Positive* effects
      - release of nutrients from humus during mineralization;
      - increased nutrient mobilization from mineral reserves through increased production of acids and chelates;
      - intensification of soil life during decomposition.

   b) *Negative* effects through lowering of
      - sorption capacity (capacity for storing nutrients);
      - water capacity (capacity for storing water)
      - structural stability and thus resistance to erosion.

3. **Consequences of Deterioration of the Structure**
   - inhibition of root growth and thus reduced accessibility of nutrients;
   - poorer soil aeration and thus reduced intake of nutrients;
   - reduced water capacity and poor conduction of water in the soil;
   - more undesirable surface runoff and thus greater losses through erosion on slopes;
   - reduced activity of soil life with its many consequences;
   - altogether poorer plant growth and increased susceptibility to pests.

- Exploitation of easily available nutrients in the soil, until they are exhausted;
- additional mobilization of reserve nutrients in the soil, e.g., by intensifying the natural mineralization of humus or even burning the uppermost soil layer;
- mobilization of nutrients in the natural vegetation by burning (fertilizing effect of ash).

The fertility of the soil, accumulated over centuries, is thus utilized through exploitation cropping. The nutrient cycle is open and considerable losses are caused by removal and leaching.

Controlled exploitation cropping may be quite useful economically and may even be ecologically acceptable as a stable form of utilization, despite all objections to exploitation cropping as such, as long as no irreparable damage is done. This, however, presupposes the availability of large land reserves, 10 to 20 times the area worked at any one time, as is the case with original forms of *shifting cultivation* [174, 176].

However, exploitation cropping causes not only considerable damage to the farm, at least in the long-term, when land reserves are scarce but also damages the national economy.

On the other hand, it may be useful on a given farm to combine exploitation cropping of some areas with fertilization of others, e. g., in order to supply the nutrients removed from an unfertilized meadow to a field (the meadow is then the "mother" of the field).

This is essentially a problem of farm economics, as long as the damage done by exploitation cropping is reparable and can be made good by an increased nutrient input. Exploitation cropping is thus "an apparent evil carrying its cure within itself" [170]. Seriously diminished yields will soon force the farmer to stop exploitation cropping.

Only the occurrence of irreparable damage due to intensive exploitation cropping is a problem to the national economy. Farmers may not be fully responsible for maintaining the "status quo" of the earth's crust (its natural wealth can and should serve humanity), but soil is a *nonrenewable asset* that should not become the victim of short-sighted striving for profits.

Irreparable damage is:

- *Deterioration* of soil fertility (Synopsis 6–7): yields decrease considerably and cannot be normalized even through the supply of nutrients, since essential fertility factors have been disturbed or destroyed. Yields finally remain approximately constant at a very low level.
- *Destruction* of soil fertility by wind or water erosion; although the soil as such cannot be destroyed, the fertility of the A-horizon, the supporter of fertility, is lost.

The results are *devastation of land or soil destruction,* e. g., where the subsoil is stony or hardens after removal of the A-horizon (as with laterite in tropical zones).

Fertile areas are thus transformed into wastelands. Even such heavy damage is not completely irreparable in the final analysis, but the costs of regeneration bear no relation to the short-term "gain" achieved by exploitation cropping.

Historical examples are the losses caused by soil erosion in the USA, failures due to exploitation cropping of relatively poor soils of tropical rain forests, in savannah zones, etc. Growing of monocultures accelerates this process, since the soil in such cases is exposed to the attacking forces for longer periods without protection.

A tillage system of exploitation cropping is *shifting cultivation*, e.g., in the form of forestburning agriculture in humid tropical zones and formerly also in Europe. It consists, e.g., in working arable fields for 3 years, followed by 20 years of regeneration as forest. This also includes shifting cultivation of steppes (prairies) with or without burning off the grass.

Exploitation cropping accompanied by irreparable damage represents destruction of a naturally available potential, which humanity with its continuously shrinking living space cannot afford. Some earlier human civilizations may have perished through the complete exhaustion of their soils, e.g., the *Mayas* in Central America and the *Khmers* in Cambodia. In any case, *famine* was often the result of soil impoverishment due to exploitation cropping.

Utilization of Soil Nutrients

Utilization cropping consists in cultivated plants removing part of the nutrients from the reserves in the soil, without (significant) impoverishment or reduction of the fertility taking place. This creates the impression of the *permanence* of agricultural production [166]. An analogue is a well which always contains the same amount of water irrespective of the number of times it is drawn upon. Such cropping systems are based on a *high soil fertility*, with small removals and other losses, i.e., on

- either large contents of available nutrients (relatively infrequent, at least with regard to all nutrients);
- or large supplies of available nutrients from reserves.

Ideal systems of removal without impoverishment are in general relatively rare and limited to soils of maximum fertility (Chap. 6.3). Special fertilization, at least as regards most nutrients, is not required even over longer periods, if the unavoidable supply of harvest residues is disregarded.

The *basic principle* is to maintain soil fertility without human intervention, starting at a high fertility potential. This should not be considered as exploitation cropping, since the soil is not significantly impoverished and yields remain constant. Fertilization would rather be "exploitation of the purse" [143]. Typical examples are *shifting cultivations* in which the soil is utilized for various periods without the yields decreasing.

Special forms of *utilization cropping* occur in modern agriculture when some nutrients suffice for years (e.g., major nutrients) or decades (e.g., trace elements) to give high yields without fertilization being necessary.

Summarizing, it may be said that utilization cropping is a suitable and recommendable cultivation system. However, it is possible in complete form only to a limited extent owing to the scarcity of highly fertile soils, and even a high fertility potential is eventually exhausted.

6.4.2 Replacement of Soil Nutrients Removed

The catastrophic consequences of exploitation cropping, which manifest themselves in severe famine, forced farmers early on to consider remedies. The principle of *replacing* what had been removed from the field is very ancient, and permitted stable cropping systems even in ancient civilizations. Examples of this are the natural replacement of nutrients by Nile silt in Egypt and careful compost management in ancient China.

Two researchers in particular are credited with the scientific formulation of the theory of replacement, i.e., *nutrient statics*:

- *Thaer* (1809) who demanded replacement of the removed plant substances by organic fertilizers [177].
- *Liebig* (1840) who suggested that minerals removed be replaced by additional mineral fertilizers [17].

The *replacement theory* is based on the fundamentally correct view that replacement of losses in an open nutrient cycle is essential for maintenance of the production potential. This may be done through regeneration during a fallow period, cultivation of soil-improving plants, or fertilization in particular. The aim is a continuous stable crop sequence with balanced nutrient supplies and maintenance of soil fertility. Measures for preserving soil fertility are soil-guarding crop sequences, especially with nutrient-accumulating plants (e.g., cultivation of clover with the accumulation of nitrogen), crop rotation (alternating demands on nutrients and structure), suitable preceding crops, etc. [144, 149].

Nutrients can be replaced by organic and mineral *fertilizers*. The use of plant waste as well as human and animal faeces is emphasized (as an extreme case, cattle are raised to produce farmyard manure). Nutrients are brought to the fields from areas under exploitation cropping (fodder from meadows, foliage from forests). Mineral fertilization with silt, marl, ash, etc., known for a long time, is supplemented by "artificial" fertilizers.

The principle of *nutrient replacement* is fully valid only in cases of a good initial situation. However, replacement is no longer sufficient if the soil has already been impoverished through exploitation cropping. On the other hand, replacement is unnecessary if the soil is well supplied with certain nutrients.

Cropping systems based on the replacement principle are therefore only rarely realised to the full extent, but are very common in an approximate form. Yields are kept at a certain equilibrium level through average replacements in the *steady state*. *Liebig's* view that exploitation cropping always exists when replacement is incomplete is therefore only valid with certain restrictions. From the point of view of farm management, it has been replaced by the *theory of fertilization sequence* [170]. *Thünen* (1850) observed that the views of these "chemists", even if correct with regard to replacements, do not affect the *statics of agriculture* [178]. We can now say that the views of the "chemists" are largely correct, and that the statics of agriculture have been replaced by the theory of fertilization sequence.

6.4.3 Soil Enrichment with Nutrients

Natural fertility is often insufficient for high yields, especially after long periods of exploitation cropping. The aim of the third *stage of fertilization* is therefore to improve soil fertility. This procedure also had its beginnings a very long time ago. Thus, replacement methods of antiquity were frequently aimed not only at replacement but simultaneously at improvement.

Enrichment should primarily extend to substances that limit production as minimum factors. The principle of enrichment should therefore be understood in several different ways:

- Enrichment with substances that are deficient;
- replacement of substances actually present in sufficient amounts;
- removal of abundant nutrients without separate fertilization.

Disregarding these differences caused the failure of *nutrient statics*, which lays the stress on the *replacement principle*. Increased soil fertility manifests itself historically by the fact that, e.g., sugar beet and wheat nowadays have high yields even on soils formerly considered as far too poor for these demanding crops.

The enrichment method is in principle an intensified replacement fertilization. Humus and nutrient contents are increased and soil life improved. This substantially raises the yield potential up to an upper limit imposed by minimum factors difficult to correct; however, this upper limit can be shifted upward by suitable measures requiring large inputs. Soil improvement through enrichment is usually profitable, as long as it is not done in an extremely one-sided manner. Improved soil always represents a higher value.

All three fertilization stages are found on many modern farms:

- *Exploitation cropping* with regard to certain nutrients whose slowly decreasing reserves are being used, as well as through the transfer of nutrients from grass-land to fields;
- *replacement management* with regard to primary nutrients (in many normal crop sequences);
- *enrichment management* in the intensive cultivation of agricultural or special crops.

Fertilization in this situation serves less to aid individual crops than the entire crop sequence (fertilization with nitrogen being an exception). An example is maintenance liming of certain plants with particularly large lime requirements (beets, wheat, rape, alfalfa) in the crop sequence. Basal fertilization with phosphate and potassium may also be expressly applied to plants with larger requirements, while the following crop uses the reserves left over.

Static "nutrient considerations" give place to more *dynamic* considerations, the more so since nutrient balances complicate the previously expected simple calculations. This is because removal values (Chap. 5.5.2) and utilization rates of fertilizers are uncertain.

A high level of nutrient content in the soil is decisive (but without unnecessary accumulation). This has to be achieved and then maintained by fertilization. The enrichment phase is followed in the steady state by a permanent replacement phase, but in precise form and only as far as necessary.

Substitution of Nutrients in the Soil

The highest fertilization stage is attained when the soil with all its imperfections is completely replaced by an *optimal nutrient system*.

This is possible in hydroponics and with optimally constituted solid nutrient substrates. They supply plants with all the necessary substances in the required amounts and at the correct time. However, such ideal systems are expensive and until now have been used in practice only in intensive horticulture.

6.5 Fertilization under Stress Conditions

The concept of *correct fertilization* mostly refers to normal growth conditions that may vary within a certain range. However, particular stresses impose special qualitative and quantitative requirements on nutrient supplies to plants. Improving the supply (sufficient for normal conditions) of one or more nutrients can in such cases increase the resistance to a special stress and thus better enable the plant to survive danger periods, if at all. Beneficial nutrients (Chap. 3.4) become important in this connection.

The following *stresses* play a special role in plant production:

- climatic effects;
- biotic effects;
- chemical (environmental) effects.

Temporary or permanent stress situations are often responsible for low yields. Fertilization measures intended to increase the resistance of plants or alleviate harmful effects are therefore of decisive importance. Fertilization does not increase resistance in general but improves deficient nutrition in a direct way.

Resistance is to be understood as the natural resistance of the plant to physical, chemical, and biotic effects.

6.5.1 Fertilization and Resistance to Climatic Stress

Extreme climatic conditions may represent serious stresses to plants in various respects, namely through dryness, heat, radiation, cold, and mechanical loads.

Dryness is common in both arid (dry) and humid (moist) climatic zones. Periods of dryness considerably restrict agricultural production. Plants begin to wilt when water becomes scarce, i.e., plants pass the wilting point and at first enter the region of reversible, temporary wilting. Continuation of the water deficiency causes the *permanent wilting point* to be passed, and the plants finally die. However, even a passing but prolonged wilting reduces production, since respiration losses are high.

The available scarce water should at least be optimally utilized through adequate nutrient supplies, if irrigation cannot optimize water supplies. *Fertilization can save water.* This is very important in regions of dry farming and also for overcoming dry periods in humid regions. The resistance of crops to dryness is increased especially through optimal potassium supplies. Potassium increases plant turgor and thus reduces transpiration, i.e., loss of water.

Improving deficient phosphate supplies can also have a decisive influence in dry farming, especially if root growth is promoted at the start. Thus, the roots of plants well supplied with phosphate can penetrate so deeply during the short humid period after sowing that they reach the still moist subsoil and can thus survive the dry period. In contrast to this, the weakly developed roots of plants suffering from deficiency are located exclusively in completely dry soil. This leads to serious yield reductions and even complete harvest failure (Fig. 6–4).

Wilting or drying of plants does not always indicate damage owing to lack of water but may be due to mineral deficiency that is intensified by dryness.

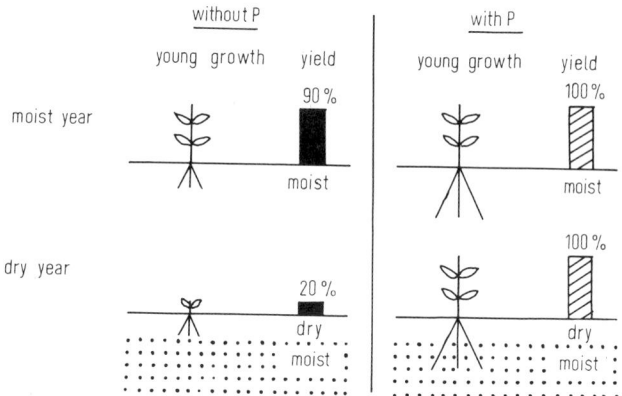

Fig. 6-4. Fertilization with phosphate and resistance to dryness (schematic [159]).

An example is manganese deficiency during drought, causing brown discoloration of plants and simulating lack of water (Fig. 6-5).

A consequence of this is that dryness stress due to lack of water is often intensified by nutrient deficiency. It can therefore be considerably alleviated or even completely prevented by precise fertilization. However, fertilization during drought must be particularly aimed, with minimum amounts of fertilizers, so as not to intensify the lack of water through unnecessary salt supplies.

Heat is frequently combined with dryness and for plants signifies increased respiration relative to photosynthesis (this may eventually cause regressive growth). Nutrition should be optimized by appropriate fertilization to prevent these growth disturbances in hot climates, so that no minimum factor in mineral nutrition unnecessarily reduces photosynthesis.

Intense *insulation* causes additional stress to plants, due to heat and dryness thus induced, if no water is supplied. Short-wave radiation causes increased decomposition of growth substances in the plant. This imposes special requirements on adequate *zinc* supplies which specifically promote the formation of growth substances (auxins) from the amino acid tryptophane.

Shading plays an important role for some crops in connection with heat and radiation, especially under tropical production conditions. A favorable microclimate exists beneath shady trees. The latter seem to be especially necessary when crops are not optimally nourished (Chap. 8.4.2).

Cold can harm plants in various ways and kill them in extreme cases. Damage by cold is frequently related to the formation of ice in cells. All fertilization measures retarding the movement of water through cell membranes thus have a positive effect on the resistance to cold.

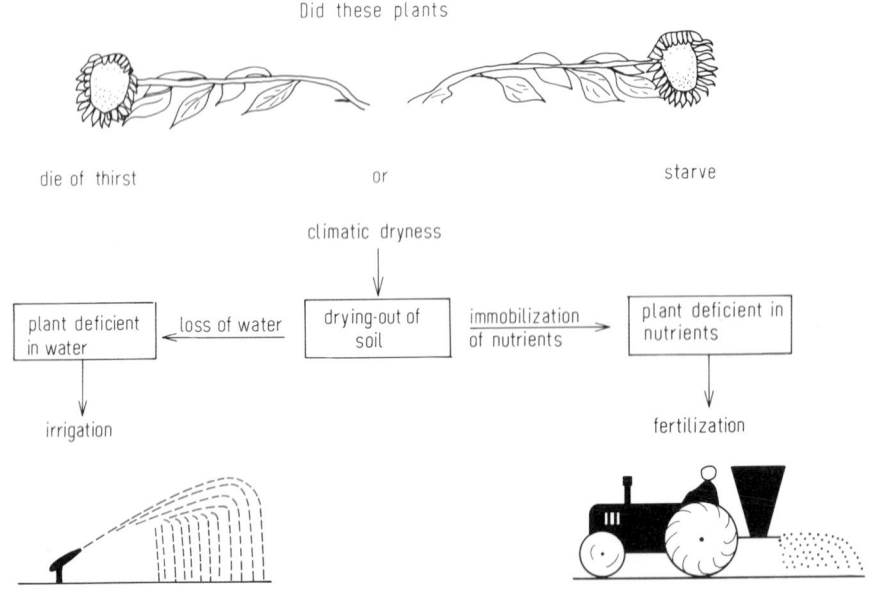

Fig. 6–5. Possible drought damage.

Proper supplies of potassium and phosphate to plants before the winter have proved to be effective fertilization measures. On the other hand, nitrogen excess is somewhat detrimental (normal optimal N-supplies are obviously always desirable).

However, it is not only the major nutrient elements that play a role in the resistance to cold. A slight deficiency of some micronutrients such as manganese or copper can occasionally considerably increase the damage caused by sudden cold at night. These elements apparently increase the resistance of enzymes to damage caused by frost.

Damage due to cold occurs not only through frost but above freezing point as well (between 0 and 12 °C) in the case of sensitive tropical plants. This *chilling damage* is also connected with disturbances of membrane and enzyme function.

Summarizing, increasing the resistance to plant-damaging low temperatures requires optimizing the nutrition. Even a relatively small fertilizer input may completely prevent or at least considerably reduce damage due to frost. However, even the best fertilization is an inadequate protection against extreme temperature effects (especially against rapidly and widely fluctuating temperatures).

Mechanical loads represent a further stress factor. The bending and breaking strength of plants is subjected to particular stresses during storms and the action of heavy weights like water or snow.

An important basic condition for plant *stability* is a balanced nutrient supply, so that abundant carbohydrates (cellulose, lignin) are available for the load-bearing parts of plants. A proper potassium supply has a positive effect on plants in this direction whereas one-sided nitrogen excess lowers the stability and *softens* the tissue. An example of this is the snow-break damage in spruce forests, when branches lack the required strength and break under the snow load.

Mechanical stresses play a role in agriculture especially with *cereals*. The long stalks require the strength elements to be particularly well formed, so as to prevent losses through lodging (especially early lodging). Silicic acid plays a role as a stability factor, in addition to the proper ratios of major nutrients (especially nitrogen and potassium). Fertilization with silicates might therefore be especially important for rice, since *lodging damage* caused by heavy rain storms (typhoons) is frequent in rice-growing areas. An important cause of cereal lodging, besides motion of the air, is the greatly increased ear weight, due to adhering water. In any case the ear weight is greatly increased when yields are higher, so that the load becomes especially large and countermeasures become important.

An important additional stabilization measure is now *shortening of the stalk,* which simultaneously strengthens its lower part. The substances used for this are discussed in Chap. 4.5.1.

6.5.2 Fertilization and Resistance to Plant Diseases

Plants often do not give the yields corresponding to their growth potential, as there are losses caused by plant diseases and pests. However, only a few fertilization aspects can be discussed from among problems in the important boundary region between plant protection and plant nutrition.

It would be ideal if correct nutrition (fertilization) could render crops completely resistant to any attack by bacteria, fungi, and insects. However, this appears to be doubtful, to say nothing of attack by locusts.

The *natural resistance* of plants to bacteria, fungi, etc. can be:

- *reduced* by insufficient or excessive supplies of one or more nutrients: acute and even latent deficiency in general promotes attack since protective mechanisms and defences have been weakened, additionally disturbance of the metabolism promotes the attack;
- *normal,* i.e., corresponding to an average supply of nutrients;
- *increased* through direct or indirect fertilization measures.

Measures to improve the natural resistance should in any case be employed as far as possible, so as to keep additional direct countermeasures for plant protection at an acceptable level.

Proper supply of all nutrients to the plants first of all increases total production, so that certain unavoidable damage can be compensated more quickly. A few examples [167] indicating the significance of certain nutrients will be given:

- *Nitrogen deficiency* often signifies greater possibilities of attack by weak parasites;
- *nitrogen excess* often causes soft, spongy tissue and promotes attack by viruses, bacteria, and fungi, via a frequently unfavorable biochemical constitution;
- *phosphate deficiency* promotes attack by harmful fungi, probably due to an unfavorable N/P ratio;
- *potassium deficiency* primarily causes reduced production of carbohydrates (thinner and weaker cell walls, weaker leaf hair), thus facilitating the entry of parasites; moreover, there are more intermediate products of metabolism, e. g., more sugar instead of starch, thus promoting attack by aphids (and virus infections);
- *calcium deficiency* weakens the strength elements and facilitates the entry of fungal hyphae, for example;
- *silicon deficiency* prevents silification of the leaf epidermis (Chap. 3.4.1), thus facilitating attack by pests.

Little is known as yet about the intake of antibiotics by plants, to increase their resistance to bacterial diseases. However, therapeutic protection of crops by antibiotics produced by fungi in the soil could play an important role in resistance. Fertilization that not only improves nutrient intake by plants but also stimulates soil life as a whole could thus have positive effects through an increased production of antibiotics (Chap. 4.3.4).

The link between nutrition and resistance of plants can be demonstrated in particular when plants are inharmoniously and insufficiently nourished. Plants are especially susceptible to attack by parasites when extreme nutritional conditions weaken the plants through deficiencies and toxic effects, for example, on highly acid soils. An example is the observed attack by parasitic nematodes on clove trees in Sumatra, causing their premature death. The primary cause of this damage was not so much attack by the parasites, as completely inharmonious nutrition (deficiencies of some necessary elements and excess of toxic aluminum). Correction of the nutritional conditions in such a case would be far more important than combating the parasites, whose influence would be negligible if the plants were optimally nourished.

6.5.3 Fertilization and Resistance to Chemical Environmental Influences

Plants are nourished by chemical substances from the environment, but harmful substances from the environment also act on the plants. Fertilization must therefore not only ensure supplies to plants but also increase their resistance to undesirable substances, if necessary.

The following *chemical environmental influences* should be considered:

- Stress on *leaves* caused by toxic substances in the air and leaf spraying in the course of cultivation measures (e.g., plant protection);
- stress on *roots* or entire plants owing to toxic substances in the soil (of either anthropogenic or natural origin).

Damage is of various kinds. Often it cannot be prevented but can be reduced by improving the nutritional situation or by precise additional fertilization.

Toxic Substances in the Air

The resistance of plants to emission damage is becoming important, especially near industrial areas. The extent of damage depends primarily on the intake of these substances, however, the possibilities of partly closing the stomata by applying shrinking agents (calcium, phosphate) are very limited. On the other hand, increasing the production of certain organic substances is very important. The more carbohydrates a plant contains, the better it is protected against sulfur dioxide and fluorine emissions. Certain growth substances seem to increase resistance to the harmful constituent of *smog*, i.e., to peroxy acetyl nitrate (PAN). A higher protein content in some plants increases their resistance to sulfur dioxide, so that proper supplies of nitrogen become important. Calcium has a direct protective function against fluorine, as this plant poison is rendered harmless by the formation of calcium fluoride (CaF_2) which is sparingly soluble [148].

The buffering capacity of the cells plays a role in the resistance to acid emissions. It can also be influenced to a certain extent by fertilization. In any case, there is a certain possibility of increasing the resistance to toxic substances in the air by fertilization. The following basic rule applies here, as to the influence of fertilization on food quality, which will be discussed later: plant resistance is increased when nutrient supplies are raised from the deficiency range to that of an optimal supply.

Substances for leaf spraying, used within the framework of plant protection, should also be mentioned in this connection. They are, in contrast to emissions, applied intentionally to protect plants or damage their competitors (weeds). Leaf spraying often represents a certain (mostly temporary and slight) stress situation for cultivated plants, despite its overall positive effects. This may be important especially when several stress factors act simultaneously, e. g., cold stress, herbicide spraying, and the existence of a deficiency. Stress due to spraying is obviously better tolerated by plants better supplied with nutrients.

Toxic Substances in the Soil

Toxic substances in the soil will be understood to be toxic excesses of substances, which either harm plants or accumulate in plant food to an undesirable level. They may be either "non-nutrients" or nutrients.

Toxic excesses in the soil can be divided into two groups according to their origin, i. e., of natural or of anthropogenic origin.

Naturally occurring excesses detrimentally affecting plant growth, food quality, or the quality of raw materials may be due to:

- Soluble salts in general (see Synopsis 6-8);
- easily soluble boron (Chap. 6.1.4);
- mobile aluminum in highly acid soils (Chap. 5.1.1);
- mobile copper, selenium, arsenic, and nickel in certain soils.

Synopsis 6-8. Effects of High Salt Concentrations in the Soil on Plant Growth.

High salt concentrations in the soil harm the plant through
- inhibition of water intake;
- disturbances of nutrition (extreme ratios of mobile ions in the soil solution (and to a smaller extent) in the plant.

Plants differ in their sensitivity to increased salt concentrations [155, 145]. *Salt tolerance* (relative, for crops) may be:
- high (barley, rape, sugar beet, cotton);
- medium (most crops);
- low (red clover, beans, sugar cane, rice, flax, many fruit trees and berries).

Characteristic magnitudes for estimating salt effects are the *salt content* or the (more easily measured) *conductivity* of a soil or nutrient substrate. *Plants sensitive to salt* still have normal yields when the salt content in a soil of medium particle size is below 0.6‰, or the conductivity is below 2 millisiemens/cm. Yields are already considerably lower when these values are doubled. *Salt-tolerant* plants still have normal yields when the salt content in a soil of medium particle size is slightly below 3‰, or the conductivity is below 10 millisiemens/cm.

It should be stated for comparison that a conductivity of 1 millisiemens/cm (at $25\,^{\circ}C$) corresponds to the following salt concentration in the solution being tested:
- for soil salts: 0.64‰
- for fertilizer salts: 1.0‰.

Anthropogenic excess substances formed in the course of environmental pollution are of various kinds. The above-mentioned examples, which may be of either natural or anthropogenic origin, are supplemented by the following possibilities:

- mobile heavy metals that are also *plant nutrients*, e.g., zinc and copper;
- mobile heavy metals without *nutrient function* in plants, e.g., lead and mercury, cadmium.

Deficiencies of nutrients in the soil can easily be corrected by fertilization, but toxic excesses are usually much more difficult to eliminate or render harmless. The best method of dealing with this problem is doubtlessly to remove the harmful substances or cause their fixation to such an extent by suitable fertilization that they cannot be absorbed.

Removal of toxic substances from the soil is only possible in some cases, for example, easily soluble salts may be leached from the root zone. In this case, displacement may be promoted by applying certain fertilizers, e.g., fertilization of saline and sodium soils with gypsum.

Another, very expensive method is to remove the entire soil layer in which toxic substances have accumulated. This procedure might be considered in small areas, e.g., if only the uppermost cm of the soil is involved. However, many toxic substances cannot be removed to any significant extent. This applies primarily to certain metals. They must therefore be fixed so strongly in the soil that the mobile fraction remains far below the limit of toxicity. A classic example of this is aluminum toxicity in extremely acid soils. Suitable liming can fix the aluminum to such an extent that it becomes completely harmless (as in most soils).

The following possibilities of *reinforcing fixation* by fertilization should be
- Liming to increase the pH;
- increased fertilization with phosphate;
- increased fertilization with humus;
- application of special fertilizers for fixation.

The mobile fractions of many substances can be diminished by increasing the pH. Liming thus reduces toxicity, if such substances occur as undesirable excesses in acid soils. Any other nutrients excessively fixed in this manner could be supplemented by leaf spraying.

Phosphate forms slightly soluble compounds with many metals. This indicates that application of large phosphate doses for immobilizing such substances might be useful. However, this should first be tested for the particular soil in a pot experiment. Large inputs are required in any case. Increased P-supplies also seem to increase immobilization of certain harmful substances in the roots, so that the stress on parts above the surface is reduced.

Fertilization with humus is a further method of immobilizing certain metals in the form of insoluble complex compounds (Chap. 4.3.4). This procedure, although not direct, is nevertheless effective, to lower undesirably high concentrations of available copper in rubber plantations, for example (a high copper concentration in raw rubber reduces its quality).

Application of special *fertilizers for fixation* is still in its infancy. Certain synthetic resins (e.g., the *Lewatit* exchange complex made by *Bayer*) are able to bind some substances so strongly that they can no longer be absorbed and thus lose their toxic properties. New possibilities for countering toxicity seem to be in this direction.

Further fertilization measures to counter toxicity consist in symptomatic treatment, e.g., by supplying iron to leaves suffering from Fe-deficiency caused by zinc excess. Such measures can obviously only alleviate the damage to a limited extent. Further, still largely hypothetical possibilities are to render plants more resistant to toxic substances, e.g., through increasing the filtering capacity of their roots.

6.6 Interpretation of Results of Field Experiments

6.6.1 Success and Failure of Fertilization Experiments

The effects of fertilization on plants via a complex substrate, as a function of climatic factors (i.e., in field fertilization experiments) cannot be calculated precisely in advance. Yield predictions are only possible in controlled systems, i.e., approximately during yield experiments with nutrient solutions in environmental chambers.

However, agriculture needs information on the effects of fertilizers under usual climatic conditions on different soils. Many *fertilization experiments* have been performed for this purpose for more than 100 years and will also be necessary in future. New fertilizers, new plant species and crop sequences, new cultivation procedures and plant-protection measures, higher yield levels, different stress situations, etc. again and again give rise to new problems that must be solved by further field experiments. Answers to some new questions are indeed provided by earlier experiments, and many effects can be predicted with a certain degree of confidence. However, the decisive test is a field experiment especially designed for the specific problem involved.

Fertilization experiments in the field are indeed expensive, but their results enjoy the aura of final and unquestionable authenticity.

The results of field experiments in cases under dispute are analogous to judgements of the last instance, in contrast to theoretical considerations or the results of rapid analytic tests of any type. Hypotheses concerning certain fertilization problems are considered to be confirmed only after passing the final test in field experiments.

Field experiments may indeed be the judgement of the last instance in principle, but there are frequently considerable discrepancies between the ideal state and reality. This is not so much a question of the well-known fact that field experiments by necessity yield variable results, since these sources of error are usually taken into account to a large extent, even if very frequently not sufficiently rigorously. Thus, there is a tendency to consider nonsignificant upward deviations as "obvious" yield increases that have to be taken into account without fail, while downward deviations are easily considered to be "insignificant". Some experimenters apparently have a greater affinity to positive results.

A special *problem* with fertilization experiments is that incorrect deductions are frequently made from correctly performed tests. There would be no need to discuss such cases separately if such inadequately thought-out and wrongly interpreted experiments were rare "exceptions". Rare errors are usually soon corrected by other experiments. Unfortunately, some common views on fertilization problems are based on inadequately interpreted field experiments. In other words: the aura of (apparently indisputable) results of field experiments as the final judgement simulates proof of many things that might better be considered not proven (since they are wrong).

Experimenters sometimes come to erroneous conclusions with the best intentions of arriving at the truth, since they disregard the complexity of fertilization. Thus, obtaining an increased yield by fertilization with nitrogen is usually interpreted as showing that better N-supplies increase growth and yield. This may be the case, but it need not be so. The mechanism of action may be quite different, and this leads to quite different consequences with regard to fertilization (Chap. 2.1.5).

The first classical example of a correctly performed but wrongly interpreted experiment on plant nutrition took place 300 years ago (*Van Helmont*, 1620). Because of the incorrect interpretation it is still highly topical.

Van Helmont supplied only water to a willow shoot in a carefully controlled pot experiment lasting for five years. The shoot developed into a tree, without the soil weight changing significantly. The interpretation was as follows: plants need only water for growth. This interpretation is wrong, since it allowed only for what was obviously added, i. e., water. Of course, the plant needs water, but also substances from the air, which fact was completely ignored, and from the soil. *Van Helmont* did find that the soil weight decreased slightly, but considered this phenomenon to be insignificant because of the slight effect.

The overall theoretical basis for correct interpretation was far too narrow. This fact is also responsible for the inadequate interpretation of many experimental results even today. An experiment can be interpreted correctly only if all possible effects, at least in principle, are taken into account. Moreover, correlations are often interpreted causally, although the factors compared do not interact but only coincide.

It might be considered to be the *success* of field experiments that these methods made it possible to uncover incorrect theories or even only inadequate hypotheses.

On the other hand, it might be called the *failure* of field experiments that despite large inputs they provide some results considered to be absolutely correct on account of the reputation of field experiments, although they may be inadequate or even useless. This statement will be explained by means of some examples. The general models employed were derived from specific experiments (Chap. 6.6.7).

In conclusion, on the basis of available experience, the following points should be taken into account or required for a correct interpretation of field experiments:

1. Clear definition of the problem and corresponding design of the experiment;

2. proper consideration of all possible connections, even if allowing for them in the experiment is not always necessary;

3. a maximum of additional information gained from the experiments, e. g., through soil and plant analyses, because only this permits exhaustive interpretation;

4. the final deductions should fit the measured results, e. g., they should *not* read as follows for an experiment on fertilization with N, in which only the yield was determined:

- the experiment demonstrated a clear N-action, but
  rather:
- the experiment demonstrated a clear N-action, provided that there were no special side effects, which were not investigated.

6.6.2 Examples of Possible Interpretations

Experiment A
Additional fertilization with N in the form of urea during the principal growth period provides a significant yield increase on a soil with pH 6.5.

Interpretation: Better N-supplies have improved growth and thus increased the yield.

Alternative: N-supplies were adequate even without additional fertilization, but manganese supplies were in the region of latent deficiency and were temporarily improved by acidification by the N-fertilizer during the principal growth period. This caused the yield increase. Direct fertilization with Mn would have been better.

Experiment B

Fertilization of rape with superphosphate provides a significant yield increase in comparison with Rhenania or Thomas phosphate.

Interpretation: Water-soluble phosphate is superior to P-forms not soluble in water, under these conditions.

Alternative: The phosphate component has approximately the same effect with all three fertilizers, but P-deficiency was accompanied by S-deficiency, which was eliminated by the sulfur fertilizer superphosphate. Sulfur supplies could also have been ensured in a different way.

Experiment C

Fertilization with Thomas phosphate increases the yield even when the concentration of available phosphate in the soil is exceptionally high.

Interpretation: The content of available phosphate in the soil was insufficient; there was probably an incorrect assessment because of the method of soil analysis. Additional fertilization with P in any case increased supplies and thus the yield.

Alternative: P-supply was already optimal, as indicated correctly by soil analysis, but Thomas phosphate with its magnesium content increased the inadequate Mg-supplies and thus the yield. Direct fertilization with Mg would have been better and probably even more effective in this case.

Example D

Potassium fertilizer provides increased yields in special experiments on heavy soils, not only with normal doses, but also when doses are several times as large.

Interpretation: K-supplies were for some reason insufficient at the normal fertilization level (perhaps due to fixation), so that much more intensive fertilization increased the yield.

Alternative: K-supplies were already optimal with normal fertilizer doses. However, yield increases on this heavy soil were due to the structure-improving (flocculating) effect of a high salt concentration. This effect is largely due to improved aeration. Application of structure-improving fertilizers would have been more effective and less problematic, in view of possible salt damage.

Example E

Potassium, applied either in normal or in double doses, provides no yield increase.

Interpretation: The soil was sufficiently supplied with potassium, and fertilization with potassium could therefore have no effect.

Alternative: Potassium was indeed deficient, but the high fixation capacity of this soil rendered both normal and doubled doses ineffective. Melioration fertilization with considerably larger doses would be indicated.

Example F

Fertilization with 50 kg/ha of P in the form of either superphosphate or Novaphos provides equal yield increases.

Interpretation: The two fertilizers should be considered as equivalent for the site conditions involved.

Alternative: Optimal P-supplies were in this case already provided by 30 kg P, so that the equivalence of the fertilizers is only apparent. Perhaps one half of the superphosphate dose would have been sufficient, whereas two thirds of the Novaphos dose would have been necessary. The design of the experiment was not such as to permit judging of any equivalence.

Example G

Fertilization of chlorotic plants with iron sulfate applied as soil dressing brings about a lush green color and better plant growth in a relatively short time.

Interpretation: An iron deficiency was eliminated by fertilization with Fe.

Alternative: Fe-supplies were optimal, but a manganese deficiency was eliminated by mobilization of manganese in the soil by iron sulfate (via reduction processes). Direct fertilization with manganese would have been more correct and effective.

Example H

Belated leaf spraying with a mixture of micronutrients causes a yield increase of only 2 dt grains; although profitable, this does not meet (the higher) expectations.

Interpretation: Deficiency in this field was only slight; opinions were probably far too optimistic.

Alternative: Spraying was much too late to effectively correct a deficiency that had already existed for weeks. Timely increase of supplies would have considerably increased the yield. The possibilities of applying certain trace elements on this field cannot be judged too optimistically.

7 Fertilization of Agricultural Crops

Practical fertilization of the various crops is a large subject. Different plant and site condition requirements have to be considered. Nutrients are abundant at some sites, and deficient at others. Some plants have modest requirements, others make particularly large demands.

However, the more plants are considered in their totality, the more the *uniformity of their requirements* becomes noticeable. The biochemical uniformity of higher plants is basically much more obvious than the many, usually slight, deviations from the norm. An apparent multiplicity, in fact, conceals a far-reaching qualitative and quantitative similarity of nutrient requirements and general demands made on the nutrient substrate. Thus, even the *least demanding* plants become demanding when suitable strains are to have high yields.

Moreover, many soils are quite similar despite their different origins. Similar substrates and similar requirements thus imply uniform fertilization in principle. Many fertilization rules apply to plants at many sites, at least as long as yield levels are average.

However, many specific requirements beyond a largely uniform basic fertilization are to be met when high yields are desired. This is high-performance fertilization.

The special features of fertilization of different plants are clearly brought out by discussing plants in groups with similar requirements. For the sake of clarity, many fertilization details are given in the form of systematically arranged recommendations. Explanations are given under the relevant reference terms in other chapters. The reader is referred to summaries published on the fertilization of plants [36, 44, 55, 194, 213, 41, 190, 200, 220].

7.1 Cereal Crops

The aim of fertilizing cereals is to produce high yields of grains containing the principal constituents (starch and protein) for food and fodder.

By way of example, the fertilization of cereals will be discussed in detail for *wheat* as the most important bread crop, as well as for *maize* and *rice*. For intensive fertilization to work, there must be optimal soil and crop management conditions, especially when high-yielding strains are grown. Sufficient water must also be available. Cereals can tolerate a slight deficiency of water fairly well, but yields are reduced. The transpiration coefficient of most varieties of grain is 350 to 400; this figure represents the number of liters of water required for the production of 1 kg of dry matter. Moreover, fertilization of cereals should be coordinated with plant-protection measures, including weed-killing, as a labor-saving measure. The quality of the grains must also be taken into account in the case of bread cereals [181, 216].

7.1.1 Wheat

Fertilization of *wheat* (triticum sp.) will be discussed with special reference to winter wheat. Modern strains have a high yield potential (at least 100 dt grains per ha), which in some fields is already fully exploited. The evolution of wheat yields is shown in Fig. 1–3.

Obviously fertilization for yields of 80 to 100 dt/ha will follow different rules than those formerly valid for yields of only half those amounts (for yield parameters *see* Synopsis 7–1). Not only must more fertilizer be applied but the rules to be followed differ. Fertilization must first of all be more differentiated and according to a more comprehensive plan (Chap. 1.1.2) [216, 204].

Most problematic of all is *fertilization with nitrogen*, this being the "principal motor" of yield increases and quality improvements [34, 193, 221, 225]. It is especially important that plant requirements be covered during the principal growth period. The available nitrogen present in the soil at the beginning of spring should also be taken into account (Chap. 5.4.1). N-fertilizers commonly in use act too quickly in comparison with the temporal demands of wheat requirements. The N-doses must therefore be divided over the individual growth stages. Such graduated dosage of the amounts required at any one time requires that the fertilizer be immediately available after application. However, dryness can delay fertilizer action, so that a deficiency may occur despite timely fertilization.

There are two alternatives for fertilization with N, depending on soil properties, climate, etc.:

1. Especially *large N-doses* at the beginning of growth, possibly followed by a late fertilizer application.
Advantage: No deficiency up to the end of the principal growth period.
Drawback: Luxury consumption of N at the beginning, i. e., inharmonious initial nutrition and fertilizer losses in spring.

Synopsis 7-1. General Data on Wheat (Winter Wheat).

Grain *yields:*   40 to 100 dt/ha
composition of grain (contents):
dry matter: 86%
starch   60 to 85%
raw protein: less than 12%
                  (medium to low)
                  more than 13%
                  (high, up to 20%)
gluten: 21% average
grain/straw ratio: 1:1.3 to 1:1.5
dry matter in straw: 90%

Yield parameters:
stalks: 500 to 800 per m^2
grains per ear: 30 to 40
weight of 1000 grains: 40 to 55 g

Development stages of grain (FEEKA-scale in numbers, BBA-scale in letters [35, 181])

1 { A = emergence
    B = first-leaf stage
    C = second-leaf stage
    D = third-leaf stage
2 = E = beginning of tillering
3 = F = main tillering
4 = G = terminated tillering
5 = H = beginning of shooting
6 = I = one-node stage
7 = J = two-node stage
8 = K = appearance of last leaf
9 = L = ligula stage

10 = M = opening of last leaf sheath
10.1 = N = beginning of ear appearance
10.5 { O = end of ear appearance
       P = beginning of blossoming
       Q = end of blossoming
10.8 { R = formation of grain
       S = milk ripeness
       T = wax ripeness
11 { U = yellow ripeness
     V = full ripeness
     W = dead ripeness

2. Moderate fertilization before the beginning of growth, with two or three supplementary applications.
Advantage: In theory, better matching to the demand curve.
Drawback: Risk of deficiency periods if fertilizer does not act immediately.
Properly dosed N-fertilizer increases the yield via the yield-determining factors:

• Tillering (stalks per plant or per unit area);
• number of earlets per ear;
• number of grains per ear;
• weight of 1000 grains (weight of 100 liters).

However, wheat has a high compensating capacity and can in part substitute one factor by another. The influence of nitrogen can be summarized in general by stating that 1 kg N produces above 15 kg grain if N is applied in spring but only half this quantity when it is applied later on (Chap. 5.6.2).

The dependence of N- and K-requirements on the growth is shown in Fig. 7-1. About 50% of the phosphate is required in the second half of the growth period. Cereal nutrient removals can be indicated only as average values (Chap. 5.5.2) (Table 7-1).

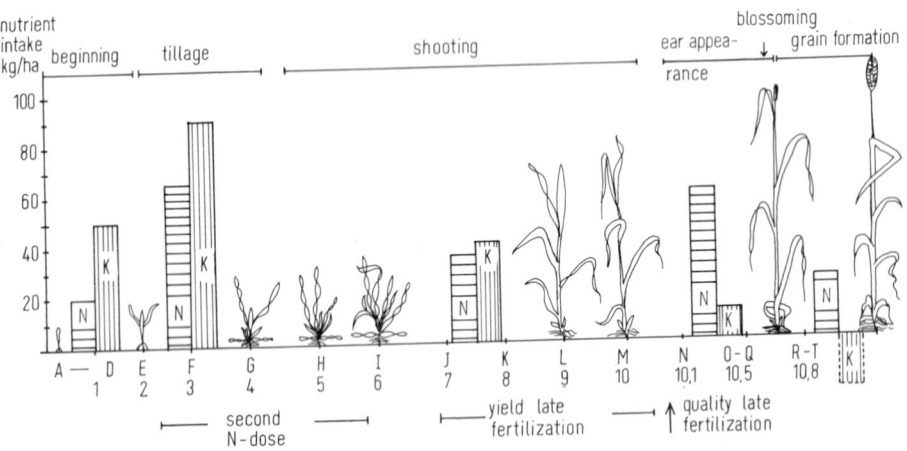

Fig. 7-1. N- and K-intake of wheat during development (separate intakes during 5 growth periods, according to [182]).

Recommendations for the Fertilization of Wheat

1. Modern wheat strains are *demanding* plants; their high yield potential can only be realized if abundant mineral nutrients are available to them without interruption.

2. A precondition for rapid supplies of nutrients is, first of all, a proper state of liming, i.e., a slightly acid to neutral soil reaction (according to soil type). Liming need not be undertaken directly for wheat but may be carried out within the framework of crop rotation (best with Mg-containing limes).

3. Additional *organic fertilization,* e.g., with farmyard manure, is not required. Fertilization with straw, within the framework of crop rotation, necessitates an N-dose corresponding to 1% of the straw weight as nitrogen compensation to avoid the detrimental N-blockade (Chap. 4.3.4).

4. The required amounts of *phosphate* and *potassium* should be applied before sowing; i.e., in autumn for winter wheat. However, the doses need not be the total amounts. The quantities required are indicated in Tables 5-11, p. 213, and 5-14, p. 215. The fertilizer form is of minor importance.

Table 7-1. Mineral contents and removals of wheat and maize (average values according to [203], partly modified and supplemented).

plant	yield (85% dry matter)	content in ‰ of fresh substance (= removal in kg/ha per 10 dt)						content in ppm of dry matter (= removal in g/ha per 10 dt)					
		N	P	K	Ca	Mg	S	Fe	Mn	Zn	Cu	B	Mo
wheat grains		20	4	5	0.5	1.5	(2)	70	(40)	(50)	5	(4)	(0.3)
straw		4	0.4	7*	2	1	(1)	70	(40)	10	5	(4)	(0.3)
maize grains		15	3	3	0.2	1	(1.5)	(40)	(30)	(20)	4	(4)	(0.2)
straw		10	2	10	4	2	(1)	(40)	(30)	(20)	(4)	(4)	(0.2)

removal per harvest unit of 10 dt grains and corresponding quantites of straw (grains/straw ratio for wheat = 1:1.3, for maize = 1:1.5)

plant		kg/ha						g/ha					
		N	P	K	Ca	Mg	S	Fe	Mn	Zn	Cu	B	Mo
wheat		25	4.5	14	3	3	(3)	160	100	65	11	(9)	(0.7)
maize		30	6	18	6	4	(3)	100	75	50	10	(10)	(0.5)

* values sometimes twice as high

5. *Requirements* of mineral nutrients are *maximum* during the period from tillering to shooting, and their totals are definitely higher than removals. One half of the requirement is absorbed after blossoming, when fertilization with N is intensive. The potassium content is maximum at the wax-ripe stage and decreases thereafter.

6. The following guidelines refer to *fertilization* of winter wheat with N: Application in autumn during sowing should be at the low level of 0 to 20 kg N per ha, depending on the N-residue from the previous crop.

Two- or three-part doses are indicated during the actual growth period (Fig. 7-1, p. 293) [235, 181]:

- First dose before or at the beginning of vegetation at the end of winter; 30 to 100 kg N, depending on nitrate reserves in the soil (Chap. 5.4.1);
- second dose in the period between tillering and the beginning of shooting (at a plant height of 10–20 cm): 10 to 40 kg N applied as top dressing or leaf spraying, possibly in combination with CCC-treatment;
- third dose: 40 to 70 kg N, either as late yield fertilization towards the end of shooting, or as quality late fertilization shortly before ear appearance.

The choice of the N-form is not crucial for wheat. Nitrate acts most rapidly in late fertilization, but the difference between it and, e.g., lime ammonium nitrate is slight. Liquid and solid N-fertilizers are equivalent but the former are sometimes simpler to apply. However, spraying of urea should be avoided after the appearance of ears (corrosion).

7. Large doses of N-fertilizers should be avoided because of the risk of *lodging*. As a preventive measure, the stability of wheat should be increased with stalk stabilizers (CCC). CCC is applied at the time of tillering and can be combined with N-fertilizers (Chap. 4.5.1).

8. The *trace elements* manganese, copper, and sometimes also zinc easily become deficient because of the high pH-values required. This limits yields, irrespective of the amounts of N-fertilizer applied. During the period of highest requirements, it may therefore be useful to reinforce natural mobilization through acidifying fertilization with N (Chap. 2.1.6) or provide direct replenishment by fertilization.

9. Further, correct fertilization (balanced nutrition) of wheat is required to increase *resistance* both to diseases that cause yield reductions and to stress effects, e.g.,

- cold causing winter killing;
- storms resulting in lodged grain;
- spraying of weed-killers that might cause spottiness.

The resistance of wheat depends on proper supplies of all nutrients and the avoidance of one-sided oversupplies of N.

10. The quality of wheat grains for use in bread (baking quality) is improved by late fertilization with N. This increases the content of raw protein. The better baking potential is chiefly due to the increased content of gluten but also to its improved quality (Chap. 9.3.1).

Fertilization of Summer Wheat

Summer wheat has a short vegetation period and is essential in the crop sequence, if winter wheat cannot be sown in time before winter. Summer wheat can also have high yields which, however, are 10 to 20% lower than those of winter wheat.

Fertilization is largely the same as for winter wheat; the fertilizer quantities may be correspondingly smaller.

7.1.2 Rye, Barley, Oats

Fertilization of Rye

Rye (secale sp.) is used for different purposes in crop rotation:

a) Rye, as a *low-demand* plant for poor sites, has a particularly high capacity for absorbing nutrients, and is therefore suitable for light, dry, and acid soils. Requirements with regard to water supply, liming state, and mineral fertilization are small, but yields are also correspondingly low.

b) Rye as a *high-yielding cereal:* rye attains high yields (50 to 80 dt/ha) approaching those of wheat, when the soil is fertile. Fertilization should therefore be similar to that of wheat (winter rye like winter wheat). Late fertilization of rye with N has a particularly desirable effect on the yield (for fertilizer quantities see Chap. 5.4.1).

Fertilization of Barley

The determining factor for the fertilization of *barley* (hordeum sp.) is whether it is to be used for fodder or brewing. Fodder barley is grown as winter or summer barley with the primary purpose of producing starch and protein. Fertilization is largely the same as for wheat. However, barley depends to an even greater extent on good soil structure, and thus a correct liming state, since an insufficient air supply in the soil easily causes yellow discoloration.

Summer barley, like oats, is susceptible to manganese and copper deficiencies. N-supplies should be abundant during later growth stages, too. However, for this reason the risk of lodging should particularly be taken into account, since stalk stabilizers cannot yet increase the stability of barley to the desired extent. Winter barley requires more nitrogen before winter than wheat.

Brewer's barley, grown in summer, yields grains rich in starch but containing not more than 9 to 11% protein (raw protein), and having a high extract yield. The proteins should consist mostly of enzymes and as little as possible of reserve protein.

Brewer's barley can largely be fertilized like fodder barley. However, fertilization with N should not exceed 40 to 50 kg per hectare. N-reserves remaining from the previous crop should be taken into account and, late fertilization with N is contraindicated for this reason.

Fertilization of Oats
Oats (avena sat.) serve mainly for fodder and, demanding little, can also grow on poor sites, but then yields are low too. On the other hand, oats are also desirable as a *restoration crop* in one-sided intensive crop sequences. However, oats yield only, e.g., 40 dt/ha in this case, despite better growth conditions. Nevertheless, modern strains are quite capable of yielding 60 to 80 dt/ha under optimal growth conditions.

The problem of insufficient oat yields in crop sequences with wheat/sugar beet/rape should be considered in detail. Particularly progressive farmers were already complaining around 1900 that they were forced to stop growing oats after their fields had just been improved through marling. Oats suffered from manganese deficiency and gave only low yields. This was not yet understood at that time.

The present-day problem seems to be similar. Usually, there is no acute deficiency with obvious symptoms, but manganese, copper, etc. are often in short supply, and this apparently limits the yield. Certain nutrients can be absorbed to only a small extent, so that the deficiency region is easily reached. This drawback should be compensated by fertilization (see recommendations). Oats are well suited for diagnosing deficiencies, since they show relatively clear symptoms as indicator plants.

Oats are sensitive to dryness, which causes the growth of empty earlets, but much so-called drought damage is in fact due to a deficiency of some mineral nutrient, which becomes acute only because of dryness, and can be prevented with better supplies (Chap. 6.5.1).

Recommendations for the Fertilization of Oats
1. The recommendations for wheat largely apply to oats, as well, but some special features should be taken into account.

2. Oats prefer (moderately) *acid* soils, since they have a low absorption capacity for the trace elements manganese, copper, etc. The pH of light soils should therefore be about 5.5, and that of medium soils not much above 6. This is quite possible with a suitable crop sequence (potatoes, oats, etc.). However, a deficiency of trace elements may easily limit the yield if oats are combined with intensively cultivated crops like wheat and sugar beet, which require higher pH-values, at least on medium soils.

3. Supplies of some *trace elements* must therefore be increased to obtain high oat yields. Two methods are best combined for this:

- Increasing the natural supplies of manganese, copper, and zinc by applying acidifying fertilizers. This creates localized "nests" of acid reaction, from which the roots can absorb mobilized nutrients (Chap. 2.1.6).
- Additional supplies of trace elements that are not present in sufficient amounts (according to diagnosis), e.g., by leaf spraying (repeatedly if necessary).

4. Oats are also sensitive to *magnesium deficiency*. Magnesium supplies should therefore be ensured by means of Mg-limes (only if liming is necessary) or other fertilizers containing Mg.

5. Oats require much N for high yields; division of the dosage is indicated. Luxury supply should be avoided, even if only to eliminate the risk of lodging. The use of stalk-shortening agents (CCC) is also advisable for the same reason. Late fertilization increases the protein content, which is advantageous when the oats are used for fodder.

7.1.3 Maize, Sorghum

Fertilization of Maize

Worldwide, *maize* (zea mays) is the most important cereal crop after wheat and rice. It also plays an important role as green fodder for silage. Maize is a plant with a high yield potential, i.e., with high photosynthesis performance because of its effective CO_2 assimilation as a C_4-plant [19]. Grain yields of more than 100 dt/ha have been obtained [179, 208].

Water requirements are relatively low, at a transpiration coefficient of 300 (300 l water per 1 kg dry matter). However, they are still considerable in view of the high yields. In particular, maximum requirements in June and July must be covered to prevent growth interruptions.

Precise fertilization into the root zone may be advisable with maize, in view of the low plant density (7 to 10 plants per m², or up to 16 plants per m² with silage maize). It is indicated especially when there are:

- large distances between rows (better possibilities for mechanized weed-killing);
- only moderate nutrient supplies.

Placement of at least part of the fertilizer in the vicinity of the roots is effective particularly for nitrogen and phosphate (Chap. 2.2.3). The same should hold true in theory for the other nutrients when they are in short supply.

Maize has large nutrient requirements corresponding to its high yield (Table 7-1). Plant analysis may be useful for diagnosing requirements and estimating minimum removals of silage maize (Table 5-7).

The quality of the maize grains plays an important role in food for human beings in large areas around the world. Maize grains contain 85% dry matter, of which 10% is raw protein. However, the nutritional value of this protein is only moderate. Larger and late doses of N-fertilizers in particular increase the content of the prolamine *zein*, which is poor in the essential amino acids lysine and tryptophan. Fertilization with N thus raises the nutritional value only to a limited extent. New strains (opaque maize) contain about twice as much lysine.

Recommendations for the Fertilization of Maize

1. Maize is a *warm zone* plant and is grown as grain maize or silage maize, both of which have similar fertilizer requirements. Optimal nutritional conditions must be provided in cooler zones, especially during initial growth, namely:

- Proper supply of P for rapid root growth and a higher resistance to cold;
- proper supply of N for a rapid formation of leaf mass;
- proper supply of K for better water status and a higher resistance to cold.

2. *Organic fertilizers* are well utilized by maize, but should be applied in autumn. However, precise application of farmyard or green manure is not necessary for maize.

3. Maize demands little in the way of *soil reaction*, but the respective optimal pH-value should always be maintained (Chap. 5.1.1) for reasons of soil structure.

4. To improve the all important *initial development* requires either a soil that is generally abundantly supplied with nutrients or placement of the initial fertilizer dose in the root zone (about 5 cm beside or below the seed). This applies mainly to nitrogen and phosphate, which should be supplied in watersoluble form. Fertilizer doses of 40 kg N and 60 to 150 kg P_2O_5 have proved effective [188].

5. *Nutrient requirements are maximum* in the month that blossoming occurs. The daily intake of a maize field at the peak of this critical period is up to 5 kg N per ha (for the fertilizer quantities see Chap. 5.4).

6. N-supplies may and should be abundant, especially as there is hardly any risk of lodging. Additional later fertilization is indicated when the plants are knee-high, with the fertilizer being placed beside the rows. Fertilizer grains should not enter the leaf axis because of the corrosion risk (for N-quantities see Chap. 5.4.1).

7. Supplies of *magnesium* and *trace elements* (especially zinc) should also be abundant, both to increase the resistance to cold and because of the large requirements, which in the case of grain maize considerably exceed removals (Table 7-1).

8. The *fertilizer form* of most nutrients is of minor importance, as long as good availability of the nutrients in the soil is ensured. However, water-soluble phosphate is best utilized if placement is in the root zone.

9. *Cold damage* due to night frost, sometimes even at temperatures above freezing point, may be considerable in regions with cool spring seasons. This manifests itself in growth interruptions and damaged leaves. Abundant initial supplies of phosphate and potassium, and other nutrients too (probably especially of trace elements), play an important role in increasing the *resistance of cold*. The stress phase induced by weed-killer sprayings should also be overcome more quickly when nutrients are more abundant.

10. The *quality* of the grains, for food, can be improved by intensive or late fertilization with N, since the content of raw protein, which is 10% on average, is thus significantly increased. However, the quality of the protein (content of vital amino acids such as lysine) is barely improved.

Fertilization of Sorghum

Sorghum (panicum, sorgum, etc.) is a collective term for several plant species grown in warm zones mainly as cereals. Some strains are also used as green fodder because of their large plant mass, e.g., sudan grass.

Sorghum plants are thermophilic, but demand little water and nutrients. Their high resistance to dryness and heat makes them popular crops in tropical and subtropical regions. Their resistance to drought is the result of a deep root system and a low water consumption. They also have a high nutrient mobilization capacity. Photosynthesis in these plants is particularly effective (like maize). The grains contain 85 to 90% dry matter, with 10 to 14% raw protein and relatively large amounts of vitamin B_1 [288].

Fertilization should be matched to water supplies; it should be limited (it may even have detrimental effects) when water is scarce or the supply is unreliable (e.g., in case of nonirrigated crops with long dry periods):

- Fertilization with N is worthwhile only if doses are small (e.g., 0 to 40 kg/ha);
- fertilization with P promotes the root growth that is so important when supplies are insufficient;
- fertilization with K is usually unnecessary, since many soils in arid zones contain sufficient potassium for low and medium yields.

Special sorghum strains and newer types are indicated when the water supply is adequate. Such crops have a high growth performance and an appreciable yield potential with regard to grains and green mass. Fertilization of such sorghum strains should be the same as for intensively grown maize.

7.1.4 Rice

Rice (oryza sat.) is an important food for one half of the earth's population. Low rice yields of 1 to 2 t/ha have been obtained since ancient times, but the average yields are 4 to 6 t/ha ("brown" rice) in countries with a proper cultivation technology. Certain high-yielding strains have already given yields of 10 to 12 t/ha on some fields [199] ("indica" rice in tropical and "Japonica" rice in temperate zones).

Fertilization has contributed much to the yield increases possible with the new strains. In principle, the rules for wheat apply to rice: nutrient supplies must match the large requirements, fertilization should include the entire range of nutrients, etc. (Chap. 7.1.1). However, the specific cultivation conditions require some particular features to be taken into account with regard to paddy rice, which will be primarily considered.

Paddy rice grows best on soil covered with water, and therefore requires large quantities of the latter (600 liters per kg dry matter). Those fertility properties of rice soils that are linked with the special requirements of water management are therefore particularly important, namely, permeability and redox potential.

Rice soils should, on one hand, be only slightly permeable so as to retain water, but on the other hand, should provide sufficient aerated root space for the plants. Compactions are therefore desirable in the subsoil, but not in the upper layer (at a depth of 10 to 20 cm). They may be due to natural processes of soil formation (bonding by iron oxides), or, intentionally or not, to soil-working measures (plough pans). Flooding of the mostly acid rice soils increases the soil reaction up to the neutral point. This contributes to the mobilization of silicic acid.

Reduction offers considerable advantages, as long as reducing conditions in the rice soil are pronounced no toxic hydrogen sulfide accumulates: iron and manganese are mobilized by conversion to the bivalent form (rice requires much Fe and Mn). Part of the phosphate is set free and thus becomes available when iron trivalent phosphate is reduced to the bivalent form. Hydrogen sulfide is detoxified by precipitation as iron sulfide. However, nitrates are easily denitrified, so that ammonium fertilizers should be preferred. Rice itself contributes to a certain oxidation of the upper soil layer by conveying air via a pipe system to the roots.

The principal problem is undoubtedly the N-supply [299, 331]. Some natural N-supply of rice fields is ensured through the fixation of N by blue algae which flourish under "paddy" conditions.

High-yield strains react positively to increased doses of N-fertilizers. Rice requires much N during shooting, and after blossoming it also needs abundant N-supplies for the particularly active upper leaves (Fig. 7–2). Many fertilization

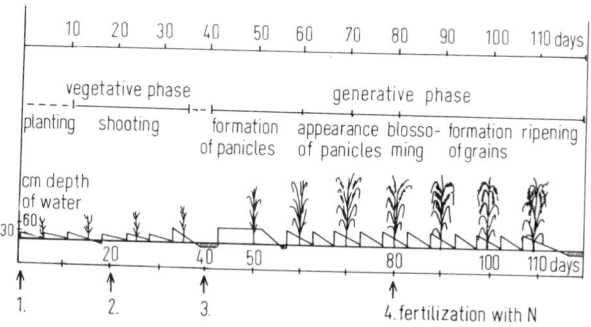

Fig. 7-2. Irrigation and fertilization of rice (according to *HSU*, cit. in [199]).

methods have been tested in attempts to achieve higher yields. This is illustrated in Fig. 7–3.

However, in future the trend will have to be towards more precise fertilization, rather than the application of large amounts of organic fertilizers. As in the case of wheat, this is likely to be possible only through improved diagnosis. The International Rice Research Institute (IRRI) has already elaborated some limit values for the plant analysis of rice [299]. Some toxicity limit values are indicated in Table 5–8, p. 207.

Recommendations for the Fertilization of Rice (Paddy Rice)

1. Modern strains can realize their high yield potential only if the preconditions with regard to supplies of water and nutrients are met. *Irrigation* must be

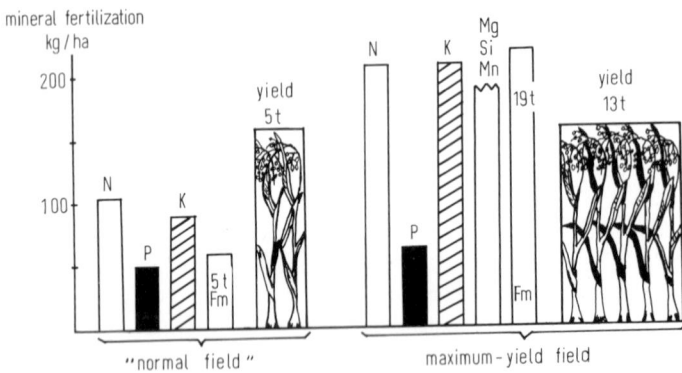

Fig. 7-3. Increased rice yields in Japan, due to more intensive and better fertilization (fm = farmyard manure; according to [231]).

matched to the strain and site conditions involved. The soil must have a well aerated (oxidized) zone on top of a reduced zone.

2. The *soil reaction* should be approximately neutral. Rice soils represent quite a heterogeneous group of soils, ranging from acid to alkaline, but timely irrigation before planting often ensures that flooding automatically leads to an almost neutral pH-value.

3. *Organic fertilizers* (farmyard manure, compost, or green manure) often substantially increase rice yields. They act both as a slow-flowing source of nutrients and through their soil-improving effects [226]. Large doses of compost, of 10 to 20 t/ha, apparently improve several minimum factors through complex effects. This, however, does not prove that organic fertilization is essential.

4. *Phosphate* has high mobility in rice soils and is not lost to any significant extent, so that adequate supply can be ensured by basic fertilization. Many P-forms have positive effects on rice.
The common potassium salts can be used for K-fertilization. A single application during planting is sufficient if the storage capacity of the soil is adequate. A later additional application of potassium may be indicated on light soils when requirements are particularly high.
Fertilizer requirements per 10 dt rice yield are:

• 10 kg phosphate (P_2O_5) per ha;
• 15 kg potassium (K_2O) per ha.

5. Fertilization with nitrogen poses many problems. Ammonium fertilizers or fertilizers yielding ammonium after reaction (e. g., urea) are primarily indicated

as N-forms. Ammonium should be preserved in the soil as much as possible. Nitrate is denitrified to a large extent. This could be the principal reason for the relatively low utilization rate of N in rice soils (less than 50% when applied as soil dressing). N-fertilizer is therefore best introduced into the (lower) reduction zone of the soil, to preserve the ammonium form. This is possible through ploughing-in before planting. However, special equipment is necessary for supplementary fertilization of the growing crop. Fertilization with urea in the form of *mud balls* (small spheres made by the farmer with soil silt and compost), placed at a depth of 10 cm, is effective, but also very expensive. N-action is doubled in comparison with that obtained by application as soil dressing.

6. The required *amounts of N* vary within wide limits, depending on the utilization rate (50 to 150 kg/ha for medium yields). Division of the N-dosage is advisable in order to better match requirements and prevent losses:

- The principal dose should be applied at the time of planting in the case of early ripening strains; when reserves in the soil are small;
- a second, large dose should be given at the beginning of panicle formation for late ripening strains, and a third dose later if requirements are large.

A possible distribution scheme with four N-doses is shown in Fig. 7–2.

7. The *risk of lodging* often limits the amounts of N applied for maximum yields. Fertilization measures for maintaining the stability (especially in view of the possibility of typhoons) are therefore important. Abundant K-supplies should be ensured together with a sufficient intake of silicon (for which separate fertilization may be necessary).

8. Iron and manganese among the *trace elements*, are abundant in most rice soils because of the reducing conditions. This means that there may even be the danger of harmful excesses (Table 5–8). Zinc deficiency might be more important than has been thought until now on the basis of an occasional acute deficiency. Boron deficiency seems to be unimportant, as with cereals in general.

9. Increasing the *resistance* of rice to diseases is very important. The often extreme growth conditions (combination of deficiencies and excesses of some nutrients) cause several "physiological" diseases in rice. There is an obvious relationship between nutrition and resistance [227]. One-sided N-excess is harmful, but abundant supplies of potassium and silicic acid (Si) are desirable.

10. Optimum *quality* of the rice grains is achieved when nutrient supplies are balanced. Late fertilization with N, in particular, increases the protein content under these conditions. This is of considerable significance for protein supplies in certain regions.

7.2 Root and Tuber Crops

These plants serve primarily for the production of carbohydrates (starch and sugar). They, in addition to cereals, thus provide important basic foodstuffs and fodder. The following classification is indicated with regard to cultivation and fertilization:

- Root crops: sugar beet ⎱ beta beets
               fodder beet ⎰

               turnip   (brassica-beet)
- Tuber crops: potatoes, Jerusalem artichoke (in temperate zones)
               tropical tuber crops.

Agriculturally, these are root crops that put very high demands on the soil structure. Good crumb structure and aeration of the soil can indeed be promoted by soil working (hoeing, etc.). However, the basic preconditions are more important, i. e., a proper liming and humus state of the soil.

7.2.1 Sugar Beet, Fodder Beet, Turnip

Fertilization of Sugar Beet *(beta vulg.)*
The aim of *sugar-beet* fertilization is to achieve high yields of beet and especially of sugar. Fertilization (in particular with nitrogen) must therefore aim at a balance between large amounts of beet and a large sugar content. A knowledge of the nutrient requirements of sugar beet [209] and numerous fertilization experiments permit the derivation of general fertilization recommendations. However, some problems also become apparent.

Fertilization experiments with sugar beet provide results that are in part significantly different. Thus, different amounts of farmyard manure are recommended, depending on the site. No useful rule can be derived from this. In fact, the quantity as such is quite unimportant, but not the effects achieved. Farmyard manure does increase the yield but can be substituted by other means.

The large nutrient requirements of sugar beet in the third month should be noted. They may lead to temporary deficiencies in dry periods, since the content of available nutrients in the soil is then reduced. High yields presuppose an adequate water supply, so that no stress situation arises in this respect. However, a smaller rather than an abundant water supply seems to be indicated towards the end of the growth period.

Fertilization with N poses particular problems. The generally recommended total amount is 140 to 200 kg N per ha (Table 5-9). N-supplies should:

- not be greatly exceeded (lowering of sugar content and recovery from the sap);
- not drop below critical nitrate content (yield reduction) (Fig. 7-4).

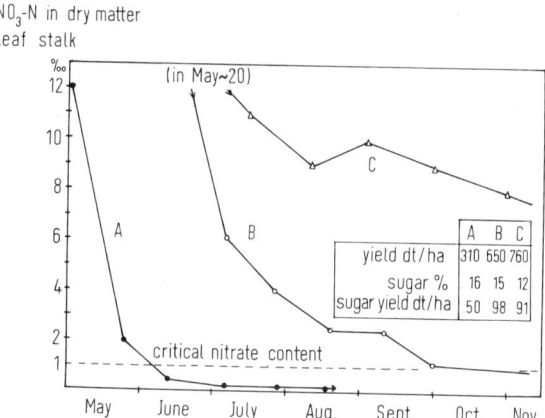

Fig. 7-4. Nitrogen supplies to sugar beet and yield (field B with precise fertilization with N has highest sugar yield, field A suffers from N-deficiency, field C from N-excess; analyses of young, fully developed leaves; according to [230]).

The nitrogen contained in organic fertilizer must definitely be included in the nutrient amount. Diagnosis of N-supplies should be based on the nitrate content (NO_3-N) of leaf stalk and central vein (leaf vein without leaf blade). The leaves tested should be young but fully developed.

Fertilization is less problematic the better the structure and nutrient state of the soil. Neither particularly large fertilizer doses nor special "placement" in the root zone are then required. High beet yields and a large sugar content in the beets presuppose favorable climatic conditions. However, the "favorableness" of the climate can be partly replaced by correct fertilization and other measures.

The question of the *quality of leaves* as fodder will only be touched upon here. A high content of nutrients, especially protein, is desirable for the animals, as well as a low content of harmful substances (e.g., oxalic acid).

Removals are indicated in Table 7-2. The following should be noted with reference to these data, which are used as a basis for calculations of nutrient supplies: There is sometimes a considerable difference between minimum and average removals. Intensive fertilization thus manifests itself in larger removals, as is to be expected. However, this implies unnecessary, large removals of nutrients from the field (Chap. 5.5.2). The indicated removals of trace elements are supposed to be only slightly above the minimum removals.

Table 7-2. Mineral contents and removals of sugar beet (contents partly modified according to [212, 230], removals partly according to [209].

		N	P	K	Ca	Mg	S	Na	Fe	Mn	Zn	Cu	B	Mo
		contents in % of dry matter — minimum contents (min) and average contents (∅)							contents in ppm of dry matter (optimum nutrient level for leaves; 3/4 of optimum nutrient level for beets)					
		min. ∅	min. ∅	min. ∅	min. ∅	∅	∅	∅						
	beets	· / 8	1.5 / 3	· / 8	· / 5	2 / 5	(1) / (2)	2 / 6	(40)	(30)	(15)	(4)	(25)	(0.2)
	leaves	15 / 20	· / 2	20 / 40	2 / 10	2 / 5	∅	∅	50	40	20	5	30	0.3
		removals with fresh substance in kg/ha							removals with fresh substance in g/ha					
com-puted data	beets 100 dt	(18)	(3.5)	(18)	(4.5)	(4.5)	(2)	(4.5)	90	70	35	9	60	0.5
	leaves 100 dt	23	3	30	8	3	3	9	75	60	30	8	45	0.5
according to Lüdecke	100 dt beets	36	6	42	11	7	(4)	12	150	120	60	15	95	1
	100 dt beets and 80 dt leaves	35-60	6-7	40-60	12-16	9-12	·	15-22	·	300-500	·	75-125	300-500	·

yield (beets = 23% dry matter; leaves = 15% dry matter)

General Data on Sugar Beet

Yields: 350 to 650 dt beets per ha.
Desirable number of plants per ha: 60000 to 80000.
Beet/leaf ratio: 1:0.8 (fresh substance).
Dry-matter content of beets: 20–26%.
Dry-matter content of leaves: 12–18%.
Sugar content of beets: 14–19% saccharose in fresh substance.
Sugar yield from sap: about 80% of sugar content.
Detrimental N = "harmful" nitrogen, i.e., N-compounds that in part prevent crystallization of sugar (Chap. 9.3.1).
Net sugar yield =

$$\frac{\text{beet yield} \cdot [\% \text{ sugar} - (\% \text{ soluble ash} \cdot 5 + \% \text{ detrimental nitrogen} \cdot 25)]}{100}$$

Recommendations for the Fertilization of Sugar Beet

1. Rapid start of growth, necessary especially in regions with short vegetation periods, presupposes optimal supplies of water and air to the young plants. The crumb structure required for this can be achieved, e.g., by

- Correct liming in autumn, or with a small additional dose of quicklime (about 3 dt/ha) before sowing;
- proper organic fertilization, e.g., with 200 dt farmyard manure per ha or correspondingly with green manure. The farmyard manure should be decayed and properly distributed in the topsoil in autumn, since otherwise the beets have a tendency to develop several roots.

2. The high yields of sugar beet require *abundant supplies* of easily available nutrients. The following primary nutrients are required (for amounts of N see Table 5–9):

- P for rapid initial root development;
- N for intensive leaf-mass production as a precondition for a high sugar yield;
- K for a stable water status and high photosynthetic performance.

3. *Nutrient requirements* of sugar beet are 20–30% larger than removals (Table 7–2), since a considerable part of the leaves and fine roots die before the harvest. Requirements are maximum during the third month, when about one half of the nitrogen and potassium and one third of the phosphate is absorbed. N-supplies should be abundant in the beginning and reach the critical nitrate content towards the end of growth (Fig. 7–4). Division of N-supplies into two doses (before sowing and at the third leaf respectively) is advisable. Late fertilization should be undertaken only in special cases.

4. The *fertilizer form* plays a minor role in the case of sugar beet:

- Different N-forms are approximately equivalent; the differences sometimes observed are usually due to side effects (e.g., boron in Chile saltpeter).
- The choice of the P-form is also of minor importance; P should be given either in autumn or in spring, depending on the solubility.
- Fertilization with potassium should be in the form of chloride, some weeks before sowing.

5. Other nutrients important for fertilization are:

- Magnesium, whose removal is at the level of P-removal;
- boron, of which beta beets require 6 to 8 times as much as cereals. Boron was formerly supplied automatically with Chile saltpeter. Precise B-supplies at a level of 1 to 2 kg/ha are provided most simply with boron-containing P- or N-fertilizers (Chap. 3.1.5).

6. Beta beets can tolerate salt (halophytes), but in the seedling stage are at first sensitive to salt. *Salt damage* in germinating plants can be prevented if the necessary large amounts of easily soluble fertilizers are applied a few weeks before sowing, or sometimes only after thinning in the form of top dressing (e.g., one third of N-supply).

7. Being halophytes, beets react favorably to additional supplies of *sodium* (Na) especially when potassium is in short supply, and probably also of chloride. However, precise Na-supplies are unnecessary when potassium fertilizers of medium concentration, containing sodium chloride, are applied.

8. There is a discrepancy on medium and heavy soils between the required high soil reaction (pH 7) and adequate supplies of *trace elements*, which in this range are more strongly immobilized. This refers mainly to manganese and boron. Additional mobilization through soil-acidifying fertilization with N (top dressing) may therefore be indicated during the principal growth period, even if this implies some slowing-down of the N-action. Direct additional supplies of these, and if required, of other nutrients too, are otherwise indicated in the form of top dressing and leaf fertilization (Chap. 6.1.4).

9. N-supplies at the end of the growth period should for reasons of *quality* be sparing rather than abundant. "Late" nitrogen increases the leaf mass and its protein content but lowers the sugar yield and its recovery from the sap because of detrimental nitrogen.

10. Intense luxury consumption of N- and P-fertilizers increases the *content of soluble minerals (ash)* in the molasses. This also lowers the sugar yield from the sap and should therefore be prevented.

Fertilization of Fodder Beet (Common Beet, *beta vulg.*)

Fodder beet can be fertilized much like sugar beet. However, the aim is to produce large amounts of carbohydrates and protein in beets and leaves [209]. Fresh-substance yields of fodder beet are approximately twice as high as those of sugar beet (600 to 1 200 dt/ha), but fodder beet contains only 10 to 14% dry matter. The dry-matter yields of the two kinds of plant are therefore approximately equal. The beet/leaf ratio is about 1 : 0.3 (varying from 0.25 to 0.5, according to the type). The content of dry matter in the leaves is about 11%.

N- and P-*removals* per 100 dt beets with leaves are about half as much as for sugar beet. However, removals referred to the total yield per ha are similar. Potassium removal per 100 dt is only slightly less (than for sugar beet). This implies about 50 to 100% greater overall removals during the harvest. However, what was said about N and P also applies to potassium, if only the necessary minimum removals are used as a basis. The fertilizer quantities are indicated in Chap. 5.4. Dosage of N-fertilizers is simpler insofar as later fertilization with N is possible. However, it is only the amide content (and less the protein content) that increases significantly in the leaves. Excessive fertilization with N has a detrimental effect on the durability of beets.

What was said about the fertilizer form with reference to sugar beet essentially applies to fodder beet, too. A certain difficulty is due to the fact that fodder beet is slightly more sensitive to salt but requires larger fertilizer doses. Division of N- and K-doses is therefore indicated at least for high yields.

Fertilization of Turnip (Swede Turnip, *brassica nap.*)

Turnip greatly resembles fodder beet as regards fertilizer requirements. Beet yields vary between 400 an 1 000 dt/ha, depending on whether turnip is grown as secondary or principal crop. The content of dry matter is 11–12% [186].

The optimal soil reaction is slightly acid. Proper liming is also an important preventive measure against clubroot.

Special attention should be paid to *boron* among the trace elements. Boron deficiency causes *glassiness* and thus reduces fodder value and durability. Fertilization with B at a rate of 1 to 3 kg per ha may therefore be indicated if natural supplies are deficient. Molybdenum supplies must sometimes be increased on more acid soils.

7.2.2 Potato, Topinambur

The potato occupies first place amongst tuber crops of the temperate zones. Topinambur = Jerusalem artichokes (*helianthus tub.*) can be fertilized similarly.

Fertilization of Potatoes *(solanum tub.)*

The general aim of fertilizing *potatoes* is to obtain a high yield of tubers and thus mainly of starch but also of other valuable substances like protein, vitamin C, etc. However, there are certain differences in fertilization, depending on the specific production aim:

- Food potatoes, whose tuber quantities and quality should be matched (see recommendations);
- fodder potatoes, where the mass production of starch takes priority;
- seed potatoes, whose fertilization should particularly promote the health and germinating power of the seed material (only moderate N-supplies, abundant P-supplies, etc.).

The following discussion refers to food potatoes [224]. The water supply should be abundant particularly during the principal growth period at the time of blossoming. Any, even only temporary, lack of water lowers production, especially at the time of large mineral intake. The transpiration coefficient is 300 to 400, i. e., this is the number of liters of water necessary for the production of 1 kg dry matter of potatoes. Fertilization should be especially directed to the quality of the external appearance, (scab), storability, taste, etc. Removals are indicated under the general data. They refer only to the tubers, since the leaves are often left on the field.

Leaf fertilization, if necessary, can often be combined with measures for plant protection. Thus, the formerly common treatment with copper preparations to prevent fungal diseases contributed to abundant supplies of Cu.

General Data on (Food) Potatoes (see also Table 7-3)

Tuber yield: early potatoes: 200–300 dt/ha
late potatoes: 300–450 dt/ha

Composition of tubers (contents):
dry matter: 20–22%
starch: 12–15%
protein: 1.5–2% (raw protein)

tuber/leaf ratio = 1:0.5 to 1:1
dry-matter content of
fresh leaves: 15–20%

Table 7-3. **Nutrient contents and removals of potato tubers** (partly modified according to [212]).

	N	P	K	Mg	Ca	S
content in ‰ of dry matter	15	3	20	1.2	0.6	1
removal in kg/ha (per 100 dt tubers)	30	6	40	2.4	1.2	2

Recommendations for the Fertilization of Potatoes

1. Potatoes, more than any other root crop, require a *loose soil structure* for root development and tuber formation. This may be achieved, e.g., by abundant organic fertilization (200 dt farmyard manure per ha) or with corresponding amounts of green manure. Farmyard manure should, if possible, be well mixed with the topsoil in autumn.

2. Potatoes prefer an acid to slightly acid *soil reaction*. Liming to obtain a good crumb structure should therefore be limited. "Structure liming" should be undertaken only shortly before sowing or as top dressing (with slaked or mixed lime) if the pH were to rise above 6 (Chap. 4.1.2).

3. High tuber yields presuppose abundant nutrient supplies, especially of *nitrogen* and *potassium*:

- N for intensive production of leaf mass as a precondition of intensive starch production;
- K for high photosynthetic performance (starch content) and high tuber quality (for fertilizer quantities see Chap. 5.4). Nutrient intake is maximal during blossoming in June and July.

4. *N-supplies* should be abundant (about 50 kg N per 100 kg tubers) but not one-sidedly excessive, since leaf mass should not be produced for its own sake. Division of the N-dose is indicated (for N-quantities see Chap. 5.4.1).

5. Potatoes prefer slow-acting *acidifying N-fertilizers* (see recommendation 8), e.g., ammonium fertilizers or fertilizers yielding ammonium after reaction (e.g., urea) (Chap. 2.1.5).

6. Only sulfatic forms of *potassium fertilizers* should be considered for potatoes, since these plants are sensitive to chloride (potassium sulfate, potassium-magnesium, NPK-fertilizers containing little chloride). Chloride inhibits the displacement of starch from leaves to tubers. If chloride-containing fertilizers must be applied, this should be done in autumn, so that the chloride is largely leached out.

7. *Mg supply* is particularly important because many potato soils are deficient in Mg, and in view of the large potassium doses required. Fertilization with Mg may be effected either with Mg-limes or with K- and N-fertilizers.

8. Of the *trace elements*, manganese, especially, becomes easily deficient in the pH-range between 6 and 7. This not only lowers the yield but also promotes the undesirable incidence of scab. Additional fertilization with Mn may therefore be indicated, e.g., as leaf spraying, but primarily as Mn-mobilizing, acid fertilization with N (also in the form of top dressing) (Chap. 2.1.5).

Sometimes, the large boron requirement can also be covered only by additional fertilization, since tubers tend to develop *fissures* in cases of deficiency.

9. The *resistance* of potatoes to plant diseases can be improved by correct fertilization. Thus, the spreading of viruses is promoted by luxury supplies of nitrogen but reduced by abundant P-supplies.

10. Fertilization should take important *quality properties* into account, i. e., durability, cooking properties, and taste. Proper K-supplies, jointly with overall optimal nutrient supplies (especially of copper) reduce sensitivity to shocks ("blue-spot symptoms"), darkening of the potato paste during processing, or the black spottiness of boiled potatoes. Excessive N-supplies increase the content of amides etc. that spoil the taste.

7.2.3 Cassava, Batata, Yam

The starch-rich tubers of various plants form important basic foodstuffs in tropical and subtropical regions [180, 209].
     The following tropical tuber crops are most important:

- Cassava or manioc (manihot, etc.);
- batata (ipomaea bat.) or sweet potato;
- yam (dioscorea sp.).

     These tuber crops are often grown without separate fertilization. Their yields then vary from low to high according to the natural supply of nutrients. Good growth is often achieved on recently cleared land within the framework of "shifting cultivation" (Chap. 6.3.1).
     Cassava, being a shrub plant, forms thickened roots as tubers. These contain 15 to 30% starch and many other nutrients but also contain the glucoside linamarin (from which hydrocyanic acid is obtained). It can be removed, however, by "watering".
     Cassava is highly sensitive to frost and therefore requires a warm, humid climate and a well aerated soil. Dryness is tolerated well but yields are then at the lower end of a wide range between 50 and 800 dt tubers per ha. Nutrient requirements resemble those of potatoes. N-doses of 50 to 100 kg/ha seem to be indicated for medium yields. They should be suitably supplemented with phosphate and potassium, possibly also with other nutrients.
     Batata, being a vegetable plant, also forms thickened roots as tubers. These contain about 25% starch and sugar in varying proportions, sometimes in approximately equal amounts. This plant is therefore also called *sweet potato*. Batata also requires a warm climate, but is not as sensitive to cold as cassava. The soil should have a good structure and be well aerated. The soil reaction

should be about pH 5.5 to 6. Proper K-supplies are important for high yields, as with potatoes. Phosphate also improves the durability.

Nutrient removals largely resemble those of potatoes. Reported values of 40 kg N, 10 kg P, and 60 kg K per 100 dt tubers [218] seem to be slightly exaggerated and probably indicate some luxury removals. The amounts of fertilizers may be based on tentative values for potatoes.

Yam is a creeper and occurs in various types. The tubers contain approximately 20% starch and other nutrients, but they also contain the toxic alkaloid dioscorine, which, however, is destroyed during cooking. Yam also requires a warm and humid climate. Nutrient requirements resemble those of potatoes, so that fertilization may be similar. Yields vary widely between 50 and 500 dt tubers per ha.

7.3 Oil Crops and Grain Legumes

Oil crops yield fatty substances in either liquid or solid state, oils and fats respectively. Botanically, they belong to different families; but according to their growth, the most important oil crops can be divided as follows:
Vegetable and shrub plants: rape and rape-seed   flax (oil flax, Chap. 7.4.1)
                      sunflower   sesame and castor-bean
                      soybean and peanuts
                      (see Chap. 7.3.2)
Tree plants: olive (olive tree), oil palm, coconut palm

Chemically, the fatty substances (oils) obtained are carbohydrate derivatives. Proper potassium supplies are therefore as important for these plants as for starch plants. The main role in fertilization is therefore played by nitrogen (for the production of plant mass) and potassium (for intensive photosynthesis).

Grain legumes are also termed *protein plants*, since they are important suppliers of vegetable proteins. These are peas, beans, vetch, lupins, lentils, soyabeans, peanuts, etc.

7.3.1 Rape and Rape Seed

Fertilization of Rape (Oilseed Rape)
Rape (brassica nap.) serves for the production of edible oil, etc. The yield potential of modern strains is about 50 dt grains per ha. This is approximately one half of the potenial yield of wheat. High yields of 20 to 40 dt per ha or more are in practice already obtained through optimal nutrition of the plants, if correct

measures are taken to prevent diseases and attacks by pests. Important preconditions are a fertile soil, early sowing (mid-August), and fertilization in autumn,
mainly intended to increase winter hardiness.

Water supplies must be ensured for rape, as for all high-yielding plants,
during the period of intensive vegetative development. Abundant growth forms
the basis of high grain yields (kernel/straw ratio of about 1 : 2).

Nutrient removals vary according to the supply level. Average values for
10 dt grains with corresponding quantities of straw are as follows: 50 kg N, 10
kg P, 40–50 kg K [222]. Most of these might be due to straw, and necessary minimum removals must be determined still more precisely.

Rape grains contain about 90% dry matter, of which about one half is rape
oil. The *quality* of rape grains is primarily determined by fertilization with N.
More intensive fertilization with N lowers the oil content (by the dilution effect)
but raises the oil yield through the higher grain yield. On the other hand, larger
N-doses increase the content of essential unsaturated fatty acids.

With regard to undesirable constituents, it should be noted that only
strains poor in erucic acid have been grown in the FRG since 1975.

Recommendations for the Fertilization of Rape

1. For sowing, rape (winter rape grown as oil crop) requires a *crumbly* soil such
as exists, e.g., after root crops, because of the smallness of the seeds. On the
other hand, rape is required in cereal-crop sequences; this also permits the necessary early sowing at the end of August. A good liming state is therefore of
principal importance for rape, especially after cereals. Liming should be performed up to the given optimal pH-value (mostly 6.5 to 7).

2. Rape must develop well before winter, and for high yields has *large* early *nutrient requirements:*

- Proper supplies of P and K for rapid initial growth and cold resistance;
- N-supply (about 30 to 60 kg per ha after cereals) should not cause luxury consumption with its detrimental effects on the resistance to frost.

3. *Organic fertilization* is well utilized by rape because of its lengthy vegetation
period. However, it is difficult to arrange this in time and is not absolutely essential for high yields.

4. *Nutrient intake is maximal* in the 3 to 4 weeks before blossoming. Hence, the
principal N-dose (one half to two thirds of the total) must be applied at the start
of vegetation. Some of the nitrogen can be applied as top dressing before the
plant cover is closed (for the quantities of N see Chap. 5.4.1). P and K need not
be applied in spring if they are supplied abundantly in autumn (Chap. 5.4).

5. The *form* of the nutrients N, P, and K is of minor importance for rape. However, important side effects like acidification, sulfur content, etc. should be taken into account, particularly in the case of N-fertilizers.

6. Supplies of the other major nutrients must be optimal if high rape yields are to be attained. The large *calcium* and *magnesium* removals should be stressed in this connection. Therefore magnesium lime should always be used for liming.

7. Rape, being a crucifer, *requires much sulfur* for the production of mustard oils. Requirements are approximately twice those for phosphorus. Sulfur is applied most simply as a secondary fertilizer constituent (Chap. 2.5.2). Strains poor in mustard oil require correspondingly less sulfur.

8. The trace elements *boron* and *manganese* in particular become deficient easily, because of the required high soil reaction:

- Natural Mn-supply should therefore be increased through acidifying N-fertilizers or, if necessary, supplemented by additional supply, e.g., by leaf spraying;
- the boron requirement of rape is particularly large (approximately like those of sugar beet); boron deficiency causes "fissuring" at the stalk. Application of 1 to 2 kg boron per ha is indicated, if necessary (Chap. 3.1.5).

9. *Resistance* plays an important role in rape precisely when yields are high (resistance to cold, fungal diseases, etc.). Fertilization should always have the aim of optimal nutrition (Chap. 6.5). This necessitates both balancing the supplies of major nutrients and providing enough other nutrients. Thus, boron deficiency could detrimentally increase the sensitivity of plants in various ways. Rape tolerates salt relatively well.

10. The *oil content* as a quality characteristic is slightly lowered by (large) N-doses; however, the total yield of rape oil per unit area is significantly increased because of the higher grain yield.

Fertilization of Summer Rape and Winter Rape Seeds
Both plants may replace winter rape as oil crops, but their yields are lower. Winter rape seeds can be sown later as winter rape in autumn. Summer rape, being a summer crop, has a place in certain crop sequences or serves as a replacement for omitted winter rape. Fertilization requirements resemble those of winter rape. N-doses should be divided, e.g., one half during sowing and the second half 3 to 4 weeks later. Overall fertilization may be less than for winter rape, in view of the lower yield expectations.

7.3.2 Sunflower, Sesame, Castor Bean

These annual oil plants from warm zones were until now either not fertilized at all or fertilized only slightly over large areas. However, fertilization may lead to considerably higher yields if carried out precisely to improve the minimum factors rather than in a general way. Nutrient contents and removals are summarized in Table 7-4 [190, 218].

Sunflower (helianthus ann.) prefers a soil with a pH of 6 to 7 and requires much potassium as it has shallow roots and a good absorption capacity for nutrients, it needs only little fertilizer for low yields (e.g., 30 kg N per ha). However, suitable strains attain higher yields when they receive improved nutrient supplies. The seeds contain about 50% oil of the linoleic acid type.

Sesame (sesamus ind.) also prefers an almost neutral soil. Under extensive cultivation, it is often the first crop after clearing within the framework of shifting cultivation (Chap. 6.3.1), and is not fertilized specially. However, newer strains are worth abundant nutrient supplies [215]. Zinc supplies are often deficient and should then be increased. The seeds contain about 50% oil of the oleic- and linoleic acid type.

The *castor bean* (rhizinus communis) prefers a soil reaction of about pH 6 and is sensitive to stagnant moisture. It has large nutrient requirements and a tendency to zinc deficiency. The seeds contain about 50% oil, which consists almost completely of ricinoleic acid and is a high-grade lubricant [207].

7.3.3 Olive, Oil Palm, Coconut Palm

Fertilization of Olive Trees
Olive trees (olea europ.) occur mainly in the Mediterranean region. The tree grows only slowly, often on poor soils. Yields vary accordingly between 10 and 100 kg olives per tree. They contain 15 to 40% oil of the oleic acid type. Olive trees prefer sandy soil with a neutral soil reaction, and grow well on calcareous soils. Water is an important yield-limiting factor despite the deep roots. A mulch cover may be indicated for improving water supply.

Fertilization should be divided to ensure more uniform supplies and prevent unnecessary salt loads. The first dose should be applied before blossoming, and the second after the harvest. Placement of the fertilizer in a large ring around the tree is indicated. The effects of better nutrient supplies manifest themselves only in the second year through increased fruit setting.

Olive trees requires much N and K in particular, and primarily boron from among the trace elements [190, 195]. Diagnosis of the nutritional state is increasingly based on leaf analysis. Tentative values for this purpose are given in Table 5-6. For nutrient removals and fertilization see Table 7-4.

Table 7–4. Nutrient removals and fertilization of oil plants in warm zones (compiled after [190, 195]).

Plant	Yields per ha	Removals (kg/ha) with 1 t seeds or fruit yields indicated				fertilizer amounts required for medium yields		
		N	P	K	Mg	N	P	K
sunflower	1.5–4 t seeds	30	3	60	(10)	50–100	20	120
sesame	0.2–2 t seeds	(35)	(3)	(15)	(3)	20–80	20	(40)
castor bean	0.5–3 t seeds	40	3	15	3	30–120	20	(40)
olive	15 t fruit	25	7	20	(5)	30–50	15–25	35–50
oil palm	20 t bunches of fruit	60	10	85	10	100	20	150
coconut palm	7000 nuts	90	15	120	25	100	20	130

Fertilization of Oil Palms

Oil palms (elaeis guin.) grow in humid tropical zones. Their fruit pulp yields palm oil (30% of the fresh fruit, of the oleic acid type). Their kernels yield palm-nut oil (auric acid type). Full yield of 10 to 20 t fruit bunches are attained after 10 to 20 years.

Oil palms prefer moderately acid soils. Nutrient removals with the fruit are high, to which the requirements for wood growth must be added. Growth in youth is accelerated by placement fertilization during planting. The nursery bed should also be generally rich in nutrients [202].

The effects of fertilization are slow to appear, as might be expected, and manifest themselves only in the second and third year after fertilization. Proper K supplies should be stressed, but magnesium supplies should also be taken care of precisely for this reason. On acid soils, very high Mn concentrations sometimes occur in the leaves. This may cause excess damage. Boron require-ments are large. Fertilization of older trees is best based on leaf analysis. The leaves of *frond 17* are often used. These are 8 to 9 months old. Their dry matter should contain: 25‰ N, 1.8‰ P, 13‰ K, 3‰ Mg, 5 ppm Cu [190]. The value for manganese, given as 150 ppm, is probably greatly exaggerated and should rather be 40 ppm. The considerable chloride requirements should be noted. The Cl content of leaves should be at least 4‰, since otherwise reduced yields must be expected.

Fertilization of Coconut Palms

Coconut palms (cocos nucif.) grow in humid tropical zones. They provide full yields from the 15th to the 60th year. A tree then bears 60 to 100 nuts or 10 to 20 kg copra (air-dried endosperm, coconut pulp) per year. Copra contains 60 to 70% oil of the lauric acid type.

Coconut palms prefer neutral soils, but also grow relatively well on mod-erately acid soils. The soil should in any case be deep and permeable. Like oil palms, growth in youth, especially, can be promoted through proper nutrient supplies.

Fertilization action is delayed in older trees, as might be expected. Re-transportation of nutrients from the fibre shells to the field is beneficial, since these account for a large part of the removals. Proper K supply is particularly important, but (as with oil palms) no Mg deficiency should be caused by this. Fertilization of older trees is best based on leaf analysis diagnosis. No. 14 leaf should have the following contents: 20‰ N, 1.3‰ P, 10‰ K, 5‰ Ca, 3‰ Mg [190]. The Na content should not exceed 4‰.

7.3.4 Peas, Beans, Soybeans, Peanuts, etc.

Grain legumes (pulses) have relatively uniform nutrient requirements. Fertilization problems will therefore initially be discussed jointly. The grains of legumes provide food rich in protein and, in part, of fatty substances, too. Accordingly they play an important role in human nutrition and as a basis for the production of animal protein in worldwide agriculture, besides serving as green-fodder plants (Chap. 7.5.1). Their ability to supply nitrogen to themselves (Synopsis 7–2) and to mobilize soil nutrients well makes it possible for them to have medium yields on many soils even without mineral fertilization [219].

However, precise additional fertilization often permits considerably higher yields. This applies also to the N supply, which normally poses no problems. However, natural N fixation is seldom sufficient for maximum yields, so that additional fertilization may be necessary. Legumes are even extensively nourished with fertilizer nitrogen in extreme cases.

Recommendations for the Fertilization of Grain Legumes

1. Legumes require *no fertilizer nitrogen*, since they supply nitrogen to themselves through their nodule bacteria. However, a small starting dose of 10 to 20 kg N per ha is advisable to promote rapid initial growth. The seed must be inoculated with suitable bacteria when a legume type is first cultivated on a given field.

2. Legumes mostly prefer *a neutral soil reaction*, since they require much calcium. Some legumes (types and strains), however, are sensitive to lime, since they tend to suffer from chloroses (iron deficiency, etc.) at higher pH-values.

3. Legumes prefer soils with *good structure*, but their strong and deep-reaching root systems also improve the structure.

4. Legumes have good *mobilization capacity* for nutrients in the soil, and with their deep-reaching root systems also utilize nutrients well in the subsoil. However, abundant supplies of phosphate, potassium, etc. promote high yields.

5. *Sulfur requirements* almost equal P requirements. S supplies must therefore be taken care of in particular. Some yield increases through superphosphate or ammonium sulfate are mainly due to the S supplies.

6. *Calcium requirements* are large and, for some legumes, equal potassium requirements. At least Ca supplies should be improved by fertilization (e. g., with gypsum) if such legumes have to be grown on acid soils. P fertilizers sometimes act through their Ca components. Ca supplies are in any case more important than the pH-value.

Synopsis 7–2: Nitrogen Fixation by Legumes.

Legumes (and some other plants) can supply nitrogen themselves by symbiosis with N fixing bacteria, and thus do not normally require fertilization with N. The bacteria live in the root nodules. The nodule bacteria belong to the genus *rhizobium*, which consists of at least six species with some still unknown subspecies (races) [172]:

1. *Rhizobium leguminosarum* in peas, vicia beans, lentils, flat peas.
2. *Rhizobium phaseoli* in phaeolus beans (e.g., kidney beans).
3. *Rhizobium trifolli* in clover, with two or three subgroups:
   a) red and white clover;
   b) crimson clover, Egyptian clover.
4. *Rhizobium meliloti* in alfalfa, melilot, etc.
5. *Rhizobium lupini* in lupins, serradella, etc.

Inoculation of the seed with nodule bacteria is necessary when legumes are grown for the first time on a soil (inoculant *Nitragin*).
Plants of the next crop are infected from the soil, in which spores (bacteria in a state of rest) are preserved for more than 10 years.

Symbiosis process: The spores form mobile swarms, which penetrate into the root bark, and there form proliferations. The bacteria multiply in these nodules and fix nitrogen (N_2) from the air, i.e., convert it via hydroxylamine into bacterial protein with the aid of red leghemoglobin, enzymes, and heavy metals (Fe, Mo, Co). This protein can be utilized by the legumes after decomposition of the bacteria. In their turn they put carbohydrates (sugar), active agents, etc. at the disposal of the bacteria.
Nodules are occasionally *ineffective*, i.e., for some reason they do not fix nitrogen. Requirements as regards soil reaction correspond to those of legumes. The pH-value as such is less important than nutritional conditions for the bacteria (e.g., there is a tendency to Ca deficiency at pH 4). Neighbouring plants can also profit from N fixation by legumes, e.g., after root parts rich in N have died, sometimes also from excretions. The amounts of usable nitrogen for the following crop are increased, especially by leaf legumes.
A small *N-dose* is indicated for overcoming the first infection phase. Larger amounts of N fertilizers tend to reduce the self-supply, since legumes utilize N fertilizer preferably and reduce the fixation of N. In intensive cultivation, this may cause a stoppage of natural N fixation in practice.

7. *Boron requirements* of all legumes are particularly large. Additional fertilization with B may be indicated, especially at higher pH-values.

8. Chloroses due to *iron* or *manganese* deficiency are common at high pH-values. However, the susceptibility to these chloroses seems to be strain-specific, so that correct selection of the strain can obviate such deficiencies. Leaf spraying is otherwise indicated.

9. *Requirements of nutrients* are *maximum* during the phase of abundant vegetative development, which sometimes lasts for a long time.

10. A high *quality* of the legume grains, e.g., large contents of protein and fats, can be achieved only by optimal nutrition. Precise fertilization based on diagnosis is therefore indicated.

Fertilization of Peas and Beans

Peas (pisum sat.), *broad beans* (vicia faba), and *kidney beans* (phaeolus sp.) prefer a neutral soil reaction. However, liming should not exceed the requirements of the particular soil type (Chap. 5.1.1). Liming should be carried out in autumn. Yields and removals of major nutrients are indicated in Table 7–5. The nitrogen removed obviously need not be replaced. Large N doses, besides the starting dose, are sometimes applied for high yields in intensive cultivation.

Kidney beans require more nutrients than broad beans. Fertilization may be based on removals (for tentative values see Table 7–5).

Table 7–5. Yields and removals of grain legumes (modified and supplemented according to [219]).

plant	good yields dt/ha		nutrient removals per 10 dt grains with straw, kg/ha					
	grains	straw	N	P	K	Ca	Mg	S
peas	30	45	60	8	30	25		
broad beans	40	60	60	8	40	25		
vetch	20	20	60	(12)	30	30		
kidney beans	20	30	80	8	40	40	6	(7)
yellow lupins	20	35	70	8	40	20		
lentils	15	15	70	8	25	30		
peanuts	35	60	70	8	30	30	8	
soybeans	35	60	80	(12)	40	40		

Fertilization of Vetch, Lupins, and Lentils

Vetch (vicia sp.), *lupins* (lupinus sp.), and *lentils* (lens cul.) prefer a neutral soil reaction. An important exception is the lime-sensitive yellow *lupin*. Lentils are grown primarily on calcareous soils. Lupins (in particular yellow lupin) are suitable as first crops on *cleared areas*. They demand little, but abundant nutrient supplies are nevertheless advisable if higher yields are desired. Little starting nitrogen is required by vetch and lupins, but lentils react positively to N doses of 40 kg/ha or more. Fertilization may be based on removals (for tentative values see Table 7–5).

Fertilization of Soybeans

Soybeans (glycine max.) are the most frequently grown grain legumes. High yields may attain 40 dt grains per ha, containing 35% protein and 20% fat. Soybeans grow on various soils, but for high yields a balanced water supply and a pH of approximately 6 are required.

A starting dose of 10 to 20 kg N per ha is usually worthwhile. As with all legumes, N fixation increases inversely to the content of available nitrogen in the soil. Proper supplies of phosphate and calcium to the soil are indicated; a balanced K : Ca : Mg ratio is essential, as for all legumes. Fertilization is best based on leaf analysis [190]. Average fertilization may be based on removals (Table 7-5).

Fertilization of Peanuts

Peanuts (arachis hyp.) are an important oil and protein crop of the tropics. High yields are approximately 50 dt unshelled fruit per ha (35 dt kernels, nuts) [192, 197].

Kernels contain 40 to 50% oil. The oil consists of oleic acid and linoleic acid. A neutral action is desirable, but the optimal pH for the soil type considered should not be exceeded (Chap. 5.1.1). A characteristic property is the *sensitivity* of some peanut strains or kinds to *chlorosis*. This problem can usually be circumvented by selecting the correct kind. Ca requirements are large but are partially covered through direct Ca intake from the soil by the fruit, which grows in the ground. Additional fertilization with Ca, e. g., with gypsum, may therefore be indicated on acid soils. S requirements are 5 to 10 kg/ha. Fertilization may be based on removals (Table 7-5).

The following limit values should be reached in the leaf for optimal nutrient supplies: 35‰ N, 2.3‰ P and S, 10‰ K, 12‰ Ca, 5‰ Mg [190]. The required amounts of trace elements should be based on those for alfalfa (Table 7-8, p. 331).

7.4 Fibre Plants and Other Industrial Crops

Fibre plants yield fibres with different properties, which can be used by industry. Other industrial crops yield raw materials for industry, e. g., raw rubber and similar products, waxes and dyes. This chapter also includes sugar cane, which provides the "raw material" sugar.

7.4.1 Cotton, Flax, Hemp, Sisal, etc.

The fibres are obtained from different plant parts, e. g.:

- in the case of *cotton*, from the hairs of the seeds;
- in the case of *flax, hemp, jute, and kenaf*, from fibres of the supporting tissues of the stalk;
- in the case of *sisal and manila hemp*, from hard leaf fibres.

Fibres consist of carbohydrates. These are more than 90% cellulose in the case of cotton, but consist of cellulose, lignin, etc. in the case of most scleren-chyma fibres.

Nutrients should be supplied to fibre plants to promote the production of carbohydrates. This means that stress should be laid on proper K supply, besides proper basic supplies of all nutrient elements. However, fertilization with N [185, 190] is often especially important, since fibre plants are frequently grown on soils poor in N but relatively well supplied with potassium.

Fertilization of Cotton

Cotton (gossypium sp.) is the most important of the fibre plants. The aim of fertilization is to produce large amounts of fibres of maximum length (up to 4 cm), as well as seeds whose oil and protein is utilized. Yields vary between 0.5 and 4 t raw cotton per ha, of which ⅓ are fibres and ⅔ seeds.

Cotton is a plant from warm zones. Its cultivation in arid zones requires a good water supply (irrigation) for high yields. Cotton prefers a slightly acid to neutral soil reaction, due to its susceptibility to acid damage (excess of soluble aluminum and manganese) [190, 195, 210, 211].

Nutrient removals with fibres are slight (about 1 kg N, 0.2 kg P, and 3 kg K per t of fibres). However, total removals, a large share of which is accounted for by the seeds, are considerably larger (Table 7-6). Nutrient intake is already intensive at the beginning of growth, with abundant initial P supplies being particularly important.

The necessary *fertilization with N* depends on the yield level and N supply in the soil, ranging from 20 to 200 kg/ha. Low values correspond to abundant N supply in the soil and low yields, while large doses are required for high yields on soils poor in N. A record harvest of 1.5 t fibres per ha requires about 200 kg N per ha, if the data in Table 7-6 are used as a basis, and even then fertilization should exceed this quantity. The larger the necessary N doses, the more important is their distribution over the vegetation period. The smaller the N doses, the more important is placement in the root zone beside or underneath the seed. Undesirable side effects (e. g., weed growth) can be prevented if fertilization with N (except on very poor soils) is begun only after the plants have been thinned.

Fertilization with N according to diagnosis is gaining in importance. Requirements can be deduced from the nitrate content of the soil (in principle as for cereals) or of the leaves, especially of the leaf stalk. The necessary or maximum nitrate content in the leaf stalk varies greatly with the age of the plant. It should be about 1 to 2‰ (nitrate N) at the time of blossoming [190].

Plant requirements of *phosphate* and *potassium* are indicated in Table 7-6. The required fertilization depends greatly on supplies in the soil. Some soils

Table 7-6. Nutrient removals and requirements of fibre plants (compiled from [190, 196, 201]).

plant	yield per ha (average)	removals at yield indicated (kg/ha)				fertilization for medium supplies (kg/ha)		
		N	P	K	Mg	N	P	K
cotton	0.5 t fibres + 1 t seeds	35	6	13	4	100	20	70
	1 t fibres + 2 t seeds	70	12	26	8	140	30	90
flax	6 t straw ⎫ + 7 t seeds	70	15	90	10	40 ⎫	20	100
hemp	6 t stalks ⎭					100 ⎭		
jute	2 t fibres (+ stalk mass) ⎫	70	13	130	60	50-100	20	100
sisal	2 t fibres (+ leaf mass) ⎭							

have large annual fertilizer requirements, while others produce several cotton harvests without application of P and K. Available supplies in the soil are decisive, together with removal levels, which are not very high if only seeds and fibres are removed from the field. In case of smaller inputs, it is advisable to apply P and K fertilizers by placement. *Leaf diagnosis* is also suitable for establishing requirements. Critical nutrient levels in the USA are 3.5‰ P in the leaf blade, 20‰ K and 2‰ Mg in the stalks of the youngest mature leaves at the time of boll formation [190].

Fertilization with *sulfur* may be necessary on some soils. The considerable boron requirements, amongst trace elements, should be noted (0.5 to 1 kg B per ha, if necessary). Fertilization with *zinc* may also be necessary, especially when the soil reaction is high. A slight *iron* deficiency sometimes occurs on irrigated soils in the rhythm of the varying redox processes in the soil.

Correct fertilization of cotton is also necessary to increase resistance to various diseases and pests. However, in general, fertilization can only increase yields in combination with correct plant protection.

Fertilization of Flax

Flax (linum usit.) should supply high-quality fibres, combined with high oil yields if grown in extensive farming. The fibre quality depends on several characteristics, e. g., stalk length, number of fibrecell bundles in the cross section, etc., and can be influenced positively or negatively by fertilization [201].

Flax prefers a slightly acid soil. Because of its low absorption capacity it has a high demand for mineral nutrients. Nutrient requirements therefore considerably exceed removals (Table 7-6).

Fertilization with ·N is possible only within limits, since large N supplies lower the fibre quality. Organic fertilization also seems inadvisable for fibre flax for the same reason. The necessary amounts of N are applied at sowing, divided into two doses at most. Proper potassium supply, which apparently should exceed the optimum nutrient level, are essential for a good fibre quality. They should take the form of potassium sulfate, since chloride has a detrimental effect. Proper Mg supply also seems indicated. Flax belongs to the plants with larger boron requirements. Boron should be applied in amounts of 0.5 to 1 kg per ha, if necessary.

Oil flax has greater stability than fibre flax. Its fertilization with N can and should be at a much higher level (80 to 90 kg/ha when the dosage is divided). Seed yields are about 15 to 20 dt/ha. The seeds contain almost 40% oil (mostly linoleic and linolenic acids).

Fertilization of Hemp

Hemp (cannabis sat.) is grown for the production of fibres suitable for ropes. It prefers a neutral soil reaction. This necessitates suitable liming on acid soils [201]. Organic fertilizers have beneficial effects. Abundant overall nutrient supplies are required, especially since premature shedding of leaves causes part of the nutrients to be lost again. Fertilization with N is important to hemp, because of its massed growth, but should not be one-sidedly intensive to ensure fibre quality. Potassium requirements are large, as is the case with all fibre plants. The same applies to magnesium.

Potassium chloride can be used for K fertilization. Calcium requirements are almost as large as potassium requirements. Hemp is frequently grown on fen soils, which may cause manganese and copper deficiency, and the larger N supplies must also be allowed for.

Fertilization of Jute and Kenaf

Jute (corchorus sp.) yields a soft fibre used for the production of sackcloth, etc. The fresh substance contains about 5% fibres. Jute requires a high soil fertility. Young plants are highly sensitive to stagnant water. Good soil structure and slightly acid soil reaction are important preconditions for successful fertilization. Proper supplies of P and K are important for the formation of fibres, but fertilization with N is decisive for the yield on most sites (Table 7-6).

Kenaf (hibiscus cann.) has nutrient requirements similar to those of jute but is less demanding with regard to soil structure and aeration.

Fertilization of Sisal and Manila Hemp

Sisal (agave sisal) produces hard fibres obtained from the leaves (3 to 5% portion). Five thousand plants per ha each produce up to 250 leaves of 1 to 2 m length. Harvesting of leaves begins after 3 years in several annual partial cuttings. Yields are 1.5 to 4 t fibres per ha per year [196]. Sisal agaves prefer a neutral soil and are highly sensitive to stagnant water. Deep root penetration renders sisal resistant to dryness with a relatively low demand for mineral nutrients. However, growth and fibre length can be significantly increased by fertilization, if there are deficiencies.

Fertilization with *leaf residues* (directly or as ash) is important for the nutrient balance of sisal fields. A large part of the nutrients is thus restored to the fields. Application is best effected in the form of mulch on the surface (e. g., 100 t of decayed leaf waste per ha). Green manure may be applied as soil cover if restoration of leaves is difficult.

Fertilization of sisal agaves is most important in the nursery bed (quantities as for demanding agricultural plants) [190].

It should be added to the information given in Table 7–6, that sisal, although a monocotyledonous plant, *requires much B*, so that B should be applied at a rate of 0.5 to 1 kg/ha, if necessary. The required concentrations of Mn, Zn, and Cu in the leaf should be as high as for cereals, but the concentration of B should apparently exceed 30 ppm.

Manila hemp (musa text.) produces a fibre with high wet strength, which is suitable for ropes. Nutrient requirements exceed those of sisal.

7.4.2 Rubber Plants, Sugar Cane

Fertilization of Rubber Plants

The most important plant producing raw rubber is *hevea brasiliensis*, the "rubber tree" of humid tropical zones. One third of the sap tapped from the bark (latex) consists of raw rubber. *Latex* can be tapped from 6 to 30 years after planting. Yields of from 0.5 to more than 3 t raw rubber per ha are obtained from about 250 trees per ha [189].

The *demands* made on the soil by hevea are less than those of any other tropical cultivated trees. This is linked to an effective nutrient mobilization by mycorrhiza fungi (Chap. 8.3.1). However, soils of at least medium fertility seem to be indicated for high yields. They should be deep (because of the long tap roots) and moderately acid (pH 5 to 6). Cultivation is quite possible at lower pH-values, if account is taken of the extreme nutritional conditions. However, in general this might be more difficult than correcting the pH. Organic fertilization seems advantageous for various reasons (mulch for protecting the soil against erosion, as a source of nutrients, etc.). Nutrient removal with latex is only about 6 kg N, 2 kg P, and 4 kg K per t raw rubber. Other sources report even smaller amounts of P.

However, this production level requires densely planted trees. Therefore, nutrient requirements are considerable at first. Hevea plantations are more profitable, the earlier exploitation can begin after planting. Fertilization of young plants is therefore especially important. This applies first to the nursery bed, and then to the first years after planting. *Fertilization in the nursery bed* may be based on the general rules for tree growing (Chap. 8.3).

Young trees in the field at first require abundant phosphate for initial root development. This is particularly important, since many soils in the tropics are poor in available P.

Stock doses of 100 to 500 g phosphate fertilizer (raw or processed) are therefore placed in each hole prepared for planting. Supplies of N, K, and Mg should also be abundant. However, fertilization should be less for stock supply (because of the sensitivity to salt), but rather in half-yearly doses. Data on optimal fertilizer quantities differ considerably for the various growing regions [190]. This is not surprising, in view of the differences in soils. Fertilizer doses of 30 to 40 g N, 20 to 30 g P, and 50 g K per tree during the first year should be correct in the principal growing region, i. e., Malaysia. Twice these amounts should be applied in the second and third year, and thereafter three to four times these quantities. Initial supplies of nitrogen can be increased by applying green manure in the form of legumes. The proper dosage of fertilizers can be assessed only from the vegetative growth during the first year. Latex production is later a further step, though only conditionally.

Therefore it might be best to use *leaf analysis* for diagnosing requirements. Not only have tentative values for optimal nutrition been established, but *Beaufils* has developed an evaluation system, allowing for the nutrient ratios N/P, N/K, and K/P [190].

Leaf analysis is performed on 3 to 8 month old leaves. Optimum nutrient levels are approximately 35‰ N, 2.5‰ P, 15‰ K, and 2.5‰ Mg. There are small fluctuations, depending on whether the leaves concerned were taken from the sunny or the shady side of the tree. The possibilities of diagnosis increase, the more sampling is precisely narrowed.

Trace elements become more important as minimum factors when yields increase, as is the case with all plantation trees. Manganese supplies should be fairly abundant on acid soils, but Mn deficiencies frequently occur on soils with higher pH-values. Zinc deficiency might also play a role on certain sites. Boron is mostly rather in excess, with detrimental effects. Copper supplies can be quite abundant in certain regions. This is desirable for growth, but not for the raw-rubber quality, which is lower at higher Cu-concentrations. Recommended countermeasures are the application of large amounts of organic fertilizer, to immobilize or fix excess Cu.

Fertilization of Sugar Cane

Sugar cane (saccharum offic.) is a high-yielding plant of humid tropical zones. It requires much light and water, as well as proper soil aeration.

The plant belongs to the grasses. It grows to a height of about 3 m and after 12 months has cane (stalk) yields of 50 t/ha (average) to more than 200 t/ha. The sugar content is approximately 10% saccharose, but may attain more than 15% (about one half of the dry matter).

Sugar cane prefers loamy soils with moderately to slightly acid soil reaction. Organic fertilizers (also applied as mulch cover) have a beneficial effect, especially because they improve the structure. Cultivation is for one or several years *(ratoon crop)* [183, 198, 210].

Mineral requirements increase with the yield target. More than one half of the total area under sugar cane is still barely being fertilized (and sometimes has yields of only 15 t/ha); however, the full potential of the plant can be exploited fully only when nutrition is correct.

The *nutrient balance* depends largely on whether only the stalks (the cane) are removed from the fields, or the leaf mass, which contains most nutrients, is also removed. Removals with 10 t cane are only 4 to 7 kg N, 2.5 kg P, and 4 to 12 kg K [187, 214]. This substantial variation is primarily due to luxury removals. Although the leaf mass is lower, it contains more nutrients. Fertilization recommendations are 1 to 1.5 kg N per t expected average cane yield, i.e., about 100 to 200 kg N per ha (in distributed amounts), or 50 kg/ha more when cultivation is continued for several years. Average amounts of K fertilizers are about the same (allowing for the K/Mg ratio) [190].

Fertilization in modern sugar-cane cultivation is based not on general data gained from experience but on *leaf diagnosis*. Diagnostic systems first developed in Hawaii, called *stalk logging* or *crop logging*, in practice imply fertilization based on stalk or leaf analysis. This serves as a check on the effects of basic fertilization and as a basis for establishing the level of necessary supplementary fertilization. *Stalk analysis* consists in analyzing the eighth to tenth internode

Table 7-7. Fertilization of sugar cane according to leaf analysis (nutrient contents in dry matter of leaf blade of third leaf from top of three-month old plants, data according to *Samuels,* Puerto Rico, cit. [190]).

supply state of soil	content ‰ N	fertilization kg/ha N	content ‰ P	fertilization kg/ha P	content ‰ K	fertilization kg/ha K
very low	< 10	300–200	< 1	130–65	< 10	250–170
low	10–14	300–100	1–1.5	65–20	10–15	250–80
moderate	14–15	100–0	1.5–1.8	30–0	15–17	80–0
normal	15–20	0–100	1.8–2.5	0	17–20	0–50
high	20–25	0	2.5–3.0	0	20–30	0
very high	> 25	0	> 3.0	0	> 30	0

(from the top). The following are minimum contents for 6-month old sugar cane, for example: 3.5‰ N, 0.4‰ P, 10‰ K. Fertilization with, e.g., 100 kg N or K per ha is indicated if the above contents have not been attained.

Leaf analysis in *crop logging* according to *Clements* consists in analyzing the leaf blades of the third to sixth leaf (from the top). The data obtained are evaluated by a special procedure [190]. Another form of leaf analysis, used in many countries, is based on the contents of the leaf blades (without the central rib) of the third leaf. Derivation of the fertilization level from the data obtained is shown in Table 7–7, p. 328.

Supplies of trace elements are assessed in the same manner. Minimum contents in the dry matter should be as follows: 20 ppm Mn, 4 ppm Cu, 1 ppm B, 0.1 ppm Mo.

7.5 Fodder Crops

Fodder crops serve to feed domestic animals. The fertilization level of fodder crops thus depends not only on the amounts required by the plants for high yields, but also on the amounts of important minerals required by the animals (Chap. 7.5.2) [191].

7.5.1 Green-Forage Crops

Various plants can be used for green forage; their fertilization may be relatively uniform despite their differences. One should, however, distinguish between:

• Legumes, which supply nitrogen to themselves;
• non-legumes, which require fertilization with N (e.g., grasses and cruciferous plants).

The fertilizer quantities depend largely on the yield target when supplies in the soil are normal. More fertilizer is therefore required for principal crops with higher expected yields, than for intercrops. Sometimes intercrops can utilize the unconsumed residues of fertilizers applied to the preceding crop exclusively, e.g., when the latter were fertilized particularly abundantly with phosphate and potassium.

Many green-forage plants are at the same time also green-manure plants, which are ploughed into the soil as green manure (Chap. 4.3.2)

Fertilization of Legumes

Legumes do not depend on fertilization with N. A starting dose of 10 to 20 kg N per ha at the time of sowing is required only for the first growth phase. Inoculation of the seeds with nodule bacteria is necessary when a given legume species is first grown on any field (Chap. 7.3.4).

The aim of legume growing is to produce fodder rich in protein. Thus, green forage from red clover or alfalfa contains about 4% raw protein at a dry-matter content of 20% (or 15% protein in hay with a dry-matter content of 85%). Nutrient removals are indicated in Table 7–8. Red clover and alfalfa are particularly indicated as fodder crops when soil conditions are good:

- *Red clover* (trifolium prat.) requires medium to large amounts of water and tolerates cold relatively well. The soil reaction should be approximately pH 6 (except in the case of calcareous soils). Good yields vary between 100 and 150 dt dry matter per ha.
- *Alfalfa* (medicago sat.) also requires a good deal of water but tolerates dry phases better (although yields are reduced). A warm climate is preferred. The soil reaction should be approximately neutral. The large boron requirements should be noted. This makes fertilization with 1 to 2 kg B per ha advisable if natural supplies are insufficient. Yields may exceed 150 dt dry matter per ha. The required amounts of fertilizer are indicated in Table 7–8.

Other fodder crops on fields are *white clover, alsike clover, Egyptian clover, Persian clover, serradella, lupins,* etc. *Black medick trefoil, common birds-food trefoil, sainfoin,* and also *lupins* are indicated on poor, dry soils.

Fertilization of Fodder Grasses

Grasses grown for the production of protein are harvested at the stage of abundant vegetative growth. Yields are about 100 to 120 dt dry matter per ha, containing 12 to 15% raw protein. Various plants can be used as fodder grasses. Among those frequently grown are, e.g., Italian rye grass (lolium mult.), rye, oats, Sudan grass (sorghum).

Fodder grasses require intensive fertilization with N for high yields (50 to 100 kg per cutting and ha). However, the amide content in particular is increased by very large N doses. Accumulation of nitrate should be prevented.

Enrichment with some minerals (important for animals) beyond plant requirements is desirable, since the grasses serve as fodder (Chap. 7.5.2). Fertilization of forage grasses may be essentially like that for grassland.

Silage maize occupies an important position among fodder crops, due to its high yields. It should produce starch and protein at yields of more than 150 dt dry matter per ha, containing about 8% raw protein. Fertilization may be the same as for grain maize (Chap. 7.1.3).

Harvesting for silage is carried out at the waxripe stage, at a dry-matter content of about 25%. Silage maize has a relatively small content of some minerals, and this should be taken into account for feeds (Table 5–7, p. 204).

Table 7-8. Yields and removals of green forage (referred to dry matter, modified according to [184, 191]).

Plant	Good yields dt/ha	Nutrient removals per 10 dt kg/ha					
		N	P	K	Ca	Mg	S
red clover	50–100	22	3	20	15	4	2
alfalfa	50–150	27	3	15	20	(2)	1
silage maize	60–120	15	2.5	18	6	3	3
Landsberg mixture	40–60	25	(5)	30	13	3	.
silage rape	40–60	40	4	40	15	.	.

trace-element removals for red clover and alfalfa (in g/ha per 10 dt):
Fe and Mn = 40, Zn and B = 30, Cu = 5-9, Mo = 0.3

Fertilization of Legume-Grass Mixtures

Mixtures often have higher and more reliable yields than pure crops. Various mixtures are in use; the *Landsberg mixture* has been known for a long time. It consists of crimson clover, hairy vetch, and Italian rye grass.

Fertilization is largely the same as for the individual crops. The level of fertilization with N obviously greatly depends on the ratio of grasses to legumes. The greater the share of grasses, the more N fertilizer is required. The problem of fertilization with N is that, e.g., in a grass-clover mixture, the clover reduces its self-supply with N when the amount of N fertilizer is increased (Chap. 7.3.4).

Fertilization with boron, if necessary may be problematic because legumes require much more boron than grasses. Because of the sensitivity of the latter, boron should on no account be applied in excess. Removals are indicated in Table 7-8.

Fertilization of Cruciferous Plants

Rape, rape-seed, and *cabbage* are often grown as field intercrops and have yields of 30 to 40 dt dry matter per ha, containing 20% raw protein. The summer rape called *liho rape* is particularly well known. It is harvested when blossoming begins. Removals are indicated in Table 7-8. Large protein contents require intensive fertilization with N. *Winter rape-seed* can be sown later than rape.

Marrowstem kale has a role to play as a principal crop. It is popular because of its resistance to frost. Yields are 50 to 100 dt dry matter per ha. The large sulfur requirements of cruciferous plants should be noted (Chap. 7.3.1).

7.5.2 Grassland

Immanuel Kant wrote 200 years ago: "It cannot be expected that anyone will ever be able to explain the creation of a blade of grass". This is certainly correct in the philosophical sense, but agrochemistry has meanwhile been able to give a far-reaching idea of the constitution of grass with its many nutrient and active components. It has also permitted the definition of rules towards an ideal composition by suitable fertilization measures.

Grass is not always the same *grass*. Cows can die of hunger even with abundant grass rich in protein, if it lacks the necessary minerals. Moreover, grassland vegetation rarely consists of only one kind of grass, but is mostly composed of various grasses and herbs.

Principles of Grassland Fertilization

Fertilization of grassland has two chief goals [206, 223, 232]:

- a *high yield* of tasty fodder, and thus large animal production (milk, meat, wool, etc.);
- *stable health* and thus long life of domestic animals.

The key to an understanding of fertilization is to consider grass (or the manifold grassland fodder) as a combination of organic and mineral nutrients for animals. Grass (grassland fodder) should primarily contain:

- large amounts of *protein* and carbohydrates (energy carriers);
- optimal amounts of *active agents* (vitamins, etc.) and flavoring substances;
- optimal amounts of minerals;
- no toxic amounts of organic or inorganic substances.

The composition of an ideal fodder depends on the production target and is now largely known (see also Chap. 9.3 on quality problems). The given production target must be achieved on the basis of the site conditions involved, employing corrective measures (fertilization). Two problems should then be noted in general:

a) An optimal *mineral constitution* of the plant not only increases the supply of minerals to the animals, but usually the content of valuable organic matter (protein, vitamins, etc.) too. Missing minerals can indeed be given to the animals directly. However, the lower quality of the organic matter, due to deficient mineral supplies makes it advisable to provide most of the required minerals via the grass.

b) The mineral requirements of *plants and animals* differ in part:

- *similar requirements* of plants and animals: phosphate, sulfur, Ca and Mg in herbs;
- *larger requirements* by plants: potassium, boron, molybdenum; supplies of these elements should therefore be only up to the optimal range for plants;
- *larger requirements by animals:* sodium, chlorine, and compared with Ca and Mg in grasses;
- *requirements only of animals:* see Table 7–9.

Table 7–9. Necessary mineral contents of grassland fodder (for cows, at intake of 12 kg dry matter per day) (according to [205], modified in parts).

Major nutrient element	Contents in ‰ without milk	20 kg milk	micro-nutri-ents	Contents in ppm without milk	20 kg milk	Additional elements required by animals
P	3.3	4.3	Fe	50	60	iodine 0.3
Ca	5.4	7	Mn	50	60	cobalt 0.1
Mg	1.3	1.8	Zn	(30)	50	selenium 0.1
Na	1.3	1.8	Cu	(8)	10	chromium ·
S	(2)	(3)	Mo	0.1	(0.1)	(also beneficial elements)

The target for correct fertilization of grassland thus consists in achieving the required contents in the fodder without exceeding them too much, since very large supplies of a nutrient

- reduce plant intake of other nutrients (very large K supplies lower the Mg content of grass);
- reduce the absorption of other minerals from the fodder (a large Ca content lowers the utilization rate of copper in the fodder);
- toxic damage may occur (e. g., due to heavy metals).

With regard to the *optimal chemical composition* of the fodder, it should be noted that the contents of many elements decrease with progressing growth, due to the dilution effect. Data on contents must therefore refer to a definite growth stage. For grassland this could be the stage of pasture ripeness or hay ripeness, that is, from shortly before to the beginning of blossoming. The raw-protein content at this stage is about 15 to 20% in dry matter, i. e., the N-content is about 30‰.

Fertilization can affect the natural content of organic *toxins* positively or negatively (Chap. 9.3).

The many demands made on the quality of the fodder give a special significance to diagnosing the nutritional condition of the plants. Leaf analysis in particular has proved successful as a basis for applying fertilizers. The composition is additionally assessed on the basis of nutrient ratios, but only a few examples will be given.

Nutrient Supply to Grassland
First the problems of the individual nutrients will be discussed, and then the practical consequences will be summarized in the form of recommendations. The following data on contents are based on a consumption of about 12 to 14 kg dry matter per cow (550 kg) per day. The necessary mineral contents are indicated in Table 7-9.

Nitrogen: The amounts required vary directly according to the desired protein content of the fodder. The annual consumption of N on grassland varies around the world from between 0 and more than 1 000 kg/ha (maximum in the tropics with growth throughout the year).

Phosphate: An old rule states that grass should be fertilized with phosphate up to saturation. However, the P content of grass cannot be raised much above 5%, and this does not appear to be necessary either. The choice of the P form is of minor importance, especially since moist grassland has good mobilization capacity. P deficiency occurs less frequently in sheep than in cows, since sheep have a relatively large food intake, and thus ingest more phosphate.

Potassium: Fertilization with potassium should only be for maximum yields, since animals require relatively little of it. Luxury supplies should be avoided, since they might affect Mg supply. Much potassium is supplied in liquid and semi-liquid manure. Potassium chloride is beneficial to animals as a mineral K form.

Calcium: The large calcium content required cannot be attained easily with grasses alone, as their Ca content is often only 4‰. On the other hand, many herbs contain more than 10‰ calcium, e. g., legumes contain about three times as much Ca as grasses.

The Ca/P ratio should be 1.5 to 2 [205]. The Ca content can be increased by liming, but this should only be done up to the optimal pH-value (Chap. 5.1.1).

Magnesium: Mg is often a minimum factor in grass. Animals suffer from pasture tetany (hypomagnesaemia) if the Mg content of the grass is too low or

Mg absorption from the fodder is inhibited. The required Mg concentration in the fodder for high-performance dairy cows is 1.2 to 2‰. Alternatively, the ratio K : (Ca + Mg) must be less than 2.2 (expressed in val/kg) [217]. Fertilization with Mg must be correspondingly intensive if K supplies are large (Table 5-7, p. 204).

Sodium: Some grasses absorb only little Na and contain less than 0.1‰ (timothy is especially poor in Na). However, some herbs have large contents (more than 4‰, e.g., dandelion, white clover).

It does not seem necessary to cover all the Na requirements of animals via the grass, but a relatively high Na concentration makes the fodder tastier.

Sulfur: Little is known as yet about S deficiency on grassland. Fertilizers should be applied, if necessary.

Iron: Fe supplies usually cause no problems on grassland.

Manganese: Mn concentrations depend greatly on the soil reaction. Grassland should not be limed to above pH 6, if only in consideration of Mn supply. Sufficient Mn is supplied only under good mobilization conditions if the natural pH-value of the grassland is high (e.g., lime-containing fen). Fertilization serves primarily to replenish reserves.

Zinc: Zinc requirements for the milk yield are significantly higher than basic requirements. Natural zinc supplies in grassland vary. All Zn fertilizers may be applied but coupling with Cu fertilizers is generally advisable.

Copper: Cows require 1 ppm copper in the blood. This is achieved at a copper concentration of 8 to 10 ppm in the grass when the utilization of copper in the fodder is normal. Animals often prefer plants or plant parts with higher Cu concentrations (which obviously are tastier), if they can choose their fodder on larger areas.

Cu intake by animals depends to a large extent on the Ca content of the fodder (the Ca content should be below 6 to 8‰). It also depends on the molybdenum concentration, which should be less than 3 ppm. Copper deficiency can usually be eliminated for longer periods by fertilization.

Molybdenum: Plants require more molybdenum than animals, so that no special enrichment is indicated. Concentrations above 10 ppm are toxic. Molybdenum is also important for the activity of bacteria in the first stomach (utilization of cellulose).

Additional elements: The additional elements required by animals are often present in sufficient amounts in the fodder. Any deficiency occurring can be eliminated by fertilization, e.g., through increased *cobalt* supplies. *Iodine, se-*

lenium and *chromium* only rarely cause problems. Selenium is toxic at concentrations above 10 ppm.

Elements beneficial or possibly beneficial to animals, such as *fluorine, vanadium, nickel, bromine, silicon,* etc. have practically no role to play in fertilization. The silicic acid in many grasses occurs in the form of needles and may cause injury to the digestive tract.

Recommendations for the Fertilization of Grassland

1. The aim of grassland fertilization is an optimally composed fodder for *high performance* and *good health* of the animals on the pasture. This optimal composition is now essentially known, at least with regard to minerals.

2. *Nutrient removals* represent the reference basis for fertilization. Removals are

- small on *pastures,* because of large-scale restoration of nutrients;
- large on *meadows,* since nutrients are removed from the field with the hay. Removals of potassium and magnesium are approximately 20 to 30 times as much as from pastures; removals of nitrogen and phosphorus are about five times as much as from pastures.

3. Fertilization also serves to control the *plant population.* The share of grass increases with increasing amounts of N and K supplied, while the share of legumes decreases, which in turn affects the quality of the fodder.

4. The *soil reaction* should and can be slightly lower than on fields of the same soil type. A moderately acid soil reaction should not be exceeded by liming (Table 5.1.1).

5. Grassland often receives abundant *organic fertilizers,* but does not absolutely need it. Supplies of organic matter are large in any case, and the structure is generally good because of permanent root penetration. Supplies of semi-liquid manure should not exceed 20 m^3 (per ha) on pastures. Up to twice these amounts are acceptable on meadows.

6. Fertilization with nitrogen amounts of 50 to 350 kg/ha, depending on the intensity of utilization. Division of the N-dosage is advisable, and nitrogen should be applied at the time of individual cuttings or pasturing. Thus, 240 kg N should be applied as 4×60 kg in the case of four cuttings. Application before a cutting or pasturing is better when the fertilization intensity is high, since otherwise the salt load may be detrimental in dry weather. The N form is of minor importance. Acidifying side effects are mostly beneficial for grassland.

7. Fertilization with *phosphorus* and *magnesium* should permit maximum concentrations in the fodder. Abundant fertilization with potassium is required es-

pecially on meadows, but luxury supplies should be avoided (for fertilizer amounts see Chap. 5.4).

Fertilization with *sodium* is desirable, so that supplies may reach the animals at least partly via the grass. This is done most simply with the corresponding K fertilizers.

8. Proper supplies of *trace elements* are of decisive importance for the performance and health of the animals. The concentrations necessary in the fodder are approximately known (Table 7–9). They should be supplemented by fertilization if natural supplies are inadequate. Balanced supplies of all necessary elements should be ensured.

9. Fertilization of pastures for the production of milk differs from fertilization for meat production. The *production of milk* requires large amounts of energy carriers and protein, as well as higher mineral concentrations, depending on the milk yield. On the other hand, the *production of meat* initially requires fodder very rich in protein, such as is obtained by intensive fertilization with N (e.g., 100 kg N per ha as a first dose) and early utilization. The supply of energy carriers (starch) should be stressed more towards the end. Removals of minerals are much less when the goal is fattening the animals.

10. Knowledge of the *fodder composition* at the time of pasturing or hay-making is a decisive precondition for the production of valuable fodder. Proper fertilization of grassland is possible only on the basis of reliable diagnosis.

8 Fertilization in Horticulture, Forestry, and Special Crops

8.1 Vegetable and Ornamental Plants

8.1.1 Vegetable Plants

Growing vegetables involves a more intensive utilization of the soil than ordinary agriculture. Large harvests and high raw yields are achieved per unit area, both in open-air cultivation and, especially, in cultivation under glass in greenhouses [233, 248, 250, 258].

Abundant fertilization is therefore necessary in vegetable growing. However, as fertilization is a minor cost factor within total inputs, there is a tendency to overfertilization. This implies not only unnecessary inputs, excessive removals, and increased leaching but can also have detrimental effects on the harmonious nutrient supply, if only the primary nutrients are supplied abundantly, and thus on yield and quality, not to speak of induced deficiencies of trace elements and salt damage.

Nutrient supplies are usually abundant in fields frequently under vegetable crops. The contents of available nutrients are often three to ten times as large as in other ploughed land. The problem of maximum desirable nutrient contents has not yet been solved definitively, especially with reference to the upper limit at which fertilization should be stopped completely. However, in open-air vegetable growing the target should probably be contents in the upper part of range C or in the lower part of range D (Chap. 5.2.2). Concentrations may be 50 to 100% higher in greenhouses. Extremely high concentrations, which are not infrequent, are detrimental; this should again be stressed.

A special problem is the *fertilization of young plants* in nursery beds or at transplanting [250]. There are many plants per unit area in the nursery bed. This necessitates intensive fertilization and corresponding irrigation.

Fertilization at sowing: 2–3 g complete fertilizer per liter of earth (this corresponds to between 40 and 60 dt/ha);
3–4 g complete fertilizer per liter of peat,
or 100–150 g complete fertilizer per m^2 of soil
(10–15 dt/ha).

Supplementary fertilization: every 2–3 weeks, 3–4 g complete fertilizer per liter of earth.

Potted plants receive 50 g complete fertilizer per m^2 on two successive days before transplanting. This requires correspondingly intensive watering to prevent salt damage.

Fertilization in *greenhouses* (under glass) is even more abundant than on fields because of the very high utilization. This necessitates:

- Maintaining the, usually, slightly acid soil reaction (no strongly alkaline fertilizers);
- avoiding ballast salts (use of complete fertilizers poor in ballast, in particular, no unnecessary chloride).

Leaf fertilization ensures a precise and rapid nutrient supply (Chap. 6.1.4). Fertilization by sprinkling (Chap. 6.1.3) serves the same purpose.

Fertilizer Amounts Required by Vegetables

Fertilization of vegetables may be relatively uniform, despite the multiplicity of plants and their differing yields. Removals are also relatively uniform if referred to the same yield basis (Table 8–1). They are approximately 10 kg P, 50 kg K, and 5 kg Mg per 100 dt harvested mass. Only fertilization with N must be individually dosed, although even in this case differences, with a few exceptions, are not large.

Fertilization with N: precisely according to requirements, i.e., removal plus a possible safety margin of up to 25% (e.g., for white cabbage with a yield target of 600 dt: $35 \times 6 = 210$ kg N (plus safety margin) (Table 8–1).

Fertilization with P: precise fertilization with P is unnecessary if supplies in the soil are abundant; replacement of the average removal is sufficient (possibly with a safety margin according to supplies in the soil) (e.g., for 300 dt white-cabbage yield: $10 \times 3 = 30$ kg P or 70 kg P_2O_5).

Fertilization with K and Mg: replacement of removals similar to fertilization with P (e.g., for 300 dt yield: $50 \times 3 = 150$ kg K or 180 kg K_2O).

Trace elements: removals approximately correspond to those at high yields in agriculture (see sugar beet).

Table 8-1. Nutrient removals and yields of vegetables (compiled according to [238, 250]).

Vegetable	∅ yield dt/ha	(Total) Removals per 100 dt harvested mass kg/ha				
		N	P	K	Ca	Mg
cabbage vegetables						
white cabbage	600	35				
red cabbage	400	}	7–12	40–60	20–40	4–6
savoy cabbage	350					
cauliflower	350	55				
green kale	350					
kohlrabi	200					
Brussels sprouts	100	200	40	250	150	15
leaf vegetables						
lettuce	300	25				3
chard beet	350	}	5–8	40	12	8
spinach	250	45				
root vegetables						
carrots	400					
beetroot	400					
radish	250					
salsify	250	30–50	7–12	40–60	20–30	6–10
celery	250					
little radish	150					
onion vegetables						
onion	300	30	7	40	15	5
leek	300					
fruit vegetables						
cucumber	250	25	8	40	20	3
tomato	400		4			
pumpkin	900	5		3	1	1
pulses						
pick pea	100	130	20	80	100	
horsebean	120					12
kidney bean	140	90	10	60	70	
runner bean	180					
others						
rhubarb	700	40	10	50	30	5
asparagus	70	300	40	220	140	20

It should be noted that removals often exceed requirements in the case of frequently intensively fertilized horticultural soils (luxury removals). Fertilization according to (excessive) removals in this case also covers the requirements (which exceed removals).

Recommendations for the Fertilization of Field Vegetables

1. Vegetable plants are *intensively grown crops* that are to have high yields (often within minimum periods) of high-quality vegetables. Particularly abundant nutrient supplies are necessary for this. Fertilization must be intensive especially if there are several harvests per year. However, there is a general tendency to over-fertilization in horticulture.

2. Different demands are imposed on the *soil reaction*, but the pH-range between 6 and 6.5 is optimal for many vegetable plants. Liming to the optimal pH-value for the particular soil type seems to be indicated (Chap. 5.1.1). Relatively high pH-values should be lowered (temporarily) by means of acidifying fertilizers, to ensure larger supplies of some nutrients (Chap. 2.1.5).

3. *Organic fertilization* is particularly well utilized by cabbage vegetables, e.g., 400 dt farmyard manure per ha every two years. The additional humus supplies serve primarily for improvement of the structure. Peat, garbage compost, etc. are therefore also suitable.

4. *Fertilization with N* should ensure supplies to the plants in good time before the period of maximum requirement (mass development), e. g.:

- For plants with rapid initial growth: one half or two thirds of the fertilizer amount during sowing or transplantation;
- for plants with slow initial growth: top dressing at emergence;
- for early vegetables with large plant spacing: possibly placement in the root zone;
- pulses also often attain higher yields with N fertilization.

The preferred N forms for initial growth are rapidly acting nitrate fertilizers. However, ammonium forms also act relatively quickly in biotically active soils. Effects on the soil reaction are often more important than the speed of action.

5. *Fertilization with P* may be standardized, despite some differences in requirements (early vegetables require more P than root vegetables). Replacement fertilization is sufficient at the often high concentrations of available phosphate in the soil; occasionally it may be possible to utilize excessive reserves in the soil. Extremely high P-concentrations in the soil lower the availability of some trace elements.

6. Vegetables in general require a good deal of *potassium*; differences in requirements play a minor role when supplies are abundant. However, excessive fertilization is unnecessary, particularly if supplies are already abundant; and it increases the risk of *salt damage*. The correct potassium form is important, that is, plants sensitive to chloride should only be fertilized with potassium sulfate, e.g., potatoes, tomatoes, radishes, onions, beans, cucumbers, melons.

7. The *Mg-requirements* of vegetables should be covered by abundant fertilization with magnesium. Plants producing mustard and leek oils (e.g., onion vegetables) require much sulfur. The nutrition of certain plants (tomatoes, head cabbage, Chinese cabbage) with *calcium* is sometimes problematical on soils with high pH-values too, causing inner-leaf necrosis in cabbage. The Ca-intake as well as the Ca-displacement in the plant is decisive in this respect. Remedial action may to some extent be possible by leaf spraying [255].

8. Supplies of *trace elements* to the plants should be sufficient even when

- availability has been reduced through liming or intensive fertilization with P;
- absorption is inhibited by strong competition of other nutrients (*induced* deficiency).

The specifically large requirements of certain plants should be noted in this connection. Examples are the large boron requirements of cauliflower, beetroot, kohlrabi, tomatoes, and certain pulses. Cauliflower has a tendency to molybdenum deficiency, but abundant supplies in the nursery bed are often sufficient (Chap. 3.1.6). Deficiencies of trace elements necessitate appropriate fertilization or spraying (Chap. 6.1.4).

9. *Salt damage* should be prevented in horticulture even in the case of intensive fertilization with nitrogen and potassium. Salt concentrations of up to 0.5‰ in the soil are quite normal, but concentrations of 2‰ should not be exceeded for plants moderately tolerant to salt. Green cabbage, spinach, and asparagus are highly tolerant; radishes, celery, and beans have a very low tolerance.

10. The vegetable *quality* is improved by increased fertilization as long as nutrient supplies increase from deficiency to optimal (Chap. 9.3). However, excessive supplies of some nutrients (especially nitrogen) easily detract from taste, durability, etc. Excessive supplies of P and K may, e.g., cause induced deficiencies of other nutrients, with further negative effects on the food quality. Accumulation of nitrate in leaves intended for direct consumption should be prevented, e.g., spinach.

8.1.2 Ornamental Plants

Ornamental plants include a large range of plants with different nutrient requirements: potted plants, cut flowers, summer flowers, shrubs, woody plants, and lawn grasses.

Successful nursery cultivation of these intensively grown crops necessitates fertilization according to requirements. Correct regulation of all other growth factors, such as light, temperature, CO_2, etc., must also be maintained for greenhouse cultivation. Fertilization especially in this case should be considered as one measure amongst many and be appropriately coordinated with them.

However, it would be impractical to specialize in fertilization excessively. The transition from the many horticultural *special soils* to *standard substrates* is already a considerable step towards standardization (Chap. 4.4.1).

Certain differences in phosphate and potassium requirements of the various plants and during different growth phases can be covered by uniform abundant supplies, the more so since some luxury consumption scarcely causes any harm as long as no salt damage occurs. Greater differentiation is necessary only in the case of nitrogen. This manifests itself in correspondingly different optimal nutrient ratios (Table 8-2).

Ornamental plants too have periods of larger nutrient requirements. Knowledge of the correct nutrition of the many ornamental plants is still in its infancy and far from complete.

Crucial problems of nutrient supplies to ornamental plants are:

- Selection of the nutrient substrate;
- basic fertilization of the nutrient substrate, thus, the nutrient level;
- amount and intervals of supplementary fertilization (in liquid or solid form);
- nutrient ratios (especially N/P and N/K);
- demands on the soil reaction (especially allowing for a sensitivity to lime);
- different sensitivities to salt (Table 8-2).

Some tentative values for the correct fertilization of ornamental plants are given in the tables below. Some general hints are given in the recommendations and apply primarily to potted plants, cut flowers, and summer flowers.

Hints on lawn fertilization are listed separately. The fertilization of woody plants (deciduous trees and conifers) is discussed in the Chap. "Forest Trees".

A considerable simplification in the fertilization of potted plants is the use of depot fertilizers (Chap. 2.1.4). Their N component is a slow-acting nitrogen, so that the fertilizer amount placed in the pot suffices for a long time. The amount is 0.4 to 1.2 g N liter of peat substrate, depending on requirements. Because of the multiplicity of ornamental plants, the reader is referred for many details to the literature [253, 254, 244].

Table 8-2. Nutrient requirements and salt tolerance of ornamental plants (according to [253, 254, 90]).

Requirements, etc.	Group 1	Group 2	Group 3
nutrient requirements	low	medium	high
salt tolerance	high	medium	low
basic fertilization (g complete fertilizer per liter peat)	0.5	1.5	3.0
concentration of nutrient solution for liquid fertilization (‰)	0.5–(1)(2)	1–4	3–6
maximum salt concentration ‰ for liquid fertilization	2	4	6
maximum salt concentration g/liter substrate	1	1.5	3
optimal nutrient contents in substrate rich in humus (mg/100 ml) N	5–15	10–25	15–30
P_2O_5	4–10	30–60	40–80
K_2O	5–15	40–80	50–100
	adiantum anthurium scherz asparagus plumosus**) azaleas*) camellia japonica gardenia erica gracilis orchideas*) primula obconica**) vriesea splendens	aechmea fasciata anemone coronaria anthurium andreanum aphelandra squarr. cissus, cyclamen euphorbia fulgens freesias gerbera, gloxinias lathyrus odoratus monstera roses	asparagus sprengeri chrysanthemum dianthus euphorbia pulcher. pelargonium saintpaulia

* particularly sensitive to salt
** P and K requirements as for group 2

Note: methods for determining available nutrients: N in KCl extract, P and K in lactate extract.

Recommendations for the Fertilization of Ornamental Plants

1. The *nutrient requirements* of ornamental plants may *differ* considerably. This should be remembered when composing the nutrient substrate (horticultural soils or peat nutrient substrates) and for fertilization.

2. The aim of nutrition from substrate and fertilizers is, as with other plants, to achieve an *optimal content* of nutrients in the leaves, to permit an abundant and correct formation of blossoms, etc. However, experience shows that optimal contents are often exceeded in the cultivation of ornamental plants, since luxury supplies hardly have any detrimental effects.

3. *Overfertilization* and direct salt damage can be prevented by

- low basic fertilization of plants sensitive to salt;
- small and well distributed subsequent fertilizer applications (Table 8–2);
- use of N fertilizers not soluble in water.

The risk of salt damage is greater with peat nutrient substrates than with horticultural soils. Water for irrigation should contain less than 1‰ salt, or less than 0.5‰ for crops grown under glass.

4. Like many other plants, most ornamentals develop best at pH-values between 5.5 and 6. It is only the acid-loving plants, those sensitive to lime, that prefer highly acid soils, pH 4–4.5, e.g., ericaceae, azaleas, blue hortensia, rhododendron, bromeliaceae.

5. Young plants in the leaf synthesis phase require much nitrogen, but in the blossoming stage they need less. On the other hand, phosphate and abundant K supplies are required more at the time of blossoming. The nutrient ratios thus vary in the course of growth; moreover, differences in season, light, etc., have a part to play. Standardization of fertilization, at least according to requirement groups, is nevertheless possible to a certain extent (Table 8–2).

6. However, intensive fertilization with NPK can cause inharmonious nutrition of the plants, apart from the already mentioned risk of salt damage, and in particular may interfere with the supply of trace elements. This causes damage to leaves, reduced flower setting, etc., and should be avoided.

7. Ornamental plants require *trace elements* in approximately the same quantities as other plants.

At higher pH-values, plants sensitive to lime tend to suffer from chloroses caused by the fixation of iron, manganese, etc., in the soil. Trace elements must therefore be taken into account in the composition of substrates and in supplementary fertilization, e.g., in a balanced relationship with suitable complete fertilizers.

8. *Cultivation* of many plants in nursery beds at a high density necessitates appropriate nutrient supplies matched to the growth rhythm of the plants. Intensive fertilization is required before transplanting, since immediately afterwards the plants are sensitive to higher salt concentrations.

9. The fertilization of ornamental plants, especially on peat nutrient substrates (Chap. 4.4.1), can be largely standardized. It is nevertheless generally worthwhile to check the correctness of fertilization prescriptions by *plant analysis* (for all nutrients), to correct even small faults. Bigger faults manifest themselves in deficiency and excess symptoms, a knowledge of which is therefore essential. The symptoms are in principle similar to those in all other plants [253].

10. Ornamental plants grown in aqueous substrates *(hydroponics)* can be supplied with nutrients precisely according to their requirements. Examples of nutrient solutions are given in Chap. 4.4.2.

8.1.3 Lawn

The aim of lawn fertilization is to obtain a green carpet of densely growing ground cover (short grasses). Certain preconditions are necessary for this:

- a correct choice of *grasses* when sowing the lawn, i.e., hardy ground cover grasses like red fescue, short meadow grasses;
- proper *aeration* of the soil, i.e., elimination of stagnant water by drainage of the subsoil or porous vertical drainage gaps reaching the surface, loosening after compaction caused by walking;
- *soil reaction* in the moderately acid region (pH 5-6), to be achieved by suitable liming (Chap. 5.1) or acidifying fertilization;
- adequate *irrigation* (sprinkling) during dry periods, care being taken to ensure proper moistening of the entire root zone;
- frequent *mowing* (at least once a week in summer).

These preconditions form the basis for fertilization as a decisive tending measure. Fertilization should provide abundant *lush green* grass. Pale-green or a yellowish color often indicates nutrient deficiency (mostly of nitrogen, but also of potassium, and occasionally of iron, etc.). Excessive fertilization, especially with nonuniform distribution, causes "burning damage", i.e., drying of the grass because of excess salt in the soil. It is precisely the desirable short grasses that are more sensitive to salt than the longer common grasses.

Fertilization during Lawn Planting (fertilizer quantities per are $= 100 \text{ m}^2$)
Organic fertilizers (e.g., 2-3 bales of peat) and the following basic nutrients are mixed with the soil down to spade depth before sowing:

- Potassium (if necessary), e.g., 20 kg Mg-containing marly limestone;
- about 15 kg PK fertilizer (Chap. 3.2.1).

About 0.5 kg N (e.g., 2 kg lime ammonium nitrate) is mixed into the upper layer of the soil prepared for sowing.

During the first years, a lawn requires much nitrogen not only for the grass, but also for the synthesis of humus. *Nutrient removals*, and thus requirements, depend on the use [206]:

- The rules for use as pasture apply if the mown grass is left as green manure (Chap. 7.5.2). The slight amount of removals necessitates only fertilization with N (especially for the synthesis of humus);
- the rules for use as meadow (Chap. 7.5.2) apply if the grass is removed. Removals are 1.5–3 kg each of N and K, and 0.2–0.4 kg P at a harvest of 0.5 to 1 dt dry matter per are.

Annual Lawn Fertilization (quantities per are = 100 m^2)

Only *nitrogen* is supplied in *spring* and *summer*. Complete fertilizer does not harm the grass, but especially promotes the growth of undesirable weeds. Half a kg N should be supplied at least every month from the beginning of growth onward. Acidifying N fertilizers are beneficial, e.g., 2 kg lime ammonium nitrate, ammonium sulfate nitrate, or 1 kg urea. N fertilizers with especially slow action (Chap. 2.1.4) should be preferred if fertilization is irregular or may be nonuniform. "Burning damage", caused mainly by abundant lawn fertilization, should be prevented. Fertilization is best carried out before mowing, since freshly cut grasses are much more sensitive to salt. Moreover, intensified sprinkling is indicated after each fertilizer application, unless the lawn is fertilized before the rain.

Final fertilization in autumn (October–November) is to protect the grasses against frost. About 5 kg PK fertilizer (or NPK fertilizer) are indicated. This increases the resistance to cold and ensures better growth beginning in spring.

8.2 Fruit Plants and Vines

The many plants yielding fruit can be grouped as follows:

- Tree fruit: pomaceous fruit, stone fruit, shell fruit (nuts), citrus fruit;
- berries of various shrubs and strawberries;
- tropical fruit plants, e.g., banana, pineapple.

8.2.1 Tree Fruits and Berries

Fertilization must indeed allow for the differences in fruit plants, but there is nevertheless a certain uniformity of nutrient requirements within the various groups [240, 242, 235].

There are two important trends in the fertilization of many fruit trees:

- Fertilizer requirements are increasingly determined according to the supply state of the plant by *leaf analysis*. Reliable limit values are already available for a number of plants. Correct sampling is important. Relevant instructions appear in the limit-value tables. The central leaves of one-year old long shoots of tree fruits are frequently analyzed in summer. This method has been very highly developed for citrus fruit in the USA (Table 5-6, p. 204).
- Supplies via the *leaves* are becoming increasingly important, besides the basic supply of nutrients from the soil. Supply via the leaves, when required, is much more effective and faster than via the soil (Chap. 6.1.4) [240].

Fertilization of Pomaceous Fruit (Apples, Pears, Quinces)

Specific fertilization problems are caused by the longevity of the trees. It may be difficult to supply some nutrients near the roots of older fruit trees. Other procedures have therefore been tested, e.g., the injection of nutrient salts or the insertion of corresponding nails into the stem (zinc nails in the case of zinc deficiency). This, however, remains problematic because of difficulties of dosage controls. Organic fertilization should be abundant when fruit trees are planted. It later serves less for nutrient supply than for covering the soil (mulch method) (Chap. 4.3.2).

The *apple tree* is the most important fruit tree; its fertilization has therefore been investigated most thoroughly. Fertilization should at least replace removals; however, the important points of fertilization are best determined by leaf analysis and not from the balances (Chap. 6.1.4). The large potassium requirements of apple trees should be noted. There should be annual N-fertilization. Other nutrients need not be supplied each year. Use of an NPK fertilizer with magnesium (e.g., $12+12+17+2$) at amounts of 100 kg N per ha is technically simple.

Pear trees resemble apple trees in their requirements, but tend more to react to high pH-values with chloroses.

Quince trees may also be fertilized like apple trees.

More detailed information on the fertilization of pomaceous fruit (especially apple trees) is given in the recommendations.

Recommendations for the Fertilization of Fruit Trees

1. Fruit trees are *permanent crops*. They usually stand on the same site for decades, so that the soil is slowly impoverished of nutrients in the root zone. This

is particularly the case if the roots do not further spread after the tree has attained its full size. The properties of the root stock and not those of the specially selected strain grafted on it are decisive for the supply and mobilization of nutrients.

2. The *soil reaction*, especially in the A-horizon, is less important for the trees. However, correct liming according to the soil type (Chap. 5.1.1) is indicated because of the structure improvement it involves.

3. Proper *aeration of the soil*, i.e., a sufficient oxygen content of the air in the soil down to a depth of at least 50 cm, is a precondition for good root growth. The oxygen content of the air in the soil should be at least 5%, and better, more than 10%. Measures to improve soil aeration are: liming of the subsoil before planting, preventing the accumulation of stagnant water, and loosening of the surface.

There are several methods of working the soil to ensure good basic conditions for growth (supplies of water, air, and sometimes also nitrogen). These range from frequent working of the soil with the application of organic fertilizer and green manure to the application of permanent mulch or subsidiary sowing (e.g., with grass).

4. The following should be noted with regard to *mineral supplies:*

- *Nitrogen*, and on light soils *potassium* too, do penetrate into the soil when applied on the surface, but the amounts must be sufficient for a large part to pass through the shallow root zone of the secondary seed;
- *phosphate*, and on heavy soils *potassium*, penetrate only slowly into the lower layers, even if there is no secondary seed. Special deep fertilization is indicated in such cases. This is simply done using holes or fertilizer lances (Chap. 6.1.3). Proper storage fertilization down to the subsoil is indicated before planting in every case. The fertilizer form is relatively unimportant.

5. *Removals* of major nutrients with the fruit are relatively small, except for potassium. However, determination of requirements necessitates allowing for removals with leaves and wood, as well as for nutrient fixation in additional woody growth.

6. *Calcium deficiency*, which manifests itself in "brown-spot disease" of the fruit, commonly plays a role and can be combatted by spraying leaves and fruit with Ca-fertilizers (Chap. 2.5.1).

7. Attention should especially be given to *boron* and *zinc* among the trace elements. Deficiency may limit the yield even without a clearly evident chlorosis,

disregarding the effects on resistance and quality. Deficiencies of these elements are most quickly eliminated by spraying (Chap. 6.1.4).

8. Fertilizer requirements are best determined in fruit growing by *leaf analysis*. The contents to be aimed at are approximately known for the most important kinds of trees (Table 5-6). Continuous checks of the supply state, followed by supplementing the missing substances by fertilization, is therefore indicated. Fertilization according to diagnosis is much more reliable than an estimation of fertilizer requirements on the basis of removals.

9. Expenditure on combatting pests and diseases may be considerable in fruit growing. However, correct fertilization (better nutrition) increases the *resistance* of trees to infections and enables them to tolerate climatic stress better (less frost damage), although even the best nutrition can hardly ensure complete immunity.

10. The *quality* of the fruit is improved by an adequate and balanced nutrition. This refers to the important ingredients such as vitamins and fruit acids that contribute to the nutritive value and taste, as well as to the external appearance (absence of scab and tears) and to storage properties.

Fertilization of Stone Fruit (Plums, Cherries, Peaches, Apricots)

Plums, damsons, yellow plums, and *greengages* utilize organic fertilizer well. Nutrient supplies should be about 100 kg N, 25 kg P, and 80 kg K per year and ha.

Sweet and sour cherries are sensitive to higher soil reaction: their nutrient requirements are about a quarter more than plums. Fertilization should therefore be on the basis of 120 kg N per ha and year.

Peaches are fruit from warm zones and are frequently grown on light soils. This makes division of the yearly N doses advisable. Lime chloroses are frequent; acidifying fertilizers are therefore indicated on neutral or slightly alkaline soils, so as to mobilize important trace elements. Fertilization is best based on leaf analysis. General tentative values are 125 kg N, 40 kg P, and 150 kg K per ha.

Apricots should be fertilized similarly to peaches.

Fertilization of Nuts (Shell Fruit)

Relatively few nutrients are removed from the soil by *walnuts* and *hazel nuts*. Walnut and hazel nut trees often receive only organic fertilizers. (Complete) fertilization on the basis of 50 kg N per ha is otherwise sufficient.

Yields of *cashew nuts* (anacardium accid.) in warm zones often exceed 20 dt/ha. Fertilization of grown trees with 0.5 kg N per tree and corresponding amounts of P and K seems indicated. Suitable reductions are possible when there is organic fertilization.

Yields of *pecan nuts* (carya illin.) in warm zones may exceed 2 dt per mature large tree. The recommended fertilization is 1–3 kg N per tree (with corresponding amounts of P and K, if necessary). The pH should not greatly exceed 6, since there is otherwise a tendency to zinc deficiency.

Fertilization of Berries

Red currants, gooseberries, raspberries, and *blackberries* require much potassium, but are exceptionally sensitive to chloride. Fertilization with K should therefore only be in the form of potassium sulfate. Average amounts are 50 kg N, 20 kg P, and 120 kg K per ha. NPK fertilizers containing little chlorine are indicated (e.g., 12 + 12 + 17 + 2).

Black currants require more nitrogen (about 120 kg/ha) and are less sensitive to Cl.

Bilberries (vaccinium hybr.) are highly sensitive to lime and require an acid substrate. Soil acidification with aluminum sulfate may be necessary. Only N fertilizers with acid action should be considered. About 50 kg N per ha are required, e.g., in the form of NPK fertilizer (as for other berries).

Strawberries, whether annual or perennial, should be planted in organically well fertilized soil (farmyard manure) (e.g., 500 dt/ha). The soil must also be well supplied with phosphate and potassium, as for intensively grown vegetables. N requirements are approximately 100 kg/ha, e.g., as NPK fertilizer (as for other berries). Division of the dose is indicated if nitrogen alone is supplied (in spring and after the harvest).

Fertilization of Citrus Fruit

Oranges, lemons, tangerines, grapefruits, etc. grow in subtropical zones. Good yields of a mature tree are, e.g., 300–500 oranges. Nutrient removals per ha are approximately 150 kg N, 15 kg P, 250 kg K, 300 kg Ca [190, 236, 251]. The following recommendations are made for fertilization:

1. Citrus trees prefer a slightly acid soil (about pH 6).

2. Water requirements are large (if only for climatic reasons).

3. Citrus trees (like apple trees) require good soil aeration and therefore good soil structure. Organic fertilization is therefore beneficial, e.g., 10 to 20 t/ha.

4. Citrus trees should be provided with storage fertilization before planting. Mycorrhiza fungi play an important role in nutrient absorption.

5. High yields presuppose abundant fertilization with N, but one-sided N supplies are undesirable. Tentative values might be 100 to 150 kg N per ha.

6. Supplies of P, K, and Mg should be abundant, even though P removal by the fruit is only small. Good natural supplies of K often exist in soils of arid zones. Depot fertilization for several years may amount to, e.g., 1 kg P and 1–2 kg K per tree.

7. Lime chloroses are frequent, especially at high pH-values. They are due to deficiencies of Fe, Zn, etc. and can be eliminated at least temporarily by direct spraying.

8. Citrus trees tolerate salt moderately well. However, they are sensitive to boron excess (irrigation water should not contain more than 0.5 ppm B).

9. Fertilizer requirements are best determined from leaf analysis. Much data on the required supply levels are available, particularly for citrus trees (Table 5-6).

10. Correct fertilization is also decisive for various external and internal quality characteristics of the fruit.

8.2.2 Tropical Fruit Plants

Fertilization of Pineapple

Pineapple (ananas sp.) is a continuously growing perennial plant from warm zones. One plant permits two harvests in three years. The yield target for the first harvest is 50 t/ha. Removals per 10 t fruit are as follows (the lower values refer to one fruit harvest, the higher values to two fruit harvests and the plant mass): 6 to 25 kg N, 1 to 8 kg P, 30 to 80 kg K, 1 to 20 kg Ca and Mg [190]. Recommended fertilization is as follows:

1. Pineapple prefers a soil reaction of pH 5.5–6.

2. The relatively weak root system requires proper nutrient supplies to the soil for high yields. Single-plant fertilization may be indicated in view of the plant spacing (4 plants per m²).

3. Organic fertilizers like compost, mulch (e.g., from harvest residues) have a beneficial effect.

4. Pineapple is sensitive to N deficiency, reacting with slight to more severe chloroses. The color of the leaves may serve as an indicator for current fertilization with N.

5. Potassium is the major nutrient most required. The youngest ripe leaf should contain at least 35‰ K in the dry matter. Potassium sulfate is preferable, presumably because of the additional sulfur supply.

6. Trace elements are frequently deficient (e. g., Fe, Zn, Cu). Acidifying fertilizers are therefore indicated. Preventive leaf spraying is also carried out over whole areas.

7. P removals are small, but the soil should be well supplied with phosphate.

8. Magnesium should be stressed in its relationship to potassium.

9. Pineapple absorbs nutrients well via the leaves, so that part of the fertilizer can be scattered on the base leaves or applied by spraying.

10. Fertilization may be based on removals, but leaf analysis is much more reliable.

Fertilization of Bananas

Bananas (musa sp.) grow in humid, warm zones as shrubs of several meters in height. The fruit is formed within three months after a vegetative growth that lasts a year. The fruit-bearing stem then dies. Good yields provide up to 30 t fruit per ha. Major nutrient removals per 10 t fruit are 20 kg N, 2 kg P, 60 kg K, and 5 kg each of Ca and Mg. Requirements (to be covered from the soil and by fertilization) are about four times as much [190]. Proper nutrient supplies to young plants are important (Fig. 8-1). The following recommendations are made for fertilization:

1. Bananas prefer a slightly acid soil.

2. Water supply and soil aeration must be balanced.

3. Organic fertilizer is beneficial (in the soil and as mulch).

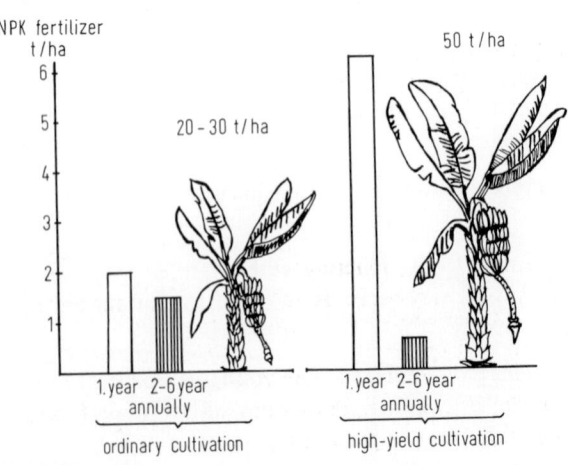

Fig. 8-1. Banana yields as a function of fertilizer distribution (Trinidad, fertilizer 9+9+35, according to *Twyford* and *Walmsley,* 1974, cited in [159]).

4. Proper fertilization (also of the subsoil) during planting is indicated. Placement of the fertilizer may also be indicated, in view of the low plant density (2000 plants per ha).

5. Fertilization with N is necessary on most soils; acidifying N fertilizers are preferable.

6. Potassium requirements are particularly large; most of the potassium is absorbed during fruit development.

7. Deficiencies of manganese and zinc among the trace elements are fairly frequent.

8. The correct fertilizer amounts lie between removals and requirements. Fertilization based on leaf analysis is better than the application of average fertilizer quantities. The following are the minimum contents in the third leaf from the top: 30‰ N, 2‰ P, 35‰ K, 3‰ Mg [190].

9. Precise fertilization is recommended especially in order to increase the resistance to fungal diseases. Deficiencies of N, K, Mg, Mn, etc. or inharmonious nutrition (too high NPK/Mg or K/Mg ratios) promote attacks by harmful fungi.

10. Direct nutrient supplies are advisable for reasons of fruit quality, too.

Fertilization of Other Tropical Fruits [190]

Avocado (persea americana) has similar, though larger, nutrient requirements to citrus trees. The optimal content is 16 to 20‰ N in the dry matter of young leaves. This content can often be obtained by fertilization with N at a level of 100 to 200 kg/ha. The optimal soil reaction is approximately pH 6.

Mango (mangifera indica) grows into big trees having the same nutrient requirements as large apple trees. The optimal pH is approximately 5.5–6. Fertilization with N is particularly important, since insufficient N supplies may cause alternating fruit-bearing. K requirements are large. Calcium deficiency causes softness of the fruit.

Guava (psidium guajava) is a fruit with a very large content of vitamin C. Fertilization may be like that of citrus trees.

Papaya (carica papaya) is a quick-growing tropical fruit with large nutrient requirements, with an especially large continuous N requirement. A pH of 6 is indicated. The sap is also used for the production of the protein-splitting enzyme papaine.

Date palms (phoenix dact.) grow in hot, arid zones and require much water. They prefer slightly alkaline soils. P and K requirements are mostly covered from soil reserves, so that only fertilization with N is necessary (1–3 kg N per tree and year). Date palms are highly tolerant to salt (up to 3‰ salt in the soil).

Fig trees utilize nutrients in the soil well and are quite tolerant to salt. About 50 kg N per ha are often advisable. K requirements are large.

8.2.3 Vines

Vines (vitis vinifera) are perennials growing in mild wine zones. At a plant density of about 5000 vines per ha they produce a yearly grape yield that provides 50 to 100 hectoliters of grape juice. The yield of table grapes may attain 40 t/ha. Nutrient removals per 100 hl yield are approximately 100 kg N and K each, 5 to 15 kg P, 25 kg Mg, 100 g B, 150 g Zn, etc. Only a small part of this is contained in the juice itself, e. g., one tenth of the N [257]. Fertilization is recommended as follows:

1. The optimal soil reaction is approximately pH 6.

2. The principal root mass is located at a depth of 20–60 cm. This requires a deep, *well aerated* soil; correct liming and intensive organic fertilization (500 to 700 dt/ha) are advantageous.

3. The *mobilizing power* of the root stock of a vine type is important for nutrient supplies.

4. *Storage fertilization* down to the subsoil is advisable when vines are newly planted: liming with Mg lime (if necessary), 200 kg P, 800 kg K, 120 kg Mg per ha.

5. Proper *N supplies* are important but should not be one-sidedly exaggerated. The usual dose, depending on the yield, is 50 kg N, in Germany it is often 100 kg N and up to 200 kg in areas with serious losses of N from steep slopes. Most of the nitrogen should be applied at the beginning of growth, the remainder after blossoming. Urea at a concentration of 0.7% is suitable for leaf fertilization.

6. The target for P supplies to vineyard soils is twice the quantities applied in agriculture. Usual amounts of P with intensive fertilization are twice the removals. This may lead to extremely detrimental interactions such as increased immobilization of zinc.

7. *K requirements* are large, to attain high yields and increase resistance. Concentrations of available potassium in the soil should be approximately twice as high as in agriculture. Average doses for high yields are 200 kg K per ha. The form of the K is of minor importance. However, potassium sulfate seems to be superior in arid zones.

8. *Mg* tends to be deficient especially when there are very large K supplies.

9. *Boron* is the most important *trace element.* The vegetative point dies if there is a deficiency. Fertilization is advisable if the boron concentration in the leaf becomes less than 10–15 ppm. Fertilization with B is effected, e. g., with 2 kg per ha every two years. Too large B supplies may cause harmful boron excess. Deficiencies of Fe, Zn, etc. are also important in viniculture [239].

10. Harmonious, adequate nutrition promotes the *resistance* of vines to cold and fungal attack, and improves the *quality* of grapes and wine. Nutrition is best checked by leaf analysis.

8.3 Forest Trees

8.3.1 Principles of the Fertilization of Forests

Forestry is a long-term production process with far-reaching utilization of nutrient reserves in the soil. Forests have already had to recede to unfavorable positions, but even the reserves in relatively poor soils suffice for a considerable production of timber, owing to the slow growth and the small removals.

Nutrition of *forest trees* is based on the following facts (Fig. 8–2):

- Relatively small requirements by trees, as compared with field plants;
- small nutrient removals by timber;
- utilization of available nutrient reserves over large areas and to considerable depths;
- adaptation to a low soil reaction, at which the mobilization conditions for most nutrients are good (especially for most trace elements);
- large-scale restoration of nutrients to the soil with leaves, etc.;
- small losses from the nutrient cycle.

Fertilization thus plays a much smaller role in forestry than in agriculture. The share of fertilizer nutrients in the total nutrient supplies of forests is probably much less than 1%. The *principal uses* of fertilization in forestry are [237, 246, 259, 261]:

- Fertilization of young trees in the nursery;
- fertilization for soil melioration (before planting trees or in older plantations) to intensify nutrient circulation, making good earlier utilization damage, etc.;
- precise fertilization to improve nutrition and thus tree growth. This not only increases the addition of timber but also renders upwardgrowth more reliable and homogeneous, and increases the resistance of the trees.

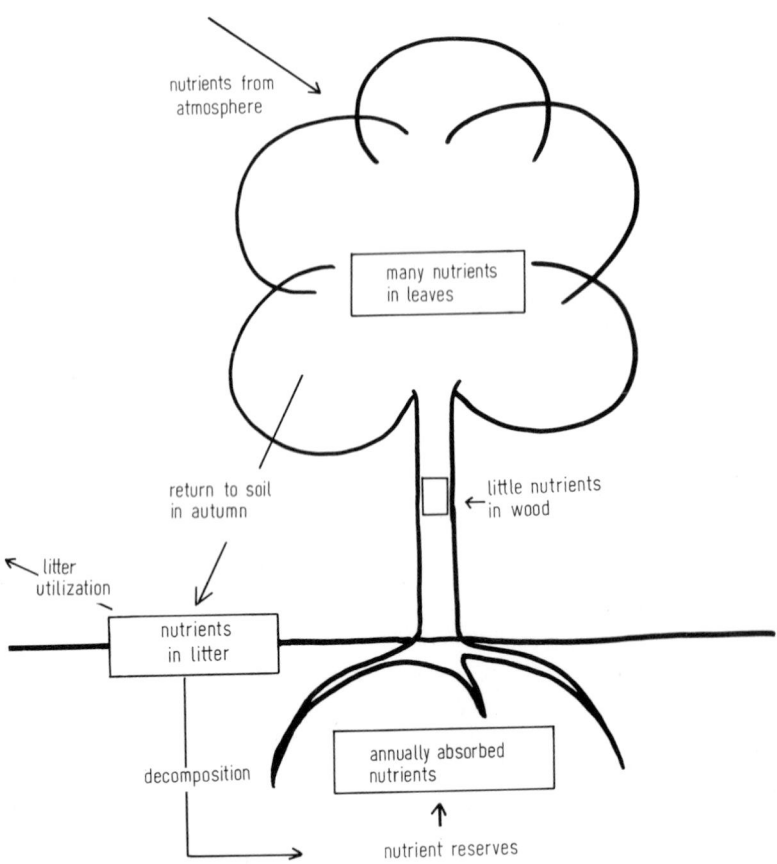

Fig. 8-2. Nutrient cycle in forests.

Nutrition of Forest Trees

The high capacity of trees for exploiting nutrient reserves in the soil is largely due to fungi living symbiotically in the roots (basidiomycetes). This *ectotrophic mycorrhiza* is common in the fine roots (instead of the hair roots) of many trees, e.g., most conifers as well as oak, beech, linden, ash, birch, etc. The mycorrhiza fungi mobilize mineral reserves from the soil and in turn obtain organic matter from the tree. This symbiosis between tree and fungus should be promoted and not disturbed by fertilization. Mycorrhiza fungi prefer a low soil reaction. This should be taken into account in liming. Nitrogen fixation by legume trees with their nodule bacteria is also important, e.g., robinias and acacias. The same applies to alders with symbiotic actinomycetes.

Fertilization in forestry can and should serve only as a (mostly slight) *supplement* to natural nutrient supplies. It must be aimed at the minimum factor [237]. Correct diagnosis of the existing nutrient state is therefore of considerable importance.

Fertilization in forestry, much more than in agriculture, should be considered as only one measure among many to increase yields. Fertilization often has multiple effects with many positive effects possibly countered by negative ones.

For example, fertilization promotes the growth of larger trees, while at the same time younger trees are suppressed by abundant soil flora, which are detrimental to natural rejuvenation.

The fertilization of forests is therefore a "wide field", and the many interdependent links are still only partly recognizable in outline. The considerable site differences only make it possible to evaluate information from fertilization experiments to a limited extent.

A factor rendering correct fertilization difficult is that different tree species and races, and possibly also single trees (clones) have different nutrient requirements and growth performances.

An unknown factor is often the supplies to trees via the leaves, e. g., with nutrients from precipitation or in fine dust (the latter may have a role to play in the case of eucalyptus trees).

The *timber quality* should remain good even when production is increased by fertilization. Effective fertilization does increase the width of annual circles, as has been established especially in the case of spruce and pine. The density (strength) is thus slightly reduced, but remains within the normal range of variation. This can therefore hardly be called quality reduction. The aim of timber production is to obtain a wide wood mantle without branches. Fertilization, or at least intensified fertilization, should therefore only begin after the natural clearing of branches or after *branch cutting*.

Nutrient Requirements and Diagnosis of Fertilizer Requirements

Correct fertilization of forests necessitates a precise knowledge of the nutrient balance of the soil and the requirements of trees. Nutrient removals in trees through storage in the wood is small. Removals at a cutting cycle of 50 to 100 years are 5 to 25 kg N, 0.5 to 2 kg P, 2 to 12 kg K, 2 to 3 kg Mg, and 5 to 20 kg Ca per ha per year (deciduous trees have larger Ca removals). These small amounts can be supplied by almost any soil. However, annual requirements are three to five times larger because of the necessary leaf formation. The nutrient requirements of forests are about 40 to 80 kg N, 10 to 20 kg P, and 30 to 60 kg K per ha per year [241].

Nutrient supplies from the soil can be determined

- approximately by estimating or establishing the mobilized nutrients in the soil;
- directly, and relatively reliably, by determining the nutrients absorbed (leaf analysis).

The potential *nutrient supply* in the root zone of the trees depends on the depth of root penetration, the amounts of nutrients mobilizable during a growth period of, e.g., 100 years, and mobilization factors like temperature, moisture content, soil reaction, etc.

The content of nutrients in the soil, mobilizable over a long period, can be roughly estimated from total reserves. Direct determination of available nutrients (as in agricultural soils) only seems useful for tree nurseries.

Apart from the diagnosis of deficiency symptoms, there remains *leaf analysis* (Chap. 5.3.3) [237, 247]. The correct timing of sampling for leaf or needle analysis is decisive, since nutrient concentrations vary in the course of time. The concentrations of many nutrients decrease because of dilution, after which there is a phase of constant concentrations. On the other hand, the concentrations of certain elements increase continuously, e.g., Ca, Mg, Mn, B. Sampling of spruce should be done between October and February, since concentrations are relatively constant in this period. For pine, the most suitable time is October and November; needles from the youngest vintage from the uppermost lateral shoots should be sampled, especially from trees of the second principal tree class. Averages of the values obtained from 10 to 30 trees are required [260].

Table 8–3. Limit values from leaf analysis for optimal nutrition of forest trees (according to [237, 263]).

Tree	Contents in ‰ of dry matter of needles (leaves)					Plant part and time
	N	P	K	Ca	Mg	
spruce, young spruce	15	2	7	(3)	1.3	needles of uppermost lateral shoots, October/November
spruce ⎫	14	1.3	5	1	1.1	
pine ⎬ older trees	18	1.5	5	(1)	0.7	
fir ⎭	13	1.6	(5)	(3)	(1.2)	
beech	(20)	(1.5)	(7)	(6)	(1.3)	leaves of upper tree-top part, end of August
poplar (black poplar)	(20)	(3)	(20)	12	2	

limit values of trace elements for spruce (in ppm): Fe = 40, Mn = 30
Zn = 30, Cu = 3
B = 15, Mo = 0.05

The nutrient contents determined by analysis, sometimes called *level values*, must then be compared with the required contents, or yield limit values.

Establishment of limit values is more difficult for forest trees than for field crops. Definite limit values can be determined only for very young trees in pot experiments, but these data can be applied only conditionally to older trees. Yield limit values of nutrient contents in trees of medium or great age were therefore mostly derived from comparisons of trees growing well and poorly, respectively. This is a useful approximation method. Tentative limit values are shown in Table 8–3, p. 360.

Leaf analysis also permits the diagnosis of possible intoxication or extreme nutrient ratios. Thus, spruce on highly acid soils contains more than 5000 ppm manganese (i. e., more than a hundred times the necessary concentrations). However, such concentrations are still not considered to be toxic, but are probably detrimental. The concentration of aluminum in pines may attain 500 ppm. This raises the question of possible toxic effects.

However, the concentration of certain substances in the leaves may also be of interest for other reasons. Thus, the silicon concentration in larch may attain 10‰. This might be important in terms of resistance to parasites.

Fertilization and Resistance of Trees

The links between nutrition and resistance of trees (Chap. 6.5) are especially important. Trees grow for many years and are also exposed to rare extreme stress situations (e. g., "once in a century" storms, particularly heavy precipitations of snow, and especially intensive attacks by parasites). When fertilization helps to better overcome such situations, its contribution to timber production far exceeds the increased accretion of wood growth. On the other hand, fertilization should not be allowed to reduce the resistance over the long term, despite possible short-term growth increases.

Some stress situations due to *climatic factors* may be snow pressure or frost damage at the beginning of growth, in which case proper K supplies are beneficial. Large quantities of wet snow constitute a considerable load on branches and may cause "snow breaks". Fertilization with N, especially of spruce, gives rise to widely branched tree tops densely studded with needles. The growth increases the nitrogen brings about thus cause a loss in the final analysis, which should be prevented by appropriate care of the tree tops.

The resistance to *immision damage* is generally increased by any fertilization that improves nutrition of the trees, and the resistance to *diseases* and *pests* plays a considerable role in forests. It should obviously not be expected that correct, adequate fertilization alone will protect the trees from every attack by parasites. However, many attacks by parasites can be limited or even largely prevented. Adequate supplies of K increase the resistance to various fungal diseases (mildew, blight of poplar trees) and reduce attacks by biting and sucking insects.

On the other hand, unbalanced nutrition of trees may be the basic cause of various kinds of damage, e.g., intensified attacks by sucking insects when there is one-sided fertilization with N, or increased heart rot of spruce, due to alterations within the nutrient supply, caused by a larger increase in soil reaction. Nibbling by game increases somewhat after fertilization, since the leaves become tastier (better fodder quality through fertilization).

8.3.2 Conifers and Deciduous Trees

Fertilization of Young Trees in Nurseries

Nutrient requirements and fertilization of seedlings can be assessed best of all. They should be assured a rapid growth start. There are large nutrient requirements per unit area in the first two years, when the plant density is high. Good nutrition in youth seems to be beneficial even when the seedlings are later transplanted to relatively poor soils. *Nutrient removals* during the first two years are [246]:

- for *conifers:* (8 million plants per ha): 120 kg N, 15 kg P, 50 kg K, 60 kg Ca, 7 kg Mg;
- for *deciduous trees* (1 to 2 million plants per ha): 150 kg N, 25 kg P, 90 kg K, 150 kg Ca, 25 kg Mg.

Fertilization may be based on these values, allowing for supplies in the soil. The concentrations of available nutrients should be the same as for intensively grown agricultural crops (Chap. 5.4).

The requirements of the various tree species must be considered separately, but the following rough classification is sufficient:

- Small requirements: most one-year seedlings;
- medium requirements: one-year seedlings of pine and more-demanding deciduous trees, several year old conifers;
- large requirements: several year old deciduous trees.

Fertilization for medium requirements, with medium supplies in the soil, should comprise at least about 50 kg N, 40 kg P, and 50 kg K per ha. These amounts should be doubled when supplies in the soil are poor.

Correct adjustment of the soil reaction is also important. Most trees prefer acid conditions; a pH of about 5 in sandy soils and about 5.5 in loamy soils is therefore indicated. The pH-target should only be neutral for calcicolous trees.

Fertilization of Young Trees in the Field

Fertilization for soil melioration (if necessary) is of primary importance. Deeply ploughed soil should also be limed deeply, if necessary. Average doses are 40 to 80 dt CaO per ha. However, liming must match the specific requirements of the trees, and should never be excessive. (Part) liming with smaller doses is recommended before the old trees are cut down.

Melioration should be in accordance with site conditions. Raw humus can be decomposed by nitrogen fertilizers and partly converted into valuable humus forms. The introduction of ammonia and the use of solid N fertilizers have proved successful. Organic fertilizers or green manure may be indicated on sites poor in humus. Soil-life activity, stimulated by all these measures, contributes much to improved soil fertility and thus provides the proper preconditions for the flourishing of newly planted trees.

Fertilization of *young plants* with primary nutrients in poor soils should be precisely applied into the seedling hole (allowing for maximum doses). Wide-area distribution is otherwise simpler, e.g., 70 kg P and 100 kg K per ha when supplies in the soil are at a moderate level.

Fertilization of young trees during planting is almost always worthwhile, at least with regard to some nutrients. Rapid growth in youth not only prevents certain afforestation risks, but also shortens the time required before utilization can commence (which improves profitability). Fertilization with potassium is preferably carried out in the second year and not in the first, to prevent high salt loads. Repetition of fertilization in the first years may be indicated.

Fertilization of Trees

Certain minimum factors in nutrition can be improved by precise fertilization at any tree age. Fertilization during the decade-long polewood and timber growth stage cannot be based on any prescription, since site conditions vary widely. Correct diagnosis is then decisive to recognize requirements. Fertilization towards the end of the utilization period (10 to 15 years before tree cutting) is especially indicated from the profitability point of view, since this capital shows returns within a reasonable time.

Fertilization for the purpose of *melioration* may also be indicated for older trees, i.e., for activating an interrupted nutrient cycle (Fig. 8–2) or for compensating excessive losses (e.g., through use as litter). Nutrient blockade is caused by the accumulation of raw humus, in which considerable quantities of nutrients may be fixed and thus removed from circulation. These may amount to, e.g., about 2500 kg N and 120 kg P per ha in the uppermost 10-cm layer.

Fertilization of Conifers

Nutrient requirements of conifers differ (Table 8-4). Forest soils supply 6 to 100 kg N per ha per year through mineralization. The N requirements of spruce and pine, approximately equal, are 50 to 100 kg/ha. The required fertilizer amounts are then obtained from the respective difference, allowing for the utilization rate in the first year after application, which in forests is only 20% However, fertilization on the basis of tentative values has proved superior, since supplies in specific cases are unknown, unless more accurate information is provided by needle analysis. The following *fertilizer amounts* (per ha per year) are recommended in case of insufficient supplies in the soil:

- Nitrogen for spruce and pine   50 to 100 kg N
- nitrogen for larch   40 to 60 kg N
- phosphate ⎫   70 kg P
- potassium ⎬ for most conifers   100 kg K
- magnesium ⎭   30 kg Mg

The form of the fertilizer is of minor importance in this case. The fertilizers are applied by scattering or blowing with suitable equipment. Fertilization obviously benefits not only the trees directly, but also the soil flora and soil life. Fertilization of forests has residual effects lasting several years, sometimes even many years. Fertilization models have been established to realize multiple fertilization aims [237, 241, 259].

Fertilization of Deciduous Trees

In considering the fertilization of deciduous trees, it should be remembered that these trees grow on the best forest areas (even if the total forest area is shrinking). Moreover, their nutrient requirements are mostly lager than those of conifers (mainly for nitrogen and calcium, but also for magnesium). *Robinia, poplar, ash,* and *elm* require particularly large contents of bases in the soil, i.e., mainly calcium and magnesium, but also potassium. Oak and copper beech, on the other hand, have smaller requirements. The Ca concentration in leaves is approximately 10 to 20‰ of the dry matter. Poplars require particularly large amounts of potassium.

Contents of trace elements vary widely according to site. Thus, manganese concentrations are high in acid soils but may drop to the lower supply limit when the pH increases. The following sequence represents the reaction to fertilization, especially with N [237]: Alder > hornbeam > oak > copper beech > birch.

The nutrient requirements of important deciduous trees are summarized in Table 8-5. Eucalyptus has been included as a deciduous tree from warm zones.

Table 8–4. Nutrient requirements of conifers (according to [237]).

Tree	Required soil reaction	Large requirements for	Good utilization of nutrients in soil	Remarks
spruce (*picea abies*)	highly to moderately acid	N + P	K and Ca	N supply frequently deficient, annual requirements about 50 to 70 kg N per ha, with 100 kg N accretion of about 10 cubic meter timber per ha over entire period of action;
pine (*pinus sylvestris*)		N + K	(N), P, Ca	N supply frequently deficient on raw-humus sites, with 100 kg N accretion of 4 to 7 (up to 10) cubic meter timber per ha over entire period of action
fir (*abies alba*)	moderately to slightly acid	P + K	K	N and P supplies sometimes deficient
Douglas fir (*pseudotsuga*)		N, P, K, Ca trace elements		N and P supplies sometimes deficient; fertilization with N as for spruce
larch (*larix*)		P and Ca	N and K	Larch demands little and reacts to fertilizer better than other conifers; N and P supplies sometimes deficient

Table 8-5. Nutrient requirements of deciduous trees (according to [237]).

Tree	Required soil reaction	Large require- ments for	Remarks
copper beech (fagus)	5.5–7.5	N, Ca	natural supplies often sufficient
oak (quercus)	5–7	N	grows well also on acid sites, durmast oak has larger requirements than English oak
alder (alnus)	about 6	P, K, Ca	fixation of N through symbiosis with actinomycetes
robinia (robinia)	about 6	Ca	as legume supplied with N through root nodules
birch (betula)	about 5	—	undemanding deciduous tree whose growth is often stimulated by fertilization
poplar (populus)	7	all nutrients	large fertilizer requirements simi- lar to agricultural crops, depending on supplies in soil
eucalyptus	(some- times) 7	Ca, K	small nutrient requirements, but growth is stimulated by proper sup- plies of P and K at time of planting

Recommendations for the Fertilization of Forests

1. Trees produce timber over long periods and are nourished almost solely from natural nutrient reserves in the soil. Supplementing these supplies by fertilization plays only a minor role, and only affects yields significantly if aimed at the minimum factor.

2. Any fertilization measure for the purpose of improving soil fertility or for direct supplementation of nutrient supplies must be judged in conjunction with all growth factors, especially climatic and biotic ones.

3. Temporary benefits of fertilization are less important than long-term posi- tive effects on timber yield, resistance, etc. Short-term benefit may cause failures over long periods.

4. Most trees, in particular conifers, prefer highly to moderately acid soils (ap- proximately pH 5) in the root zone. Liming of the more deeply located root zone is therefore not advisable, except in cases of extremely acid soils.

5. The significance of frequent important liming lies in the increase of biotic activity in the humus cover or uppermost soil layer. This reactivates the partly blocked nutrient cycle.

6. Fertilization with N serves the dual purpose of direct, increased N supplies to the trees and general improvement of the humus, so that supplies of several nutrients are increased indirectly.

7. Fertilization with the other major nutrients plays a minor role except in tree nurseries and possibly during planting. Precise fertilization is indicated, and thus also profitable, only on sites where supplies of a nutrient element limit production.

8. The decisive precondition for the correct fertilization of forests is determination of the minimum factors. This can be done by long-term fertilization experiments or relatively reliably through leaf analysis.

9. Any fertilization improving the nutrition of trees generally also increases their resistance to climatic stress and attacks by parasitic organisms. Correct nutrition is an essential precondition for tree health.

10. Precise fertilization raises yields over long periods and to an approximately constant timber quality and is therefore profitable.

8.4 Stimulants, Spices, and Medicinal Plants

8.4.1 Stimulant Plants

The aim is to produce plant parts containing, besides aromatic substances and nutrients, *stimulants*, i.e., substances stimulating the body and mind (caffeine mainly in coffee and tea, theobromine in cocoa, nicotine in tobacco).

Fertilization of Coffee Trees
Coffee trees (coffea arabica or robusta) are perennial plants growing in tropical highland forests. They attain maturity after 10 years and remain productive for 20 more years. The time between blossoming and ripening of the fruit is 6 to 10 months. Yields of 1 to 4 t marketable coffee per ha are obtained at a density of 1 000 to 3 000 trees per ha. Arabica coffee is more valuable, and robusta coffee has higher yields. Coffee beans contain about 1.3% caffeine. Coffee trees prefer deep, permeable, well aerated soils rich in humus, and have large nutrient requirements [190, 243, 210].

Nutrient removals per 1 t marketable coffee beans with fruit pulp are about 40 kg N, 3 kg P, and 40 kg K. Fertilizer requirements are three to four times as much. An example of fertilization based on leaf analysis is given in Table 8-6.

Table 8-6. Fertilization of robusta coffee trees, based on leaf analysis (Ivory Coast, according to *Loué*, cited in [190]).

Content in dry matter of third leaf, ‰	Symptoms and fertilization per tree
N 15–18	leaves yellowish to pale green; fertilization with compost and 50 g N
18–25	fertilization with 40–50 g N; 24‰ are already sufficient in dry period
25–30	fertilization with 40–25 g N
30–33	leaves dark green to bluish; good yields, little or no fertilization
>33	N-excess
P 0.6–0.9	serious P-deficiency; fertilization with 20 g P
0.9–1.1	fertilization with 15 g P
1.1–1.3	fertilization with 13–9 g P
1.3–1.5	optimal supplies; good yields, little or no fertilization
>1.5	P-excess
K 3– 8	serious K-deficiency; fertilization with 80 g K
8–15	fertilization with 80–50 g K
15–25	fertilization with 45 g K
25–30	optimal supplies; good yields, little or no fertilization
>30	K-excess; danger of Mg and Ca deficiency

The following recommendations are made for fertilization:

1. The optimal soil reaction is moderately to slightly acid (about pH 6);

2. Organic fertilization is beneficial particularly as mulch cover on the soil. Elephant grass (*pennisetum purp.* which sometimes has a large K content) is frequently used.

3. Coffee plants have large nutrient requirements and therefore need abundant supplies of available nutrients. The less the coffee tree is shaded, the faster it grows and the larger its nutrient requirements.

4. Young plants primarily require N and P during the first years. For example, 70 g P are applied in each planting hole as stock fertilization.

5. Fruit-bearing coffee trees require N and P in particular. Thus, N requirements are large four weeks after blossoming (for yield and reserve nutrients).

6. Branches may die when N and K are deficient. This can largely be prevented by abundant supplies.

7. Fertilization for high yields should comprise 100 to 200 kg N and K each per ha.

8. Deficiencies of B, Fe, and Zn among the trace elements are especially important. Mn toxicity may occur on acid soils at Mn concentrations above 500 ppm in the leaves.

9. Fertilization based on leaf analysis is the most reliable (Table 8–6). Younger mature leaves of fruit-bearing branches should have the following contents in the dry matter: 30‰ N, 1.5‰ P and S, 25‰ K, 10‰ Ca, 2‰ Mg, 80 ppm Fe, 40 ppm Mn, 20 ppm Zn, 25 ppm B.

10. A precondition for good-quality coffee beans is abundant and balanced nutrition of the coffee tree.

The problem of shading trees requires special discussion. Shading trees (legumes are best, because of the additional N supplies) reduce the intensive insulation, provide a temperate humid microclimate (and thus help the trees to survive dry phases better), supply organic fertilizer, etc. By reducing photosynthesis they slow down growth and thus lower nutrient requirements. This prevents or reduces stress damage, e. g., due to an imbalance between intensive photosynthesis and deficient mineral supplies. Shading is especially important for young trees, but later only at low yield levels. Intensive photosynthesis is necessary for high yields with correspondingly large nutrient supplies, and shading is then detrimental.

Fertilization of Cocoa Trees

Cocoa trees (theobroma sp.) resemble coffee trees in their ecological requirements and development. However, cocoa trees grow best in tropical lowlands at a slightly higher soil pH (about 6.5). Yields are usually less than 1 t/ha, but more than 2 t dry beans per ha can be obtained. A particular problem is wilting of a large part of the fruit setting (*cherelle wilt*). This is sometimes due to lack of water and, mainly, to deficiencies of mineral nutrients. It can therefore be prevented, at least in part, by fertilization.

For fertilization recommendations, the reader is referred to coffee trees, in view of the great similarities [190, 243]. This also applies to the problem of shading. A B-concentration of about 10 ppm in the leaves is sufficient (this could also be true for coffee, instead of the 25 ppm indicated).

Fertilization of Tea Plants

Tea plants (camellia sp.) grow in tropical highlands. The plant attains its full size after about 10 years and remains productive for many decades thereafter. Yearly cuttings trim the plant to a height of about 1.20 m. Regular picking of the youngest shoots (tips and the two youngest leaves) stimulates intensive vegetative growth. Annual yields are about 1 to 1.5 t marketable tea per ha with "fine" picking. Very high yields (more than 4 t/ha) are mostly based on "coarse" picking, in which the third and fourth leaves are also harvested. This, however, lowers the quality. Tea leaves contain about 3% caffeine.

Like coffee trees, tea plants require a deep, permeable soil and possibly shading. Water requirements per unit area are very large. Nutrient removals at 1 t tea (yield) per ha, including waste, are about 65 kg N, 7 kg P, and 30 kg K [190, 243]. The following recommendations are made for fertilization:

1. The optimal soil reaction is in the acid region (pH 5–5.5). Higher pH-values may have to be lowered by fertilization with sulfur.

2. Organic fertilization considerably promotes growth.

3. Nutrient requirements are large despite relatively small removals, since storage in the wood and losses through cutting have to be allowed for.

4. N supplies are most important for tea plants. About 100 kg N should be supplied per 1 t tea (yield). There should be two applications: a few months after cutting and before the next cutting.

5. All acidifying N fertilizers may be used, but the pH of the soil should not be allowed to drop below 4.

6. Supplies of P and K should be abundant especially for young plants, e. g., the amounts of K should be equal to those of N; fertilization with P especially may be considerably reduced later.

7. Supplies of trace elements are hardly a problem, at least with regard to the heavy metals iron to copper, because of the acid soil reaction. Mn concentrations are sometimes extraordinarily high (far above 1000 ppm in the leaves), but have no toxic effects.

8. The high aluminum concentrations of 500 ppm and more, due to the pH, are remarkable, but they apparently have no toxic effects on tea plants, in contrast to many other cultivated plants, and may even be beneficial.

9. Fertilization should be based on leaf analysis. The contents in the (first to third) youngest mature leaves should be: approximately 40 to 45‰ N (25‰ already signifies acute deficiency), 3‰ P and Ca, 20‰ K [190].

10. Correct fertilization of tea plants is of the greatest importance, especially for quality, since the latter and not the yield is decisive.

Fertilization of Tobacco

Tobacco (nicotinana sp.) can be grown in a wide range of climatic zones despite its heat requirements, since its vegetation period is only 2 to 5 months. Yields vary between 5 and 30 dt/ha dried leaves, where quality as well as quantity is of the utmost importance. Dry tobacco leaves contain 1 to 4% nicotine, amongst other substances. Removals per t dried leaves are 40 to 50 kg/ha N, 4 to 7 kg P, 40 to 70 kg K, and 70 kg Ca [249]. These values are doubled if the total yield is considered. The following recommendations are made for fertilization:

1. Soil reaction is optimal in a fairly wide range around pH 6.

2. Organic fertilization is important for structure improvement and as a slow-flowing N-source.

3. Nutrient requirements are maximum during the elongation phase of the tobacco plant and depend strongly on the yield target.

4. *Cigarette tobacco* should only have a moderate protein content in the leaves: tentative values are less than $15\%_{00}$ N; this permits only a low N fertilization, i. e., up to 50 kg/ha at most, 20 to $30\%_{00}$ K (obtained by moderate fertilization with K), about $2\%_{00}$ P.

5. The target for *cigar and pipe tobacco* is 20 to $30\%_{00}$ N in the leaves (fertilization with N is about 100 kg/ha), 40 to $50\%_{00}$ K, required for a higher smouldering capacity, necessitates intensive fertilization with K; high P concentrations are also desirable.

6. Mg concentrations should be high to provide white ash (especially in cigars) (more than $5\%_{00}$).

7. Large chloride contents are undesirable. Fertilization with K should therefore be in the form of potassium sulfate.

8. Boron is a trace element that tends to be deficient; concentrations of about 20 ppm in the leaves are required.

9. The resistance of the tobacco plant to certain fungal diseases is increased by abundant potassium supplies.

10. The great importance of tobacco quality necessitates aiming fertilization measures primarily in this direction.

8.4.2 Spices and Medicinal Plants

These plants are grown for the purpose of producing drugs. Drugs are dried plant parts: leaves, blossoms, fruits (kernels), and roots. They contain certain active agents, mostly secondary plant substances, required for medicines or food seasoning. The aim of fertilization is to produce high drug yields with large contents of active agents. However, fertilization is still only carried out to a limited extent. Most data on fertilization are empirical and very generalized, tentative values [256].

The active agents in spices are chiefly *essential oils*, in medicinal plants often *alkaloids*. Their concentrations are basically determined genetically, but can be influenced positively to some extent by fertilization. Correct fertilization in any case increases the yields, so that the amount of active agents produced per unit area is increased even at a constant content. Even a certain reduction of active-agent concentrations would be unimportant, provided that the required minimum contents are maintained.

Fertilization of Spices

We distinguish between herb, blossom, fruit (kernel), and root drugs according to the plant part used. The production of herb drugs can be increased by fertilization with N and K, as would be expected. Proper P supplies are also important for the yield of drugs from generative organs (blossoms, fruits).

Liming depends mainly on the soil reaction required by the plants concerned. The form of the N fertilizer depends (as for all agricultural plants) on whether simultaneous effects on the pH-value are desired. Division of the N dose into two thirds at the beginning of growth, and one third as top dressing is usually desirable. However, certain plants are sensitive to "top dressing" insofar as no fertilizer should reach the leaves. The choice of P and K forms is of minor importance and is based on the general guidelines for agricultural plants. Fertilization with magnesium should also be considered, since occasionally the requirements are larger than average (especially in the case of peppermint). There is still relatively little known about the requirements for trace elements, but it may be expected that they resemble those of agricultural plants.

Table 8-7 contains data on nutrient requirements and fertilization of spices grown in Europe. The fertilization of hops is discussed separately.

Fertilization of Hops

Hops (humulus lup.) may be included among spices, since they produce the spice required for brewing beer. The umbels yield bitter essences etc. for brewing. The *bitter value* depends on the content of humulon (α-humulic acid) and lupolon (β-humulic acid). Hops grow from the rootstock to shoots of 7 m length. About one quarter of the nutrients in the shoots return to the root in autumn; therefore, the shoots should not be cut off too early.

Yields are about 10 to 40 dt dry hops (umbels) per ha. Removals per 10 dt dry hops (allowing for leaves) are 60 kg N, 15 kg P, 90 kg K, 100 kg Ca [262].

Table 8-7. Spices and their fertilization (according to [44, 256]).

Plant	Yield of drug or active agent	Required soil reaction etc.	Removals kg/ha	Fertilization kg/ha
A leaf and herb drugs				
marjoram (majorana hort.)	30 dt herbs	neutral, sensitive to salt	⎫ 50–70 N	⎫ 80–100 N
peppermint (mentha pip.)	25 dt herbs with 1% oil	acid, sensitive to salt, intensive fertilization indicated, sensitive to top dressing	⎬ 6–10 P	⎬ 20–30 P
balm mint (melissa off.)	30 dt herbs		⎭ 50–80 K	⎭ 70–100 K
sage (salvia off.)	30 dt herbs	neutral, prefers loamy soil, fertilization with N decisive		
thyme (thymus vulg.)	30 dt herbs			
B. blossom drugs				
true camomile (matricaria cham.)	10 dt blossoms with 0.4% oil	slightly acid, fertilization primarily with N and K	⎫	⎫ 60– 80 N
lavender (lavandula ang.)	5 dt blossoms, 20 dt herbs	neutral, fertilization with P to be stressed	⎭	⎬ 20– 30 P ⎭ 70–100 K
C. kernel drugs (fruit drugs)				
caraway (carum carvi)	15 dt with 4% oil	neutral, abundant N	(including straw) ⎫ 100 N	150 N ⎫ 30 P
fennel (foeniculum vulg.)	15 dt with 5% oil	neutral, nitrate-N best	⎬ 20 P ⎭ 100 K	100 N ⎭ 100 K
coriander (coriandrum sat.)	10 dt with 0.5% oil	neutral, sensitive to top dressing	30 N, 8 P, 30 K	50 N, 30 P, 70 K
D. root drugs				
valerian (valeriana off.)	20 dt roots	slightly acid, moderate fertilization best	70 N, 10 P, 80 K	80 N, 30 P, 100 K

The fertilization of hops, on the basis of about 4500 plants per ha, depends largely on the yield. Hops prefer a slightly acid to neutral soil. Organic fertilizer applied every two years is beneficial. The following remarks concern fertilization with individual nutrients:

- *N supplies* should be abundant but not one-sidedly extreme, as there is otherwise the danger of mildew attack and a reduced bitter value. Recommended doses are 100 to 150 kg per ha for moderate yields, and 200 to 250 kg for high yields. Doses should be staggered, the third dose being given approximately at the time of blossoming.
- *P supplies* should be about 60 kg per ha for medium requirements, but smaller doses may be indicated because of the frequently large contents of available phosphate in some hop soils.
- *K supplies* should be abundant, e.g., 150 to 200 kg per ha for medium requirements.

Mg requirements of hops are also considerable and should be taken into account. Boron is the most important trace element in fertilization, since hops require much of it: the leaves contain 30 to 90 ppm B in the dry matter (boron does not return to the root in autumn). High yields remove 300 g B per ha. Fertilization with boron may be as for sugar beet. Copper sometimes accumulates considerably in hop soils, due to plant-protection measures, but there have been hardly any reports on toxicity damage. Leaf analysis is becoming important for the correct fertilization of hops.

Fertilization can influence the quality of hop umbels since moderate supplies of P and K increase the bitter value, which is reduced by very large doses of N and K.

Fertilization of Important Tropical and Subtropical Spices

The *paprika plant* (spicy paprika, capsicum sp.) has a spicy fruit pulp containing various amounts of the pungent capsaicine and considerable amounts of vitamin C. Paprika plants produce 70 to 100 t fruit per ha and have large nutrient requirements. Fertilization with 60 kg N per ha and corresponding amounts of P and K is usual. Acid N fertilizers and potassium sulfate are preferred.

The *pepper plant* (piper nigrum) produces kernels as spices. The pungent taste is due to the alkaloid piperine. Pepper plants require much nitrogen (fertilization with 200 kg/ha) and abundant supplies of P and K.

The *clove tree* (clove, syzygium arom.) has flower buds that are used for their flavor. Fertilization may be based on that of fruit trees. Excess aluminum on highly acid soils sometimes harms clove trees before the principal production period. Suitable liming of the soil is therefore indicated (if possible before planting). The pH should be approximately 5.

The *vanilla plant* (vanilly sp.) produces fruit containing aromatic vanillin. Fertilizers can only be applied to a limited extent, due to the epiphytic character of this orchid.

Fertilization of Scent Plants

Essential oils are important not only in spices but also as raw materials for perfumes. They are extracted for this purpose from suitable plant parts. To obtain 1 kg orange-blossom oil (orange-flower oil) it is necessary to distill 1 000 kg orange blossoms. More than 1 million flowers are required for 1 kg rose oil. The quantities of raw material produced can be increased by suitable fertilization of the plants concerned (orange trees, roses, jasmine, violets, acacias). However, the contents of the desired aromatic substances can only be increased to a very limited extent. Some information on fertilization is given in the corresponding chapters (e. g., for roses under Ornamental Plants).

Fertilization of Medicinal Plants

Cultivated medicinal plants mostly contain alkaloids whose active agents are used for the production of medicines. Medicinal plants should also be abundantly supplied with nutrients when planted [44].

The *deadly nightshade* (atropa bellad.) produces 10 to 20 dt leaf drug per ha, and after three years about as much root drug. The alkaloid content (atropine, etc.) of about 0.5% can be increased by fertilization with N. Fertilizer amounts applied are 100 kg N (acid forms) and K (sulfate), as well as 30 kg P per ha.

The *thornapple* (datura stram.) produces 15 to 30 dt leaf drug per ha. The alkaloid content of at least 0.2% can also be increased by fertilization with N. Fertilizer amounts applied are 100 to 150 kg N (acid forms) and K (sulfate), as well as 30 kg P per ha.

The *black henbane* (hyoscyamus niger) produces 15 to 25 dt leaf drug per ha and can be fertilized similarly to the thornapple.

The *poppy* (papaver somn.) produces a milky juice in the pod, from which raw opium is produced; about 10% of the latter is the alkaloid morphine. Moderate fertilization with 60 to 80 kg N per ha and corresponding amounts of K generally increase the morphine content.

Pyrethrum (chrysanthemum cin.) may also be considered to be a medicinal plant in a certain sense. Its flower tips contain pyrethrins (up to 2%). These are highly toxic to insects but harmless to man and other warm-blooded animals. Yields can usually be increased by fertilization (e. g., 60 kg N per ha).

8.5 Lower Plants

Higher land plants play a predominant role in food production, but the share of lower plants may nevertheless be considerable in some branches and might in future become even larger.

The following are some possibilities:

- Lower water plants (algae, etc.) for fish breeding (fertilization of ponds);
- edible mushrooms (e. g., champignons);
- green algae for the production of algal food (fodder);
- yeast and bacteria for the production of protein.

8.5.1 Water Plants (Fish-Pond Fertilization)

The *fertilization of fish ponds* implies intentional eutrophication of shallow bodies of water. The purpose is to stimulate the growth of plants living below the water surface, especially the green algae of plankton. Dead plant parts (including those of submerged higher pond plants) form *detritus* that serves as food for lower water animals, which in their turn are devoured by useful fish (e. g., carp). Fertilization thus reinforces the nutritional chain at a decisive point in bodies of water poor in nutrients. Mineral fertilization in this case acts exclusively through increased plant growth, while organic fertilizers (farmyard manure, green manure, liquid manure, waste water) also provide food directly to lower water animals (see scheme). Stimulation of the growth of swamp plants in ponds is undesirable because of their poor decomposition [234].

Intentional eutrophication of fish ponds may appear to be paradoxical at the present time of constant danger of excessive, unintentional eutrophication of bodies of water. This need not be so. Additional fertilization is in any case unnecessary if fish ponds obtain sufficient nutrients through increased supplies of the latter. On the other hand, bodies of water poor in nutrients do not react negatively to an increased production of matter from the environmental aspect, since the detritus is not slowly oxidized with a large oxygen consumption, but serves as food for fish and thus requires no oxygen for its decomposition. Successful fertilization of fish ponds should aim at the *minimum factor*. Application of multiple-nutrient fertilizers is therefore less indicated than that of the deficient single nutrient. Nitrogen is the decisive minimum factor in some regions, whereas in Europe it is most commonly phosphorus.

The natural P content of many ponds is very small and often limits plant growth. This is due to the small supplies and strong immobilization of phosphates in the pond mud. Fertilization with P is worthwhile if its concentration in the water is below 0.7 ppm in spring, or below 0.3 ppm in autumn.

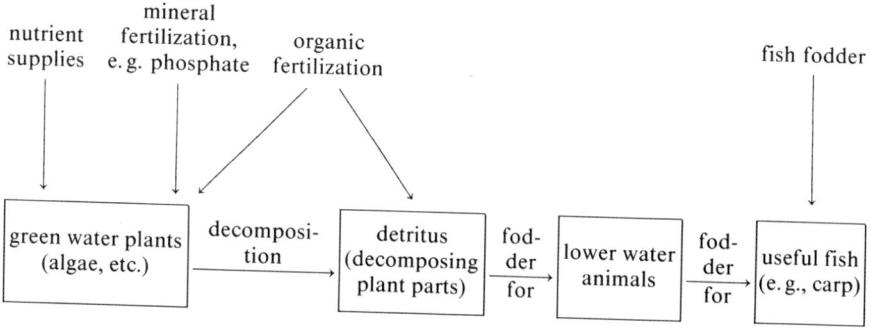

Nutritional chain and fertilization in pond management

Application of 10 to 15 kg P per ha ensures an accretion of useful fish by 30 to 120% (average value about 70%) per year. Application of 1 kg P in case of P deficiency of the water thus ensures an accretion of 5 kg carp. Division of the fertilizer dose is indicated because of the continuous fixation of P at the bottom of the pond. Application is by scattering either on the surface of the water or on the pond bottom when dry.

Fertilization with N may be as worthwhile as with P, when nitrogen is deficient. Potassium usually has no role to play in fertilization. The pH-value of ponds should be above 5. It can be adjusted by suitable liming when the ponds are dry.

8.5.2 Fungi, Algae, Bacteria

Fertilization of Champignons
Champignons (agraricus sp.) have white fruit bodies containing 3 to 4% protein, vitamins, etc. Yields of about 10 kg/m^2 are obtained in 40 days, with 3 to 5 harvests per year [245]. The mushroom lives on organic matter in the nutrient substrate (compost). It uses pentosanes and cellulose for synthesizing its own carbohydrates, as well as amino acids, primarily from the decomposition of bacteria, for protein synthesis. The presence of many active agents in the compost is also important for the nutrition of champignons.

The temperature should be between 10 and 20°C, the relative air humidity should be about 90%, the CO_2-concentration in the air should, if possible, be less than 0.1%. The substrate must not become excessively wet when watered.

Successful growing of mushrooms depends on the correct *compost*. Far-reaching decomposition of all easily decomposable substances is not desirable, in contrast to horticultural substrates (Chap. 4.4.1), since it is precisely these materials that should nourish the mushrooms. The best compost still appears to be based on horse manure (oats and hay used as fodder). The fresh manure, containing about 50% dry matter, is subjected to hot fermentation for 2 to 3 weeks. The following additions of mineral fertilizer per kg fresh manure are recommended: 2 g urea, 5 g superphosphate, 15 g each of gypsum and lime.

Partial substitution of horse manure by a similar material, e.g., refuse-sewage sludge, seems to be possible. A large number of major and micronutrients must be added if only straw compost is used [245]. The application of a 2-cm thick covering soil layer, consisting of lime-containing sandy loam or peat, for regulating the water and air balance is absolutely necessary. Champignon compost can be used after the harvest in horticulture as a substrate rich in nutrients.

Fertilization of Algae Cultures

Algae permit considerable amounts of food or fodder to be produced. This applies in particular to the green single-cell *chlorella* and *scenedesmus*. They are grown in hydroponics and with correct illumination and CO_2 supplies have yields of 5 to 7 kg dry matter per m^2 per year, compared with 1 kg wheat at most. Algal dry matter is about 50% protein, 10 to 20% fat, etc. Until now it has been used primarily as fodder. High yields can be obtained only if the nutrient solution (Chap. 4.4.2) has the correct composition and is continuously or frequently replenished. Nitrogen and potassium requirements are particularly large (200 to 1 000 ppm nitrate-N, 400 to 2 000 ppm K). On the other hand, calcium is hardly needed at all. All trace elements including vanadium must also be supplied [252].

Fertilization of Yeast and Bacterial Cultures

Yeast or bacterial cultures can be used to convert nitrogen-containing wastes or raw materials into food protein. Certain yeast strains (e.g., *torula*) synthesize their protein (about one half of the dry matter) from sugar as the energy source and mineral nitrogen. The sugar needed for this can be produced from wood waste, straw, etc.

Certain organic substances needed in food can also be produced in this manner (e.g., vitamins).

A further possibility is to use petroleum products for the production of protein by certain bacterial cultures. This necessitates complete fertilization.

9 Fertilization and Quality of Vegetal Food

Fertilization should ensure not only high yields per unit area but also high quality produce, either by

- improvement of low initial quality, caused by insufficient nutrient supplies, or
- maintenance of high quality even with increased yields.

High quality is important in almost every harvested product, whether food, fodder, or industrial raw materials of every kind. Many hints on improving the quality of vegetal raw materials by fertilization have already been given with reference to the crops involved. Because of the great importance of food stuffs and fodder for the health of man and domestic animals, the links between plant nutrition and the quality of vegetal foodstuffs will be discussed in conclusion in greater detail.

The need to produce food of high quality exists both in regions where food is in short supply and in regions with abundant supplies. Food, when in short supply, should at least have maximum nutrient content. When supplies are abundant, foodstuffs should have the properties desired by consumers, to increase sales.

Both producers and consumers are equally interested in high quality vegetal products, although for different reasons. The producer wants to sell his produce at a high price and therefore aims at quality aspects affecting the selling price. The consumer desires enticing, wholesome, full-value food, free from harmful substances.

Full-value food is an essential precondition for the health of man and animal. Almost one half of all human diseases are caused directly or indirectly by incorrect or inadequate nutrition. Special attention should therefore be given to the contents of vital and beneficial substances in the food.

9.1 Basic Facts about the Influence of Fertilization on Quality

9.1.1 Concept and Factors of Quality

The many quality aspects can be divided into two groups [300]:

a) *Commercial quality* (market quality), i.e., the value as a commercial product on the market. The commercial quality of foodstuffs for direct consumption chiefly depends on external organoleptic (sensed) characteristics, i.e., attractive appearance without blemishes, good taste, inviting smell ("partaking value"). Durability also plays an important role in storage. The product should not deteriorate or go bad. The contents of important ingredients are decisive for industrial processing (sugar, starch, protein, fat, oils, etc.). Produce is classified according to the desired properties into commercial grades that determine the monetary value (payment).

b) *Food quality* (in the narrower sense), i.e., the value for nutrition (nutritive value): It depends on the energy content and the content of value-determining ingredients, mainly ingredients essential for man and animal. The nutritive value of food is higher, the larger its content of vital and beneficial nutrients and the smaller the content of harmful substances that might be present. Food quality, which for the consumer is so important, is unfortunately not always taken into account in the price.

The quality of vegetal products depends on many factors, e.g.,

- genetic (inherited) factors that determine the basic quality specific to the kind;
- environmental factors realizing the genetic potential and modifying it in a positive or negative sense. Such factors are climatic (light, temperature, etc.) or soil-dependent (water and mineral supplies, etc.).

The level and regulation of *nutrient supplies* to plants through fertilization play a central role in determining the quality. Fertilization in mineral or organic form can considerably improve the quality, but it may also have detrimental effects.

The crucial problem with food quality is whether the food, produced formerly and even today with conventional fertilization (especially with organic farmyard fertilizers), is always of high quality, and/or whether yield increases due to intensive mineral fertilization occur at the expense of the nutritive value, even if the external quality is improved. The increased use of chemical means of production in agriculture has in part caused a crisis of confidence between consumer and producer, so there is some justification in raising the important questions:

- Are yield increases made at the expense of quality?
- how and to what extent does additional, sometimes considerable, mineral fertilization affect food quality?

The answers to these questions may be speculative or based on experiment. There are some differences in opinion, but one answer is becoming increasingly clear on the basis of many considerations and experimental results.

Properly applied fertilization, whether organic or mineral, is a suitable means of increasing the yield and also of improving the quality of the produce. However, when wrongly applied, it can also reduce the quality, especially when it causes inharmonious nutrition of the plants, e. g., one-sided excesses.

In general, quality improvement through fertilization should be considered the norm, while occasional detrimental effects due to incorrect fertilization would be the exception. This general claim will be substantiated in detail below.

The influence of "chemicals" on food quality is discussed in this book only insofar as they serve as fertilizers. The effects of toxic substances and chemicals for plant protection on the quality will not be considered [288].

9.1.2 Speculations on Nutrient Quality

Formerly, it was only possible to incompletely determine the quality of food; therefore efforts were made to bridge the gap in knowledge by speculation. The main point in this discussion is the *naturalness* of food. This implies that the tenet of the philosophical theory of the value of naturalism is assumed to apply to human food as a natural product: "What is natural, is good; what is unnatural and artificial is bad". From this point of view it is obvious that food produced with natural fertilizers (e. g., farmyard manure, compost) is considered good and valuable, while food produced with (additional) artificial fertilization (synthetic mineral fertilizers) is considered inferior (less valuable). The logic of this argument seems quite obvious and plays a role especially with consumers who are close to nature.

Against this it might be argued that naturalism belongs to the imperfect theories of the ethics of goods and is, moreover, incorrect in this form. One of the characteristics of human evolution is the striving to change nature in its original form and if possible to improve it. Neither is "natural" food always of full value; wild-growing plants may contain little valuable substance or may contain harmful, even carcinogenic, substances; nor need intervention by fertilization in principle imply deterioration.

Positive arguments for mineral fertilization can be adduced even when one largely adheres to the ideas of naturalism. Phosphate and potassium fertilizers can be clearly assessed according to their origin. They are obtained as natural products from crude-phosphate and salt deposits, respectively, and are only slightly processed or converted into forms better utilizable by plants. Even though these fertilizers are now produced in factories, they are nevertheless basically natural fertilizers from which negative effects on food quality should not be expected, even if the principle of naturalness is stressed to the limit.

The situation might be different in the case of nitrogen fertilizers. These are in fact largely totally synthetic artificial products (chemicals). Their possible negative influence cannot thus be disclaimed in general, but should be estimated on the basis of their substance.

The substance of natural and artificial fertilizers could differ in two aspects: with regard to *plant nutrients* and with regard to *fertilizer composition*. Even fully synthetic N fertilizers contain nitrogen in a form absorbed by plants at their natural sites (mostly nitrate, sometimes ammonium), or in a bonding form yielding nitrate after reaction in the soil. The substance is thus still a natural fertilizer form despite its synthetic production. It is, however, assumed from our present state of knowledge that, e. g., synthetic nitrate is completely identical with nitrate derived from humus. This need not be the case if the influence of a "life force" be postulated, that might cause changes in nutrients passing through organisms, which factor cannot be measured by present-day methods. However, there is no proof whatsoever that this is so, and the argument therefore remains completely hypothetical.

In conclusion, we shall critically consider the composition of synthetic fertilizers. We shall disregard the extreme point of view of biological-dynamic theory, which claims that natural fertilizers form a unity that cannot be replaced by artificial mixtures of substances [282]. However, an essential difference between many natural and synthetic fertilizers is the degree of their purity. Farmyard manure contains not only nitrogen but also provides all necessary plant nutrients; natural sodium nitrate (Chile saltpeter) contains many admixtures, in contrast to synthetic sodium nitrate that is essentially a pure chemical. The trend to increased purity of fertilizers is no justification at all for considering them to be harmful, but it does represent a potential danger to food quality because of a possible one-sidedness in fertilization. On the other hand, greater purity also ensures smaller amounts of possibly detrimental admixtures.

Statistical Speculations

The search for the causes of an increased incidence of certain diseases led to the inclusion of mineral fertilization in the considerations, in view of the considerable changes in food production. There is no doubt that there is a strong correlation between increased fertilization levels and the increasing incidence of certain diseases (e. g., cardiac and circulatory disturbances). Mineral fertilization could therefore be the cause of this increased incidence through inferior nutrition. Such a theoretical possibility, however, is meaningless with regard to causal investigations. This method in no way enables us to decide whether mineral fertilization has any influence on the incidence of human diseases. It is much more likely that these correlations are only apparent and not causal, since the magnitudes being compared can easily be shown to be synchronous functions of time.

Mineral fertilization may be related to both increased and reduced incidence of certain diseases. Direct deductions would in either case be a gross violation of the rules of scientific research, which are frequently neglected precisely in statistics.

The same is true for statistical connections between the contents of valuable substances in food and the level of mineral fertilization, as far as these are derived from time functions. Thus, the Reports on Nutrition in the FRG [274] states that the content of vitamin B_1 (thiamine) in foodstuffs has decreased during the last 50 years. This is surely a significant fact, since thiamine requirements are only covered to less than 80% in probably one quarter of all households. However, it is inadmissible to ascribe this mainly to increased mineral fertilization. It is precisely in the case of vitamin B_1 that it is obvious that the content in the food produced, which is in fact considerable, (e. g., in grain kernels) is reduced by "refining", i. e., separation of the flour from the bran that is rich in valuable substances. This results in a low concentration of vitamin B_1 in fine bread.

Speculations about the possible negative effects of mineral fertilizers on food quality are therefore in general largely groundless. There is no logical reason why mineral fertilizers should "in principle" have detrimental effects because of their "artificial" origin or synthetic components. On the other hand, neither can the opposite, i. e., positive effects, be deduced by speculation. Scientifically substantiated research results are required.

9.2 Food Quality as a Function of the Production System

We shall here discuss whether, apart from individual links between plant nutrition and food quality, which will be considered below, there are certain quality differences solely due to differences in the agricultural production systems. This problem is becoming more topical with the evolving discussion of an "alternative" agriculture that claims to be producing better foodstuffs and be more desirable for the environment.

This problem will be elucidated, or at least discussed factually, by comparing "conventional" and "alternative" agriculture, apart from the reference to "conventional" cropping systems discussed in Chap. 6.4.

9.2.1 Conventional or Alternative Agriculture

There are many factors involved in a comparison between conventional and "alternative" agriculture, i. e., yields, profitability, effects on the environment, food quality, etc. Here, we shall only discuss the problem of whether agriculture using alternative fertilization, sometimes at the expense of yields, may furnish products of better food quality [273, 276, 277].

Given the multiple conventional management procedures still used by more than 99% of all farms and the many forms of alternative agricultural systems, a polarizing confrontation is the wrong formulation of the problem. Actually, the two systems merge into each other and overlap in many ways. Profitability must be ensured in practically all production systems, being a function of quantity and price of the produce. In conventional agriculture, stress is placed on the yield, but in alternative agriculture, on the "special quality" that is supposed to justify the higher price of the produce. The same target is aimed at, but different production methods are employed on the basis of different scientific or philosophical concepts. Synopsis 9-1 gives an indication of the range of variation of the possibilities; groups have already been formed out of the existing variety.

"Alternative" agriculture is based primarily on the concept of *natural, biological* production, unaffected, or not significantly affected, by synthetic chemical substances. There are two main groups, i. e., organic-biological and biological-dynamic agriculture [269].

Organic-biological agriculture in its turn comprises several trends, e. g., biological cultivation according to *Dr. Müller* (Switzerland), organic agriculture by the *MIGROS* cooperatives (Switzerland), and natural quality cultivation (ANOG Paderborn) [FRG].

The common concept of these groups is to aim at a maximum of "natural" production conditions, i. e., stress on good soil fertility for quality production, moderate use of fertilizers to prevent any harmful excesses, moderate plant protection, and above all, controlled use of chemicals, to obtain suitable brand products [294, 296, 297, 298, 304].

Biological-dynamic agriculture also comprises numerous groups. It is based on *R. Steiner's* (1861–1925) anthroposophic teachings, which have not only intellectual and religious but also agricultural aspects [301]. The totality of the farm is important according to this theory, and growth forces are more important than growth substances. This has many consequences for production, far

Synopsis 9-1. Use of Nutrients and Other Means of Production in Different Groups of Production Systems.

Fertilizer or procedure	Application in agricultural production system considered				
	conventional		organic-biological	biological-dynamic	
	highly intensive without animals	moderately intensive with animals		moderate forms	original form
organic fertilizers	harvest residues	moderately	intensively	intensively	intensively
mineral N fertilizers	much	moderately	no	no	no
processed P and K fertilizers	much	moderately	extensively	no	no
crude P and K fertilizers	hardly	sometimes	moderately	little	no
rock powder	no	no	little	little	moderately
lime fertilizers	as required	moderately	moderately	little	no
growth regulators	yes	partly	no	no	
weed killing	mostly chemical	chemical and biological	chemical and biological	biological and physical	
plant protection	mostly chemical		chemical and biological	almost only biological	
special preparations for growth activation	no		no	yes (necessary)	

beyond the emphasis on good soil fertility [281, 282, 293, 306]. *Steiner's* original teachings stated that the use of "artificial" ("chemical") fertilizers should be rejected altogether, since plants obtain nourishment from the soil and organic fertilizers. On the other hand, the fields of the farm should be subjected to the influence of cosmic forces by the use of special preparations (e. g., siliceous preparations) made according to mystic prescriptions.

More moderate forms appeared within the framework of *Steiner's* teachings, as happens with every new, extremist trend. One of them is the *Demeter* group. It acknowledges plant requirements for minerals and thus permits moderate mineral fertilization in addition to the emphasis on organic fertilization.

However, the fertilizers should not be processed chemically and should, if possible, not be completely soluble in water. Lime, raw phosphate, and sometimes crude potassium salt are permitted. In this case, an exaggerated effect is ascribed to some permitted fertilizers, e.g., rock powder [307]. There is thus hardly any qualitative difference between this agricultural system and organic-biological or moderately intensive conventional agriculture, if fertilization with N, for which organic forms are used, is disregarded. The only point to be stressed is the smaller supply of available nitrogen. The main difference is thus the use of special biological dynamic preparations for the purpose of improving growth and quality.

The specific problem of the effects of fertilization on food quality can be clarified without discussing the principles and philosophies of alternative agriculture in detail, since this is neither necessary nor useful. Isolated aspects can well be separated from highly complex problems and clarified by themselves. In considering fertilization, it thus seems to be unimportant whether the term "natural" or "biological" is defined precisely, whether a plant possesses special life forces or even a "soul", whether cosmic forces influence plant growth (this happens in any case through insulation), whether all philosophical preconditions of biological-dynamical agriculture are fulfilled. The "totality" principle that is continually cited, "Everything acts on everything" is in any case wrong from the philosophical and physical aspect, and should read: "Much acts on much".

Rather, the relatively simple question of whether a certain production system, and in particular a special fertilizer, does provide more valuable food or not is the important point. This is a problem of establishing *quality properties*. Generally accepted standards and methods exist for certain quality characteristics, such as cleanliness of the merchandise, taste, storing quality, etc., but this is not quite true with regard to the actual food quality.

Accepted science adheres to standards based on medicine and the natural sciences, which can be established by exact methods. On the other hand, biological-dynamic agriculture uses different criteria and methods, e. g., capillary dynamolysis, crystallization patterns [281, 282, 279]. These methods may be of real significance in assessing food quality, but they may also simulate an imaginary value in which there is no limit to the interpreter's fantasy. Many "quality methods" can easily be developed with some imagination, and the products classified as desired according to one's own special quality scale. These methods would acquire significance only if they could satisfy the elementary requirements of all scientific methods, namely reproducibility and causal interpretability. This could also be expressed as follows: He who wants to see something good in his products, easily finds a suitable scale for this. The latter, however, remains without significance as long as it is not generally acceptable.

9.2.2 Fertilization System and Food Quality

Basing ourselves exclusively on generally accepted scales of values (Chap. 9.3), which seems to be the only sensible procedure, it becomes obvious that vegetal food of good quality can be produced with all production systems and is clearly also being produced by the overwhelming majority of farms. The quality may also be lower in any production system, but this rarely happens in the extreme form. Thus, conventional agriculture of necessity produces low-quality plants on poor, infertile soils with one-sided, extremely intensive, exclusively mineral fertilization with NPK. This is due to excesses of some and deficiencies of other nutrients. However, the same might apply on corresponding sites with alternative forms of agriculture using specific, exclusively organic fertilization. Both cases are extreme exceptions. However, products of medium to good quality (without harmful ingredients) can be expected with approximately balanced fertilization in either agricultural system on soils of reasonable fertility. Food quality may be above average or particularly high on highly fertile soils with the usual mineral or organic fertilization, but this need not be so (Chap. 9.3).

The influence of a production system on quality in any case does not so much depend on the kind of fertilization applied (whether organic or mineral), but on the sensible use of fertilizers for optimal plant nutrition: avoidance of both deficiencies and excesses. This is possible in both conventional and alternative agriculture. The problem of guaranteed quality is independent of this. A production system can claim that its produce is of reliable quality only when the application of chemical means of production is monitored. Monitoring production systems have until now restricted themselves to rough checks of fertilization to prevent excesses. On the other hand, as yet, there hardly seems to be a production system in the open in which optimal nutrition of the plants is fully monitored.

A particular problem concerning biological-dynamic agriculture is whether *special preparations*, e.g., "horn manure" and "horn silicon", sprayed on the plants in highly diluted form to improve quality and growth do improve the quality in a special way (yield increases appear to be just as questionable, but are unimportant in this context). The affirmative claims in this connection [264, 281] must be countered by the fact that unquestionable, scientifically definable, positive effects have not yet been demonstrated.

There remains the claim of a mobilization of special *cosmic forces*, caused by spraying, even if no effect of the material components supplied is really expected [301]. Undoubtedly a plant is more than just an ordered accumulation of chemicals. It also has electrical potentials, creates electromagnetic "fields" and can thus in principle be influenced by corresponding "fields" created by the planet earth or cosmic systems. However, these forces are extraordinarily weak in comparison with those that operate in molecular regions of the plant cell. Such an influence is therefore hardly possible in any appreciable degree and in any case is probably not measurable [276]. There is no reason to accept the idea of the qualitative superiority of foodstuffs produced with special preparations, as long as their influence is simply asserted without proof. In the same way, one could claim that special "spells" cast on the growing plant or even one's own belief in or desire for quality improvement are effective. This might be even less in error, since the food quality of crops also depends on the correct use of the means of production by managing man.

On the Way to an Understanding

Fortunately, an understanding with regard to correct fertilization in the different agricultural systems seems to be on the way. The necessity of certain substances for plant nutrition is now generally acknowledged. Mineral fertilizers are now used in nearly all more developed agricultural systems, even if their choice and dosage is sometimes restricted. It can no longer be claimed that mineral fertilization is harmful in principle, and this is hardly being asserted anymore. It is now recognized that fertilizers are often beneficial and it is only when they cause nutrient excesses or are not used to eliminate deficiencies that they are harmful.

The discussion of "biological" or "chemical" plant production has created much *confusion*. Plant, animal, and man are composed of chemical substances, require chemical substances, and can be influenced by chemical substances. The deeper the understanding of biology becomes, the more it must be considered as chemistry, irrespective of the fact that even today our knowledge is incomplete and that there might be still other dimensions beyond all "chemistry".

A person suffering from scurvy can be cured by additional nutrition with synthetic vitamin C, since his organism is of a chemical nature and requires this vitamin, irrespective of any possible life forces, etc. Fertilization should also be

considered pragmatically as specific, controllable intervention and should therefore be as precise as possible.

It might be the lasting merit of "alternative" agriculture to have pointed
out indisputable errors, especially those that occurred during the initial phase of
fertilization (Chap. 1.4.3), even if sometimes one error was simply replaced by
another. However, critical observations have led to a perfection of scientific theory, and this learning process could be mutual.

Such criticism can also be credited with having stressed the *special place of
food production* among the *industries*, since food is something unique among the
products made by man. Obviously, food quality is not improved by giving first
priority to maximum profits in agriculture, clearly at the expense of what should
be understood as quality. This knowledge, however, is not specific to any production system but represents age-old wisdom.

Today there is hardly any significant difference between large sectors of
conventional and *alternative* agriculture. Only some extreme cases on either side
should be recognized.

Agriculture, stimulated by scientific research, has provided amazing production results within a wide range of crop systems through increasingly more
precise fertilization. This will be so to an even greater extent when biological
principles are given more consideration. Practical agriculture and science have
learnt much from each other, and this mutual assistance will probably continue.

Results of Comparing Crop Systems
The following results are obtained from the above considerations and arguments, as far as fertilization problems are considered when comparing production systems:

- A particular production system by itself does not in principle ensure a higher
  quality of vegetal products; the quality is often high, but may be lowered by
  improper plant nutrition;
- "alternative" agriculture as such provides neither better nor worse produce
  than conventional agriculture, but many farms belonging to either trend pay
  much attention and effort to obtaining a high product quality;
- there is no indication, to say nothing of proof, that special preparations with
  unknown effects can improve the quality;
- the advantage of certain groups of production systems lies in certain checks
  on the use of chemical means of production;
- complete checks of vegetal production with a comprehensive diagnosis during
  plant growth have not yet been realized in any open-air production system;
  this, however, would be the real *alternative* agriculture of the future.

9.3 Food Quality as a Function of Nutrient Supplies

Disregarding speculations and hypotheses, we shall now discuss the influence of fertilization on food quality in detail from the scientific aspect [267, 270, 292, 300, 266]. Some basic facts should be stated first of all. Many erroneous views are based on contradictory experimental results that in one experiment show that a certain application of N fertilizer raises the quality, hardly affects it in another, and distinctly lowers it in a third. It is often impossible to draw any valid conclusions at all when many experiments are considered jointly. The difficulty lies in the lack of precision of the possibilities of interpretation. What has nearly always been investigated is the

- connection between *fertilization* and quality but hardly ever the more important and better interpretable
- connection between *nutrient supplies* and quality.

In experiments with solid substrates, fertilization is only a conditionally suitable indicator of nutrient supplies. The results of fertilization experiments with positive or negative effects on food quality need not be contradictory. In assessing fertilizer effects, one must therefore distinguish how nutrient supplies have been changed:

- Increasing supplies from *deficiency to optimum* usually also implies quality improvement;
- increasing supplies within the *optimal range* usually has no effect but sometimes causes still further improvement;
- increasing supplies from the optimal range to luxury consumption may lower the quality, but need not do so;
- extreme increases of supplies up to the toxicity range clearly lower the quality.

The side effects of fertilizers should be considered in addition to the main effect; this makes interpretation even more difficult.

It must also be asked whether yield limit values (Chap. 5.2.2) are identical with *quality limit values*. This may be so in some cases, but the quality limit values might be higher, at least for the contents of important minerals. This applies primarily to the nutrient element supplied in increased amounts. Thus, increased fertilization with copper obviously raises the value-determining copper concentration beyond the yield limit value.

The optimum content of desirable organic ingredients, as a function of the supply of one nutrient element is probably optimum at the yield limit value, but significantly higher at a quality yield limit only in exceptional cases.

For more details on vegetal ingredients and biochemistry the reader is referred to [280, 286, 287, 295, 298].

9.3.1 Fertilization with Nitrogen and Quality

Nitrogen is the "main drive" of plant production. It also influences the quality considerably and in many ways, especially through its effects on

• protein content and value;
• contents of other valuable substances (containing N or not);
• contents of quality-reducing substances.

The health (resistance) of the plant and the quality of its descendants are also influenced. Increased N supplies to plants cause the following changes in protein metabolism (Synopsis 9–2):

a) The content of *crude protein* is increased, in some cases considerably: e.g., the crude protein content of cereal grain is raised from between 10 and 15% to between 16 and 20%;

b) the content of pure protein increases up to the optimal N-supply level, because there is more than a proportional synthesis, despite the counteracting dilution effect;

c) the *albumen* content increases with the content of pure protein;

d) the content of *essential amino acids* sometimes increases up to the optimal N-supply level, sometimes hardly changes, and sometimes decreases through dilution, especially when there is luxury N consumption;

e) excess N (in the case of luxury supply) is normally stored as

• *amides* in young green plant parts, e.g., leafy vegetables, roots, tubers;
• *prolamine* in kernels, thus increasing the gluten content of grain kernels and improving the baking quality;

f) the *biotic value* of the protein usually increases up to the optimal N-supply level, and then decreases through dilution with low-value storage products;

g) other N-compounds may accumulate in vegetative plant parts when N-supplies are abundant, e.g.,

• *nitrate* in the leaves, especially when illumination is reduced: concentrations of 50 ppm nitrate N should not be exceeded in food, due to the risk of nitrite formation: nitrite can already form in the leaves under reducing conditions (e.g., when spinach is stored in the absence of air) and cause methemoglobinemia (oxidation of the iron in blood hemoglobin, which is then no longer able to transport oxygen);

- *nitrosamines,* some of which are harmful to health, but the contents of which in plants are normally insignificant;
- *betaine* as an important constituent of the "detrimental nitrogen" reducing the sugar yield of sugar beet.

Increased N-supplies cause changes in other substances, e. g.,

- the contents of *carotin* and *chlorophyll* increase up to the optimal supply level;
- the content of *vitamin B*₁ (thiamine) in cereal grains increases until supplies reach the luxury level;
- the content of *vitamin C* (ascorbic acid) decreases, especially when there are luxury supplies;
- the content of *oxalic acid,* which is harmful (in leaf vegetables for humans, in sugar-beet leaves for cows) is raised especially by fertilization with nitrate N; however, it should be remembered that even when ammonium fertilizers are used it is mainly nitrate that plants absorb;
- the content of hydrocyanic acid in grass is slightly increased; normal contents promote health (vitamin effects), while larger doses are toxic.

Important quality changes, caused by fertilization with N are illustrated in Fig. 9–1.

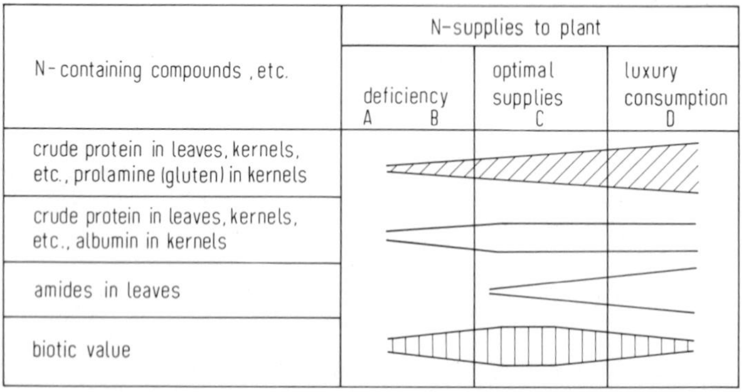

N-containing compounds , etc.	N-supplies to plant		
	deficiency A B	optimal supplies C	luxury consumption D
crude protein in leaves, kernels, etc., prolamine (gluten) in kernels			
crude protein in leaves, kernels, etc., albumin in kernels			
amides in leaves			
biotic value			

Fig. 9.1. **N-Supplies to Plants and Quality Characteristics** (widening signifies increase).

The following should be noted about the influence of N forms on quality: plants absorb mainly nitrate when N fertilizers are applied to the soil. An influence from the N-form can hardly be expected in this case. However, no accumulation of nitrate is possible obviously, when the ammonium form is preserved by special treatment, e. g., with N-Serve (Chap. 2.1.5). Other influences observed

Synopsis 9-2: N-Compounds Important for Quality Assessment in Plants.

1. *Nitrate:* mineral N-form absorbed from substrate, important initial substance for protein synthesis, contents in plant mostly insignificant.

2. *Nitrite:* may be formed from excess nitrate under certain conditions, contents in plant ordinarily wholly insignificant.

3. *Crude protein:* rough measure of protein content. Crude protein [content] = 6.25 times N content.

4. *Pure protein:* proteins subdivided into following fractions:
   a) prolamine: soluble in alcohol ⎱ low-value
   b) gluteline: soluble in alkalis ⎰ protein
   c) albumin: water-soluble ⎱ high-value
   d) globulin: soluble in salt solution ⎰ protein
   (Prolamine contains much glutaminic acid and as gluten ("glue") is important for the baking quality of grain).

5. *Essential amino acids:* 8 to 9 protein constituents vital for man, which must be contained in food. Their content determines the biotic (biological) value of the protein, expressed, e.g., by the EAA-index (Essential Amino Acid index). Vegetal proteins have values of 50 to 70, if the EAA-index of a chicken egg is assumed to be 100.

6. *Amides:* e.g., the acid amides *asparagine* and *glutamine* are storage forms of nitrogen, especially in leaves and vegetative reserve organs. They are only of little value for human nutrition, but sometimes form substances that have a bad taste: they are well utilized by ruminants.

7. *Amines:* various N containing compounds whose concentrations in plants are small. Some of them have important functions (e.g., choline), some are toxic products of decomposition (e.g., putrescine). The following are particularly important for quality assessment:

   • Nitrosamines, some of which (e.g. diethylnitrosamine) are carcinogenic and a considerable danger to health. They are formed from nitrite and secondary amines;
   • Betaine.

8. *Cyclic N-compounds*, e.g.,

   • chlorophyll;
   • N-containing vitamine, e.g., vitamin B_1 (thiamine);
   • alkaloids, e.g., nicotine in tobacco, purine derivates (theobromine in cocoa).

of different forms of N on the qualitative composition are mainly due to side effects (changes in pH, etc.).

Concerning the influence of fertilization with N on the quality of plant descendants, it should be noted that, e.g., proper N supplies to cereals increase vitality and germinating power. However, excess N supplies may promote virus incidence in the following crop when potatoes are grown.

9.3.2 Fertilization with P and K, and Quality

P-Fertilization and Quality

The many important tasks of phosphate in metabolism are the reason P-supplies play a central role in quality. Important quality indicators are:

- The P-content and the composition of the P-fraction (see Synopsis 9–3);
- The contents of other valuable substances;
- the contents of toxic substances.

Synopsis 9–3. P-Compounds Important for Quality Assessment in Plants.

1. *Inorganic phosphate:* phosphate anions are the absorbed and partly stored P-form.

2. *Phosphoric acid ester:* product of phosphorylization, i.e., bonding of phosphate anions as phosphoryl group ($-H_2PO_3$) to organic molecules (radicals = R); general formula = $R-O-H_2PO_3$.

3. *Phytin:* storage form of phosphate in organic bonding as Ca-, Mg-salt of phytinic acid, an inosite-hexaphosphoric acid (6 phosphoryl groups are combined with the ring-shaped sugar inosite).

4. *Phosphatides* = phospholipoids: important constituents of cell membranes, consisting of glycerine, fatty acids, phosphoryl groups, and amines. An important representative is, e.g., lecithin.

5. *Phosphoproteids:* products of the combination of protein and phosphate.

6. *Nucleoproteids* (nuclein-P): P-containing enzymes: complex compounds important for cell synthesis and metabolism.

The health (resistance) of the plant and the quality of its descendants are also influenced. Larger P-supplies cause the following changes in P-contents and P compounds (when increased from deficiency to the optimal level):

a) The *total P*-content grows, but this increase is limited when it is supplied via the soil and most often reaches a level of only twice the yield limit value. Extremely high, possibly detrimental, accumulation of P in the plant is difficult to reach and need therefore not be feared.

The P-content of fodder is an important quality criterion, e.g., for ruminants (Chap. 7.5.2). Insufficient P-contents are detrimental to the fertility of cows; however, relevant experiments mostly yield only correlations and thus cannot be interpreted with certainty.

b) The following changes occur within the P-fractions:

- The content of inorganic phosphate increases in green plant parts, as does the content of phytine as storage form;
- the phytine content increases particularly in the kernels, whereas the content of inorganic phosphate increases in the straw; the content of nuclein P increases only slightly, while the contents of phosphatide P remain approximately constant.

Increasing P-supplies up to the optimal level causes the following changes in other value-determining substances:

- The content of crude protein in green plant parts increases (together with all values varying with the crude-protein content);
- the content of essential amino acids in the kernels is partly increased;
- the content of carbohydrates (sugar, starch) increases;
- the contents of some vitamins, e.g., B_1, increases;
- the nicotine content of tobacco is reduced;
- the oxalic acid content of leaves is reduced;
- the coumarin content of grass first increases and then diminishes.

K-Fertilization and Quality

The N- and P-contents of harvested produce are important quality characteristics as such, but this is not the case with K. Food usually contains more potassium than is required by man or animals, since sodium more than potassium regulates swelling capacity. On the other hand, excessive potassium contents are not of themselves detrimental (the increased content of radioactive ^{40}K is disregarded as probably being harmless).

However, intensive absorption of K by plants tends to reduce the contents of Ca, Mg, and Na. Extremely high concentrations of K, e.g., in grass (in practice in the luxury-consumption range) signify deficiencies of other minerals, which must be allowed for in the assessment. Potassium has a great effect on enzyme activity through regulation of swelling capacity, and thus affects the entire metabolism and the quality of vegetal products. Increased K-supplies to plants up to the optimal level bring about the following changes:

a) The content of *carbohydrates* increases due to intensified photosynthesis; therefore there are larger contents of sugar, starch, and raw fibres (cellulose) but also of vitamin C as an essential carbohydrate; fertilization of plants sensitive to chloride with potassium chloride leads to positive effects of potassium on the production of carbohydrates, while chloride inhibits the displacement of starch from the leaves.

b) The content of *crude protein* is reduced through dilution. Through this indirect influence the relatively greater increase of carbohydrate content reduces the protein concentration. However, the content of pure protein is reduced much less than that of crude protein. The more valuable fraction of pure protein may sometimes increase, so that intensified fertilization with K may primarily cause the less valuable N reserves to be reduced. The contents of other substances, e. g., chlorophyll, are reduced in parallel with the content of crude protein.

c) The *vitamin* content increases, e. g., carotene as a preliminary form of vitamin A, vitamin B_1 (whose content, however, diminishes again in case of luxury K consumption, although this may be largely due to the chloride), and vitamin C.

d) The content of harmful *oxalic acid* is reduced, but again increases to some extent when there is luxury consumption.

e) Losses during storage of starch-containing tubers (e. g., potatoes) are reduced, since potassium reduces the undesirable decomposition of starch by enzymes.

f) "Darkening" of potatoes is reduced; this phenomenon is due to the formation of melanines and is particularly pronounced when potassium is deficient; proper potassium supplies therefore also have a beneficial effect on "black spottiness" upon cooking [291].

Better K supplies increase the quality of plant descendants by, for example, reducing virus incidence in potatoes.

9.3.3 Fertilization with Other Nutrients and Quality

Calcium is important for fluid retention regulation and as a constituent, e. g., of pectin. The calcium content is an important quality characteristic of fodder, e. g., grass. Increased Ca-contents antagonistically lower the Mg- and K-concentrations, but the Ca-content can only be increased up to a certain limit. Ca-deficiency manifests itself in fruit, e. g., in the brown-spot disease of apples, which considerably lowers the commercial and storage quality.

The influence of fertilization with lime on quality is due less to the Ca itself than to indirect effects, caused by changes in pH-value, e. g., changes in the supplies of trace elements. A typical example of this is the appearance of potato scab after liming, caused by manganese deficiency.

Magnesium is an important constituent of the plant and regulates the activity of enzymes. The Mg-content itself is an important quality criterion. It is increased by fertilization with Mg, but this is partly at the expense of the K- and Ca-contents. Larger Mg-supplies up to the optimal level also increase the contents of chlorophyll and carotene, as well as of total carbohydrates.

Sulfur is a building block required by the plant for various purposes, especially for the S-containing amino acids cysteine, cystine, and methionine. S deficiency thus lowers the protein quality (reduction of biotic value). Some plants (cruciferous, onion plants) contain sulfur in secondary plant substances, e. g., mustard oil and leek oil, whose synthesis is inhibited when S is deficient. These substances are important not only for food quality but also to increase the resistance of the plant to infection.

Fertilization with Trace Elements and Quality

A deficiency of trace elements means the failure of important metabolic functions in the plant. The quality of harvested produce is thus reduced, at least in cases of acute deficiency. Highest quality is usually only achieved with optimal supplies of trace elements. However, the quality limit value in some cases exceeds the yield limit value. This is so when the content of a trace element is by itself a quality indicator. On the other hand, contents should not be increased close to the toxicity limit. Toxic contents are not only detrimental as such, but also negatively affect the composition of organic food constituents.

The contents of many valuable substances are generally raised to the optimal level when the supplies of trace elements are increased. In particular, protein quality is improved, i. e., the content of essential amino acids, and thus of high-value protein, is increased, while the amide content is reduced.

The contents of trace elements in plants, especially in green vegetative parts, vary considerably. Average values stated in publications (e. g., the Nutritional Report [247]) give only a rough idea of actual contents. An example of the possible range of variation is given in Fig. 9–2.

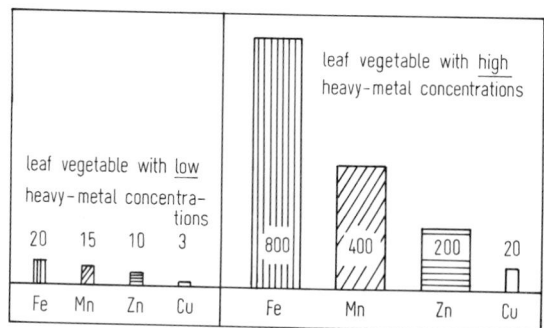

Fig. 9–2. Range of Variation of Heavy-Metal Concentrations in Leaf Vegetables (ppm in dry matter, rounded off values).

The beneficial effects of adequate trace element nutrition on the quality of plant descendants should be stressed, since abundant supplies stimulate initial development.

The *iron* in green leaves (e.g., spinach) is an important source of Fe-supplies to humans. However, contents vary greatly according to soil conditions, so that soil-reaction regulation rather than fertilization should be the mode of intervention.

The *manganese* contents of food and fodder are important quality criteria. However, fertilization increases the Mn-content only if mobility conditions in the soil are favorable. Increased Mn-supplies up to optimal level also raise the contents of some vitamins (e.g., carotene, vitamin C).

Fertilization with *copper* increases the Cu-content of plants, which is especially significant for fodder. Increased Cu-supplies raise protein content and quality in particular, as well as the contents of substances accompanying proteins. Cu-deficiency causes spottiness in certain fruit.

Fertilization with *zinc* increases the Zn-content of plants. However, plants have contents exceeding the normal values many times when supplies are very large, so that the toxicity limit may be exceeded. The Zn-content is also an important quality characteristic.

Fertilization with *boron*, increasing the B-content beyond the yield limit value, does not improve the quality and therefore makes no sense. Increasing supplies up to the optimal level raises the sugar content, for example. Moreover, the commercial quality of fruit and vegetables is significantly improved, since deficiency causes spots and fissures that substantially reduce the value of the merchandise.

The *molybdenum* content is also an important quality criterion. It can be increased by fertilization with Mo, but mobility conditions in the soil play an important role in this case. The function of molybdenum in metabolism (nitrate reduction) shows that increased Mo-supplies raise protein content and quality (in the case of legumes, too).

The far-reaching effects of insufficient mineral supplies in food are demonstrated by the example of an *influence chain* (soil – plant – man) in New Zealand. Certain soils are poor in available molybdenum, so that vegetables grown on these soils contain too little Mo, without there being an acute deficiency. Persons subsisting mainly on such vegetables have an inadequate Mo-intake. This heavy metal (in addition to fluorine, etc.) is required for strong teeth; the teeth are therefore less healthy when Mo supplies in the food are low. The incidence of caries is then correspondingly greater (Synopsis 9–5).

Combined Effects of Nutrient Elements on Quality
Maximum quality of vegetal produce is achieved not by one-sided supplies (fertilization) of one nutrient element, but by balanced, *harmonious* nutrition of the plant. Only fertilization matched to requirements, and thus corresponding opti-

mal supplies to the plant, ensure maximum contents of value-determining ingredients.

Luxury supplies improve certain quality components only in exceptional cases, but this is often accompanied by quality reductions of other kinds. Thus, intensive fertilization of cereals with N improves the baking quality but lowers the protein value.

An important problem is how abundant fertilization with NPK affects the concentrations of trace elements in the plant. It might be expected that dilution would reduce the concentrations of these important valuable substances, e. g., in fodder grass. However, there are two possibilities in this case:

- Contents of trace elements in the plant are even increased by fertilization with NPK when mobilizable reserves in the soil are *abundant* (because of the improved and more active root system);
- contents in the plant are reduced when reserves are *small*. This implies a lower value, which obviously is of practical importance only if contents are reduced below the required limit values.

9.4 Fertilization, Food Quality, and Health of Man and Animal

Connections are increasingly becoming clearer, despite the complexity of the action chain "soil – plant – animal – man". Full-value nutrition is an important precondition for the health of man and animal. Generally accepted scales of value are required for assessing the effects of different fertilizers on health, e. g.,

- The contents of value-determining ingredients in food, and the effects of deficiencies on health;
- assessment of the food according to medical indices.

9.4.1 Value-Determining Ingredients and Effects of Deficiencies

Assessment of food quality on the basis of the contents of value-determining ingredients is inviting but does not provide an adequate overall estimate. This would be possible only if all necessary, beneficial, and harmful substances were known and determined. However, even if all single components and effects were determined, their sum, being a complex effect influenced by human metabolism, might yield a different result from that expected on the basis of analytical data.

On the other hand, insufficient contents of valuable substances can be evaluated directly from the incidence of deficiency diseases in man and animal. The limitation of this procedure consists in the fact that clear diagnostic symptoms appear only when there is a relatively severe (acute) deficiency. The main problem, however, is precisely the latent (slight, hidden) deficiency, since there are many cases of latent deficiencies for every case of acute deficiency [278]. Nevertheless, consideration of deficiency diseases provides much information on the problem of correct fertilization [272, 277, 283, 284, 285, 290, 305].

Value-determining ingredients will be discussed according to biochemical groups (Synopsis 9–4):

a) Supplies of *protein*, especially of essential amino acids, at present appear to be adequate in highly developed countries with abundant food supplies. Individual cases of hypoproteinemia are in no way due to the production of low-value food. On the other hand, protein deficiency, especially of infants *(kwashiorkor)* is common in some less developed countries. The cause is both quantitative undernourishment with protein and inadequate protein quality. Better fertilization with N might largely eliminate this deficiency.

b) Supplies of *fats* (essential fatty acids) are less a problem of food production than of consumption habits.

c) Supplies of *vitamins* at present seem to be largely adequate in developed countries, if the incidence of *avitaminoses* is used as an indicator. Acute deficiency has become a rarity, but hypovitaminoses (at least of vitamin A, B, C) are common ("Deficiency in surplus"). They represent deficiencies manifesting themselves in disturbances of the cell metabolism and in nonspecific symptoms

---

Synopsis 9–4. Nutrients Necessary or Beneficial to Man.

A. Approximately 50 *necessary substances* that have to be taken in with food in addition to general energy carriers (starch, sugar):

- 8 or 9 essential *amino acids* as protein constituents, listed in the order of quantities required: leucine, valine, lysine, iso-leucine, threonine, phenylalanine, tryptophane, methionine, histidine (only for children) (daily protein requirements = approximately 1 g per kg body weight, i.e., about 70 g per day).
- 3 essential *fatty acids* as lipid constituents: linoleic acid, linolenic acid, arachidonic acid (daily requirements = about 7 g).
- about 15 vitamins, especially
  fat-soluble vitamins: A, D, E, K;
  water-soluble vitamins: B_1, B_2-complex (4 vitamins), B_6, B_{12}, C (ascorbic acid), H.
- about 20 *mineral nutrients:* minerals as constituents of salts (Table 9–1).

B. *Beneficial* substances, e. g.,

- aromatic substances (taste and scent substances);
- ballast substances, e. g., cellulose;
- special active agents, e. g., resistance substances (antibiotics).

(unwell feeling, headaches, nervousness, low resistance to infections), and are therefore difficult to recognize as such. Hypovitaminoses are common in certain population groups and could increase globally in future. The increased requirements for certain vitamins in stress situations (overwork) should be noted in this connection.

Vitamin A is produced in plants only as a preliminary stage (carotene) serving them as a photosynthesis pigment. It occurs mainly in green leaves, carrots, etc. The carotene content of plants can be increased by suitable fertilization.

Vitamin B₁ (thiamine) serves plants as a co-enzyme and occurs primarily in the germ of grain kernels (highest contents in oat kernels). The content increases with N supplies up to high fertilization levels. The main problem of supplying thiamine to the population is not the production of foodstuffs rich in thiamine but the trend to food refining. The thiamine produced is actually contained in the food only if the whole grain kernel (whether shredded or not) is used in the baking of bread as, for example, in whole-grain bread. Flour ground to a high grade (fine bread) contains hardly any vitamin B_1.

Vitamin C (ascorbic acid) serves the plants as a redox system and occurs especially in fresh fruit and leaves. Being an essential carbohydrate, its content in plants is increased by all fertilization measures that also add to the carbohydrate component. However, the normal content of vitamin C in vegetal products is quite sufficient; deficiency, common despite this, is rather a problem of consumption habits.

d) *Minerals* (including the trace elements listed separately in medical publications), except for the abundant elements (sodium, phosphorus, calcium, etc.), attracted the attention of biochemistry just a few decades ago. There is still no complete list of requirements or detailed information on the quantities needed. Much information on human mineral needs and their effects on metabolism was obtained from animal nutrition and veterinary science [268, 278, 305]. Table 9–1, indicates the requirements for minerals and some pathological effects of deficiencies (see also Synopsis 9–5).

e) *Resistance substances* should be mentioned as very important *beneficial* food ingredients. They are produced by certain fungi in fertile soils, composts, etc. (Chap. 4.3.4). Their contents are approximately as follows: 5 ppm streptomycin, 0.1 ppm terramycin, and 0.02 ppm aureomycin, etc. These antibiotics are absorbed by plants in which they occur in low concentrations in the leaves, and apparently they act as protective agents against certain light infections. This natural resistance of plants (and man and animal) might be of considerable importance, but it has still not been investigated sufficiently to permit any firm conclusions. Correct organic and mineral fertilization can promote the growth of these fungi and thus the production of antibiotics.

Table 9-1. Necessary Mineral Nutrient Elements (besides N and S); Daily requirements of an adult and effects of deficiencies (according to [268, 275, 278]).

Element	Require-ments	Deficiency symptoms in man and animal (only partly established in animals)
P	1.5 g	disturbance of bone formation
Cl Na	} 7–15 g	disturbance of kidney function, cerebral salt-loss syndrome, iso-topic dehydration
K	2 g	disturbances of growth and fertility (nutrition-induced deficiency is very rare)
Ca	0.8 g	reduced bone stability, neuromuscular disturbances
Mg	0.3 g	cardiac insufficiency, tetanic seizures, nervous disturbances of sleep, pasture tetany of cattle
Fe	10 mg	anemia (deficiency widespread on earth)
Zn	10 mg	disturbances of growth, healing of wounds, hair growth
Mn	2 mg	disturbances of growth and fertility, skeletal deformities
Cu	2 mg	anemia, disturbances of bone formation, fertility, and pigmentation, damage to coronary blood vessels
J	0.2 mg	disturbances of thyroid function (J is a constituent of thyroxine hormone)
Co	.	constituent of vitamin B_{12}
Se	.	necrosis of liver
Cr	.	disturbances of growth
Mo	.	caries (Synopsis 9–5)
Sn	.	.
V	.	.
(F) and others	1 mg	caries, necessity of fluorine and other elements is a matter of definition

Synopsis 9–5. Connections between Nutrition Deficient in Molybdenum and Incidence of Caries in School Children in New Zealand (according to Ludwig, 1962, cited in [7]).

Examination	Town of Napier	Town of Hastings Distance 15 km
teeth of school children	almost healthy	high caries incidence
fluorine concentration in drinking water	normal	normal
contents of many trace elements in vegetables	normal	normal
molybdenum content in vegetables	normal	very low
origin of soils	from recent marine sediments	from older material poor in Mo

9.4.2 Assessment of Food by Medical Indices

The effects of the quality of certain food on the consumer can be determined only on the basis of complex indicators (growth rate, fertility, resistance to diseases, etc.). This necessitates suitable nutrition or feeding experiments. They are based on the principle that food produced under different fertilization conditions is given to testpersons and animals and its effect on their health is determined. The relatively few tests carried out until now usually consisted in comparing the effects of organic fertilization with farmyard manure and/or compost to those of mineral fertilization, sometimes with small additions of farmyard manure.

The following results were obtained:

- *Rats* were rather more than less fertile over six generations when given *food produced with mineral fertilization*. This nutrition produced 5% more animals in total, but the difference is hardly significant. In any case, there is no indication that mineral fertilization lowered the food quality in this long-term experiment.
- *Infants* gained more weight when additionally fed with *vegetables fertilized* variously with *minerals*. They also had better blood-counts and increased resistance to infections (Fig. 9–3 [271]).

Both trials rather indicate that mineral fertilization improves quality and at least does not lower it. Similar results were obtained with small children in a more recent trial [300].

The following more recent trials concern the influence of fertilization on the fertility of domestic animals.

Fodder from fields fertilized with manure compost improved the quality of *bull* sperm (this is an important fertility indicator) more than fodder from fields supplied with average amounts of mineral fertilizers [265]. This trial, however, does not prove much, since the fields were not simply comparable. Like other aspects still to be discussed critically with regard to all experiments, this would have been a basic precondition for a valid comparison. Another experiment showed that rabbits are more fertile when fed with grass from unfertilized pastures than when the pastures were fertilized intensively (mostly with minerals). In this experiment too, it was not only the fertilization that was a variable (since the test areas were not statistically-distributed fertilized plots of the same soil); the validity of the results is therefore limited. The experiment nevertheless shows clearly the influence of value-determining mineral constituents of food. The better food composition is matched to requirements, the higher is the fertility of animals. The particular fertilizer form might be unimportant in this case.

Overall *critical consideration* of all these experiments is indicated for other reasons. A precondition for causal interpretation would be a comprehensive

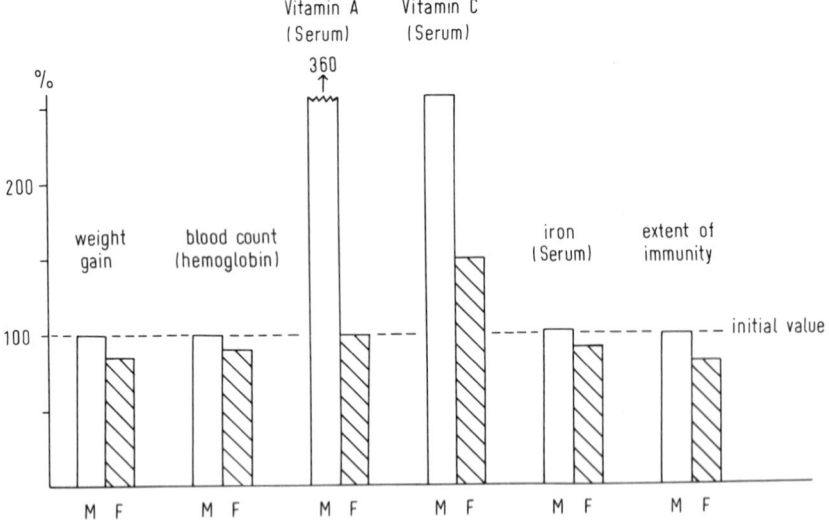

Fig. 9-3. Influence of Varying Fertilization of Food Plants on Health of Infants (M = mineral fertilization and farmyard manure, F = farmyard manure) (according to *Catel* [271, 292]).

characterization of the food used in the experiments, based on analysis for important mineral and organic constituents.

Past experiments were methodically based mainly on the fertilization practice of earlier decades. A clear distinction between influence factors is unfortunately lacking in the experiments (farmyard manure is often applied to both experimental plots). However, even testing the influence of mineral fertilization without application of farmyard manure does not exclude the soil as a source of nutrients. Comparison of plants exclusively grown on synthetic nutrient media (hydroponics) would be best. In theory there are two possibilities in a suitably designed experiment:

- Mineral fertilizers have no detrimental effects in hydroponics even in pure form, in which case no detrimental effects in the soil should be expected;
- mineral fertilizers might detract from the quality in hydroponics; these detrimental effects are either compensated or not by the soil under natural site conditions.

However, the validity of this result would be restricted to the use of mineral fertilizers for correct, sensible fertilization, even when no detrimental (or even beneficial) effects of mineral fertilizers could be demonstrated in scientifically exact trials. This leaves the problem of the applicability of these results to fertilization practice in agriculture and horticulture, which is certainly not predominantly scientifically performed.

The results of past nutrition experiments can in general be summarized as follows (as far as the basic requirements of agrochemical design of experiments were fulfilled): Mineral fertilization appears superior to organic fertilization; both fertilization methods have at most equal effects on food quality.

9.4.3 Correct Fertilization and Healthy Food

The relationship between fertilization and the quality of vegetal products is a complicated problem, but it can nevertheless be made surveyable for the most part. *Liebig* in 1850 was the first to demonstrate the action chain "soil – plant – animal and man". The significance of the circulation of matter in inanimate and living nature has since then become increasingly obvious. However, many aspects have only recently been discovered. Some theories (both within the dominant schools and by outsiders) have caused controversies and have been revised. Many connections between fertilization problems, which at first appeared to be relatively simple, have proved to be quite complex. Many farmers may long for the "good, old, simple" conditions of earlier times, but this paradise (if it ever existed) is probably lost for ever. Fertilization will never again be simple but neither will it become confusingly complex.

Mineral fertilization has gone through all degrees of judgement from highest praise to total rejection (the latter is now being concealed as far as possible, having long ago been recognized as erroneous). Agriculture in general can no longer be imagined without it (even if it is not always needed in large amounts). Mineral fertilization need not be advocated only for the reason that it is required for the production of large amounts of food, since quantity need not be at the expense of quality. There is no reason to object to high yields in agriculture because of this, especially since they make food cheaper.

General objections to "chemicalization" of agriculture are completely unjustified. Not even the use of poisons renders production "chemical" or "unbiological", since many (biogenic) poisons occur in nature. Organic fertilizers, considered to be so natural and harmless (e. g., farmyard manure and compost) may also contain harmful substances, even if these are largely decomposed in the soil. Mineral fertilizers are not poisons (except in some cases where damage to plants is intentional) but contain or supply the same plant nutrients as natural soil.

All agriculture is *biological* in the final analysis (purely synthetic chemical production of individual nutrients in factories is not considered here). The quality designation "produced biologically" cannot therefore be reserved to some special forms of agriculture. Obviously, "alternative" agriculture cannot be judged only from the fertilization aspect but this kind of fertilization must be considered as it produces neither better nor worse food in principle. Gross mistakes are possible even when organic fertilizers are used; it is the correct use of fertilizers that is always the determinant, and testing it is an important factor.

A special aspect will be discussed briefly. This is whether the aim of obtaining food of permanently high value is justified. The normal condition of man for millenia has been that periods of good and poor nutrition alternate; the cycles could last for years or days. Periods of food scarcity have been so common in the history of human evolution, that it should be asked whether they might not be somehow *advantageous*. However, there might be an advantage in a temporary nutrition deficiency of calories, but not in deficiencies in the supply of essential nutrients, even though the human body can survive periods of deficiency.

One of the values of research in the natural sciences is that it has made it possible to study the entire problem of food quality. An understanding of whole systems, in the natural sciences too, is possible only after logical dissection, if one tries not only to consider the *whole*, but also to understand and, if necessary, improve it. Only causal dissection of the complex problem can test the correlations on which the understanding of agricultural production was based in prescientific times for their validity and evaluate them correctly.

The statement that agriculture is the "guardian of health" [302] is correct to the extent that food quality, being the basis of the health of man and animal, depends on the correct use of the means of production, especially fertilizers.

There are still gaps in agrochemical knowledge of plant growth factors, but these gaps in knowledge about the production of high-value vegetal food today seem to become comprehensible or at least their possible significance can be assessed, if they are considered carefully. Science is the laborious and unbiased striving for truth. Belief that one is in possession of the truth should be left to ideologies.

Liebig correctly wrote in his *Chemical Letters*:

"Without a knowledge of natural laws and natural phenomena, the human mind fails in the attempt to imagine the magnitude and unfathomable wisdom of the Creator. Any picture that the richest fantasy, the best training of mind can imagine, is like a colored, glistening, empty soap bubble if compared with reality."

Conclusion: Fertilization, Food Supply, and Plant Quality

1. The *quality* of vegetal products depends on many factors, among others, optimal supplies of all substances required for growth. Minerals occupy a central position, but, they are sometimes supplied to an insufficient extent by natural substrates (soils).

2. This *gap* in *requirements* between insufficient natural supplies and optimal supplies for plants should be closed by the use of fertilizers, since only properly nourished plants can provide products of overall high quality.

3. Plants growing on sites without additional fertilization have greatly differing *contents of ingredients* (valuable, and sometimes harmful substances) that are important for an assessment of their nutritive qualities. This is due to the different and incomplete supplies of nutrients to plants on natural substrates. This initial state of sometimes large, sometimes small, contents of valuable substances in production without fertilization should always be remembered when fertilizer effects are assessed.

4. *Food quality* is a *complex concept* depending mainly on the content of value-determining ingredients. In practice, it is impossible to improve all value components simultaneously by any means. Quality improvement always means improvement of some particularly desirable components, not all.

5. Any fertilization *improving the supply* of nutrients to plants, i. e. , increasing supplies from deficiency to the optimal, raises the content of valuable substances in a plant (in different degrees for the individual components). The concentrations (i. e., percentage contents) of valuable substances also frequently increase in this case, but they may decrease with increasing yields because of dilution. This occasional concentration decrease of certain valuable substances (despite increases of their overall amounts and of the concentrations of other valuable substances) is unavoidable. It occurs with any kind of fertilization and mostly does not affect the overall assessment of the effects of fertilization on quality.

6. *Excess fertilization* may lower the quality. Onesidedly exaggerated fertilization (with one or several nutrient elements), implying luxury supplies of certain nutrients to the plant, further increases the contents of some valuable substances (e. g., if the mineral applied as fertilizer is itself a valuable substance), but does not increase the contents of many other valuable substances, or may even reduce them to varying degrees. It may also cause the accumulation of detrimental or even harmful substances. This quality reduction, that may occur with excess fertilization, manifests itself less the more fertile the soil. *Fertile soils* are excellent *regulating systems* largely buffering excesses of soluble nutrients undesirable for plants, using inorganic and organic (soil organisms) components. Fertilization imbalances are thus at least partly compensated.

7. The *fertilizer form* (whether organic or mineral, farmyard or commercial fertilizer with slow or rapid action) is in principle quite unimportant for the quality of vegetal products. Plants are nourished mainly by salts derived either from the soil or (directly or indirectly) from fertilizers.

There are some important differences in fertilization practice, especially if carried out by laymen, if it is empirical and not based on precise diagnosis. Thus, slow-acting fertilizers are less problematic than those that are water-soluble, with regard to undesirable, easily available excess nutrients. Fertilizers containing several nutrients (whether mineral or organic crude fertilizers, or mixtures produced from single salts) are more likely to ensure a balanced plant nutrition than single-nutrient fertilizers, if applied empirically. Some organic and mineral fertilizers combine both advantages. This, however, does not turn them into ideal fertilizers, since even a complex composition need not automatically be ideal, if nutrients in the soil are taken into account.

8. Some fertilizer components (micronutrients, organic active agents) may have *great effects* on plant production even if *applied* only in *small amounts* (dilution). However, extreme dilution of fertilizer components has a lower limit of effectiveness, long before it becomes completely senseless because of the atomic structure of matter.

It is quite unimportant whether highly diluted, special preparations of uncertain composition can improve the quality (for material or energy reasons), as long as this is only asserted but not proved by scientifically acceptable methods, and this is not yet the case.

9. Without fertilization man and animal would have to make do with food seriously lacking in valuable ingredients. They would suffer from (acute and latent) deficiency diseases, that were common in former times and sometimes are so even today. Poisoning might occur too, since fertilizers can also deactivate natural toxins.

Properly applied fertilization improves food quality through a higher quality of vegetal products (and thus indirectly of animal products, as well) so that it contributes to the health of man and animal.

10. *Highest food quality* is an ideal aim that can be achieved only approximately. Our knowledge of plant growth factors is incomplete, and even complete knowledge could not be transformed into practical results on complex substrates like soils simply by suitable fertilization.

Striving for fertilizer application based on facts, i.e., on as exact a diagnosis as possible and thus under control, is essential. This is the only way to achieve that highest quality attainable, given all the indeterminacy of *primary production*.

Appendix

Definitions of Chemical Terms

Acid: compound forming hydrogen ions (H^+) in aqueous solution.

Alkaline solution: aqueous solution of base.

Analysis: investigation of chemical compound and determination of its composition.

Anion: negatively (electrically) charged particle, e.g., NO_3^- (nitrate ion), HPO_4^{2-} (hydrogen phosphate ion).

Atom: smallest particle of chemical element.

Atomic weight: weight of atom, referred to weight of hydrogen atom.

Base: compound that forms hydroxide ions (OH^-) in aqueous solutions, in a wider sense also metals, e.g., K, Ca (see *alkaline solution*).

Cation: positively (electrically) charged particle, e.g., K^+ (potassium ion), Mg^{2+} (magnesium ion), NH_4^+ (ammonium ion).

Colloidal solution: mixture of very fine solid particles with liquid (e.g., protein colloid), either in liquid (sol) or jelly-like state (gel).

Compound (chemical): substance consisting of several elements, e.g., water (H_2O) consists of the elements hydrogen and oxygen.

Concentration: content in total quantity, e.g., for solutions: content of dissolved constituent per quantity of solution (stated in %, ‰, ppm).

Diffusion: gradual interpenetration (of liquid and gases), caused by the natural motion of molecules.

Element (chemical): basic chemical substance consisting of chemically uniform atoms; there are 92 natural elements arranged in the periodic system.

Emulsion: heterogeneous mixture of liquids (e.g., fat droplets in water).

Equivalent weight:	molecular weight of substance, divided by valence (see *eq.*).
eq.:	amount of substance, numerically equal in grammes to equivalent weight, e. g.,

for KCl (potassium chloride): 1 eq. $\cong$ 74.6 g
(univalent cation),

for $MgSO_4$ (magnesium sulfate): 1 eq. $\cong$ 60 g
(bivalent cation),

see also *mol, equivalent weight, valence.*

Hydrolysis:	reaction of a salt with ions of water, e. g., $CaCO_3 + 2(H^+ + OH^-) \rightarrow Ca(OH)_2 + H_2CO_3$.
Hydroxide:	compound containing (OH^-)-ions; hydroxides of metals are also called bases.
Ion:	electrically charged particle (cation or anion).
Minerals:	substances from mineral contents of earth's crust or soil (minerals are inorganic, mostly crystalline substances).
mol:	amount of (chemically homogeneous) substance, containing as many grammes as the numerical value of the molecular weight, e. g., for KCl (potassium chloride), 1 mol $\cong$ 74.6 g, for $MgSO_4$ (magnesium sulfate), 1 mol $\cong$ 120 g.
Molar solution:	solution containing 1 mol per liter (unit designation: M).
Molecules:	atoms bound to one another.
Molecular weight:	weight of one molecule (sum of atomic weights).
meq:	$1/1000$ eq. (see. *eq.*).
Neutralization:	combination of acid and base to form a (neutral) salt.
Normal solution:	solution containing 1 equivalent (eq.) of a substance per liter water (unit designation: N).
Oxide:	compound of an element and oxygen, e. g., CaO (calcium oxide).
Oxidation:	combination of element with oxygen or increase of its valence (loss of electrons), e. g., Ca (calcium) + O (oxygen) $\rightarrow$ CaO (calcium oxide); Fe^{2+} (bivalent iron ion) $\rightarrow$ Fe^{3+} (trivalent iron ion) + 1 electron.
pH:	(latin: *potentia hydrogenii*) negative logarithm to the base 10 of the hydrogen-ion concentration in eq./liter. For hydrogen ions (H^+) 1 eq. = 1 g.
Radical:	stable groups of atoms reacting as a whole.
Reaction (chemical):	process in which substances react with one another and are thus altered.

Reaction (of solution):	acidity expressed as pH (e.g., a solution has acid "reaction").
Redox reaction:	combined reaction of oxidation and reduction, e.g.,

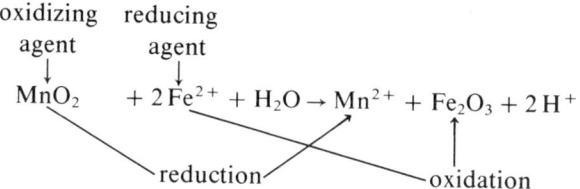

$$MnO_2 + 2 Fe^{2+} + H_2O \rightarrow Mn^{2+} + Fe_2O_3 + 2 H^+$$

Reduction:	removal of oxygen or lowering of valence (absorption of electrons). Reversal of oxidation, e.g., Mn^{4+} (tetravalent manganese ion) + 2 electrons $\rightarrow$ Mn^{2+} (bivalent manganese ion).
Reducing agent:	substance that combines with oxygen or loses electrons and becomes oxidized in the process.
Salt:	combination of one cation and one anion, sometimes easily soluble in water (common salt, or "potassium salt"), sometimes practically insoluble (e.g., Ca-silicate, apatite).
Solution:	mixture of solvent (e.g., water) and dissolved substance (e.g., salt).
Substance (chemical):	chemical elements and combinations of such.
Suspension:	suspension of solid particles in liquid, e.g., lime in water.
Synthesis:	formation of chemical compound.
Valence:	number of (positive or negative) charges of ion (univalent: K^+, Na^+; bivalent: Mg^{2+}, SO_4^{2-}; trivalent: PO_4^{3-}).

References

References to Chap. 1

[1] BELF (Bundesministerium für Ernährung): Statistisches Jahrbuch. Parey, Hamburg, from 1976 Landwirtschaftsverlag Münster.
[2] Bittermann, E.: Die landwirtschaftliche Produktion in Deutschland von 1800-1950. Quoted by Busch, W.: Die Entwicklung der Landwirtschaft seit 1800. In: Der Stickstoff. Fachverband Stickstoffindustrie. Düsseldorf 1961.
[3] Blunck, R.: Justus von Liebig. Hammerich und Lesser, Hamburg 1946.
[4] Ehrenberg, P.: Die Düngerlehre in 16 Vorlesungen. Winters, Heidelberg 1924.
[5] FAO (Food and Agriculture Organization): Annual Fertilizer Review 1960-1976.
[6] Finck, A.: Die Notwendigkeit von chemischen Elementen für höhere Pflanzen. Angew. Botanik 48, 21-38 (1974).
[7] Finck, A.: Pflanzenernährung in Stichworten. Hirt, Kiel 1976.
[8] Finck, A.: Mineraldüngung gezielt. AID-Broschüre Nr. 401, Bonn 1977.
[9] Handbuch der Pflanzenernährung und Düngung. Hrsg. Linser, H., Vol. I: Pflanzenernährung 1969, 1972; Vol. II: Boden und Düngemittel 1966, 1968; Vol. III: Düngung der Kulturpflanzen 1965; Springer, Wien.
[10] Haselhoff, E., und Blanck, E.: Lehrbuch der Agrikulturchemie, II. Düngemittellehre. Borntraeger, Berlin 1928.
[11] Haushofer, H.: Alternativen der Landbewirtschaftung? Dachverband Agrarforschung Munich 1976.
[12] Heiden, E.: Lehrbuch der Düngerlehre. Cohen, Stuttgart 1866/68.
[13] Honcamp, F.: Historisches über die Entwicklung der Pflanzenernährungslehre, Düngung und Düngemittel. In: Honcamp, F.: Handbuch der Pflanzenernährung und Düngerlehre, Vol. I und II. Springer, Berlin 1931.
[14] Kappen, H.: Die Bodenacidität. Springer, Berlin 1929.
[15] Kick, H.: Pflanzennährstoffe. A. Begriffe, B. Anorganische Stoffe. In: Handbuch Pflanzenernährung. Vol. I/1, pp. 90-122, 1969 [9].
[16] Lang, K., and Polheim, P. v.: Düngemittelverzeichnis. VDLUFA, Darmstadt 1974.
[17] Liebig, J. v.: Die Chemie in ihrer Anwendung auf Agrikultur und Physiologie. Vieweg, Braunschweig 1840.
[18] Mayer, A.: Lehrbuch der Agrikulturchemie, II. Die Düngerlehre. 7. ed. 1924, Winters, Heidelberg 1907.
[19] Mengel, K.: Ernährung und Stoffwechsel der Pflanze. Fischer, Jena 1972.
[20] Rintelen, P.: Betriebswirtschaftliche Betrachtungen zum Stickstoffverbrauch. In: Der Stickstoff. Fachverband Stickstoffindustrie, Düsseldorf 1961.

[21] Russell, E. W.: Soil Conditions and Plant Growth. Longmans, London 1973.

[22] Scheffer, F., und Welte, E.: Lehrbuch der Agrikulturchemie und Bodenkunde. Section II Pflanzenernährung. Enke, Stuttgart 1955.

[23] Schmitt, L.: Die geschichtliche Entwicklung und die Aufgaben des VDLUFA. In: Festschrift zum 75. Jubil. VDLUFA, Sauerländer, Frankfurt 1963.

[24] Schmitz, F., und Kluge, G.: Das Düngemittelrecht mit fachlichen Erläuterungen. Landwirtschaftsverlag, Münster 1978.

[25] Schneidewind, W.: Die Ernährung der landwirtschaftlichen Kulturpflanzen. Parey, Berlin 1915.

[26] Schwerz, J.: Zit. bei Franz, G., und Haushofer, H.: Große Landwirte. DLG-Verlag, Frankfurt 1970.

[27] Thaer, A.: Über die Wertschätzung des Bodens. Berlin 1811.

[28] Wagner, P.: Düngungsfragen. 3. ed., Parey, Berlin 1900.

[29] Wolff, E. v.: Praktische Düngerlehre 3. ed., Wiegandt, Berlin 1870.

References to Chap. 2

[30] Andrews, W. B.: Anhydrous Ammonia as a Fertilizer. Advances in Agronomy *8*, 62–125 (1956).

[31] Balz, O., und Hamacher, H.: Stickstoffdüngemittel. In: Stickstoff. Fachverband Stickstoffindustrie, Düsseldorf 1961.

[32] Banthien, H.: Synthetische Stickstoffdüngemittel. In: Handbuch Pflanzenernährung. Vol. II/2, pp. 1014–1110, 1968 [9].

[33] Braune, G.: Kalidüngemittel. In: Handbuch Pflanzenernährung Vol. II/2, pp. 1217–1267, 1968 [9].

[34] Buchner, A., et al.: Düngemittel. In: Ullmanns Encyklopädie der technischen Chemie. Vol. 10, pp. 201–256, Verlag Chemie, Weinheim 1975.

[35] Buchner, A., und Sturm, H.: Die Düngung im Intensivbetrieb. DLG-Verlag, Frankfurt 1971.

[36] Cooke, G. W.: Fertilizing for Maximum Yield. Granada Publ., London 1975.

[37] Fachverband Stickstoffindustrie: Der Stickstoff. Düsseldorf 1961.

[38] Gericke, S.: Thomasphosphat. In: Handbuch Pflanzenernährung. Vol. II/2, pp. 1168–1202, 1968 [9].

[39] Hauck, R. D., and Kohino, M.: Slow Release and Amended Fertilizers. In: Fertilizer Technology and Use. pp. 455–494, Soil Science Soc. Amer., Madison 1971.

[40] Höfner, W.: Schwefelhaltige Düngemittel. In: Handbuch Pflanzenernährung. Vol. V/2, pp. 1316–1320, 1968 [9].

[41] Ignatieff, V., and Page, H. J.: Efficient Use of Fertilizers. FAO, Rome 1968.

[42] Jung, J., und Jürgens-Gschwind, S.: Zur Düngewirkung kondensierter Phosphate. Landw. Forschung *28*/1, 94–109 (1972).

[43] Kaila, A.: Apparent Recovery of Fertilizer Nitrogen. J. Sci. Agr. Soc. Finland *39*, 78–89 (1965).

[44] Kundler, P., et al.: Mineraldüngung. VEB Dt. Landw.-Verlag, Berlin 1970.

[45] Lehr, J.: Chilesalpeter. In: Handbuch Pflanzenernährung Vol. II/2, pp. 1003–1014, 1968 [9].

[46] Müller, H.: Feinvermahlenes Rohphosphat. In: Handbuch Pflanzenernährung. Vol. II/2, pp. 1129–1139, 1968 [9].

[47] Olson, R. A., Army, T. J., Hanway, J. J., and Kilmer, V. J.: Fertilizer Technology and Use. Soil Science Soc. Amer., Madison 1971.

[48] Ortlepp, H., und Wagner, E.: Die teilaufgeschlossenen Phosphatdünger. Annenhof, Hamburg 1968.

[49] Rodewyk, A.: Informationen über Kali. Kali und Salz AG, Hannover 1972.

[50] Scheel, K.: Rohphosphat, Superphosphat, Glühphosphat. In: Handbuch Pflanzenernährung Vol. II/2, pp. 1110–1128, 1140–1167, 1203–1217, 1968 [9].

[51] Scheffer, F.: Die "wirksame" Phosphorsäure bestimmt den Pflanzenertrag. Phosphorsäure *16*, 105–120 (1956).

[52] Scheffer, F.: Stickstoff und Boden. Stickstoff und Pflanze. In: Stickstoff. Fachverband Stickstoffindustrie, Düsseldorf 1961.

[53] Schmidt, A.: Chemie und Technologie der Düngemittelherstellung. Hüthig, Heidelberg 1972.

[54] Schmitt, L.: Die Untersuchung von Düngemitteln. Neumann, Radebeul 1954.

[55] Schmitt, L.: Vom Segen der Düngung. DLG-Verlag, Frankfurt 1954.

[56] Sommer, K.: Nitrificide. Landw. Forschung Sond. 27/II, 64–82 (1972).

[57] Tisdale, S. L., and Nelson, W. L.: Soil Fertility and Fertilizers. McMillan, New York 1970.

[58] VDDF (Verein Deutscher Dünger-Fabrikanten): 100 Jahre Superphosphat. VDDF, Hamburg 1955.

[59] Werner, W.: Die Rhenania-Dünger. Schaper, Hannover 1967.

[60] Wehrmann, J.: Möglichkeiten und Grenzen intensiver Düngeranwendung. Landw. Forschg. 26/I Sonderh. 1–15, 1971.

References to Chap. 3

[61] Baumeister, W.: Das Natrium als Pflanzennährstoff. Fischer, Stuttgart 1960.

[62] Fritz, D., und Venter, F.: CO_2-Begasung. In: Storck, H.: Gartenbau. Ulmer, Stuttgart 1969.

[63] Katalymow, M. W.: Mikronährstoffe, Mikronährstoffdüngung. VEB Deutsch. Landw.-Verlag, Berlin 1969.

[64] Jones, L. H. P., and Handreck, K. A.: Silica in Soils, Plants and Animals. Advanc. Agronomy *19*, 107–149 (1967).

[65] Jacob, A.: Normung der Düngung durch Volldünger. Enke, Stuttgart 1955.

[66] Jung, J., and Jürgens-Gschwind, S.: The Fertilizing Effect of Condensed Phosphates. Phosph. in agric. *64*, 1–15, 1975.

[67] Löcker, H., und Rank, V.: Mehrnährstoffdüngemittel und Mischen von Düngemitteln. In: Handbuch Pflanzenernährung Vol. II/2, pp. 1345–1399, 1968 [9].

[68] Mortvedt, J. J., Giordano, P. M., and Lindsay, W. C.: Micronutrients in Agriculture. Soil Sci. Soc. Amer., Madison 1972.

[69] Reinau, E. H.: Kohlensäuredünger. In: Honcamp, F.: Handbuch der Pflanzenernährung und Düngerlehre. Vol. II, pp. 643–658, Springer, Berlin 1931.

[70] Schäfer, H. K.: Mikronährstoffdüngemittel. In: Handbuch Pflanzenernährung Vol. II/2, pp. 1320–1344, 1968 [9].

[71] Scharrer, K.: Biochemie der Spurenelemente. Parey, Berlin 1955.

[72] Schütte, K. H.: Biologie der Spurenelemente. Bayer. Landwirtsch.-Verlag, München 1965.

[73] Sluijsmans, C.: Der Einfluß von Düngemitteln auf den Kalkzustand des Bodens. Zeitschr. Pflanzenernähr. *126*, 97–103 (1970).

[74] Stangel, P. J.: Prospects of New Fertilizers, Fertilization Concepts and Bulk Blending in Asia, Africa, and Latin America. FAO, Rome 1970.

[75] Stiles, W.: Trace Elements in Plants. University Press, Cambridge 1961.

References to Chap. 4

[76] Baden, W., Kuntze, H., Niemann, J., Schwerdtfeger, G., und Vollmer, F.: Bodenkunde. Ulmer, Stuttgart 1969.

[77] Boguslawski, E. v., und Debruck, J.: Strohdüngung und Bodenfruchtbarkeit. DLG-Verlag, Frankfurt 1977.

[78] Brady, N. C.: The Nature and Properties of Soils. Macmillan, New York 1974.

[79] Fiedler, H., und Reissig, H.: Lehrbuch der Bodenkunde. Fischer, Jena 1964.

[80] Flaig, W.: Humusstoffe. In: Handbuch Pflanzenernährung. Vol. II/1 pp. 382–458, 1966 [9].

[81] Glathe, H., und Glathe, G.: Impfstoffe für Böden und Komposte. In: Handbuch Pflanzenernährung. Vol. II/2, pp. 1455–1463, 1968 [9].

[82] Hewitt, E. J.: Sand- and Water Culture Methods Used in the Study of Plant Nutrition. Commonw. Agr. Bur., Farnham Royal 1966.

[83] Kalkdienst: Düngekalk-Leitfaden. Drei Kronen Verlag, Efferen 1964.

[84] Köhnlein, J., und Vetter, H.: Ernterückstände und Wurzelbild. Parey, Hamburg 1953.

[85] Knickmann, E.: Handelshumusdüngemittel. In: Handbuch Pflanzenernährung. Vol. II/2, pp. 1400–1450, 1968 [9].

[86] Koriath, H.: Güllewirtschaft, Gülledüngung. VEB Deutsch. Landw.-Verl., Berlin 1975.

[87] Marschner, H.: Ernährungs- und ertragsphysiologische Aspekte der Pflanzenernährung. Angew. Botanik 52, 71–87, 1978.

[88] Mückenhausen, E.: Die Bodenkunde. DLG-Verlag, Frankfurt 1975.

[89] Nieschlag, F.: Der fruchtbare Boden. DLG-Verlag, Frankfurt 1969.

[90] Penningsfeld, F., und Kurzmann, P.: Hydrokultur und Torfkultur. Ulmer, Stuttgart 1966.

[91] Rauhe, K.: Wirtschaftseigene Düngemittel. In: Handbuch Pflanzenernährung. Vol. II/2, pp. 907–993, 1968 [9].

[92] Salzer, E. H.: Methoden der Hydrokultur. Franckh, Stuttgart 1965.

[93] Sauerlandt, W., und Tietjen, C.: Humuswirtschaft des Ackerbaus. DLG-Verlag, Frankfurt 1970.

[94] Scheffer, F., und Schachtschabel, P.: Lehrbuch der Bodenkunde. Enke, Stuttgart 1976.

[95] Scheffer, F., und Ulrich, B.: Humus und Humusdüngung. Enke, Stuttgart 1960.

[96] Schlichting, E., und Blume, H. P.: Bodenkundliches Praktikum. Parey, Hamburg 1966.

[97] Schroeder, D.: Bodenkunde in Stichworten. Hirt, Kiel 1978.

[98] Siebeneicher, G. E.: Neues großes Gartenlexikon. Südwest-Verlag, Munich 1973.

[99] Vetter, H.: Mist und Gülle. DLG-Verlag, Frankfurt 1973.

[100] Ziemer, F.-W.: Kalkdüngemittel. In: Handbuch Pflanzenernährung Vol. II/2, pp. 1268–1316, 1968 [9].

References 417

References to Chap. 5

[101] Barber, S. A.: Problem Areas and Possibilities of More Efficient Fertilizer Use. FAO, Rome 1976.
[102] Bayerische Landesanstalt für Bodenkultur und Pflanzenbau et al.: Die Düngung von Acker und Grünland nach Bodenuntersuchungsergebnissen. München 1975.
[103] Bear, F. E., et al.: Hunger Signs in Crops. Amer. Soc. Agronomy, Washington 1949.
[104] Bergmann, W., und Neubert, P.: Pflanzendiagnose und Pflanzenanalyse. Fischer, Jena 1976.
[105] Bergmann, W.: Ernährungsstörungen bei Kulturpflanzen. Fischer, Jena 1976.
[106] Boas, F.: Zeigerpflanzen. Verlagsgesellschaft Ackerbau, Hannover 1958.
[107] Broeshart, H.: Quantitative Measurement of Fertilizer Uptake by Crops. Netherlands J. Agric. Sci. 22, 245–254, 1974.
[108] Bussler, W.: Vergleichende Untersuchungen an Kali-Mangelpflanzen. Verlag Chemie, Weinheim 1962.
[109] Chapman, H. D.: Diagnostic Criteria for Plants and Soils. University of California, California 1966.
[110] Gram, E., Bovien, P., und Stapel, C.: Farbtafel-Atlas der Krankheiten und Schädlinge an landwirtschaftlichen Kulturpflanzen. Parey, Hamburg 1971.
[111] Grunwaldt, H. S., und Patzke, W.: Düngebedarfsermittlung nach Bodenuntersuchungsergebnissen. Landwirtschaftskammer Schleswig-Holstein, Kiel 1975.
[112] Hauck, W.: Düngung (in den Tropen). In: Handbuch der Landwirtschaft und Ernährung in den Entwicklungsländern. Vol. 2, pp. 185–204, Ulmer, Stuttgart 1971.
[113] Heller, L., und Landmann, H.: Düngungsbezugs-Preisliste 1977/78. Heller, Göttingen 1977.
[114] Henkens, Ch. H.: Bodem en bemesting. Rijkslandbouwconsulentschap, Groningen 1962.
[115] Hernando, V., y Cadahia, C.: El analysis de savia como indice de fertilizacion. Instit. Edafologia, Madrid 1973.
[116] Hesselbach, A., und Patzke, W.: Düngervoranschlag über EDV. Verband der Landw.kammern Bull. 21, Bonn 1977.
[117] Jones, J. B., and Eck, H. v.: Plant Analysis as an Aid in Fertilizing Corn and Grain Sorghum. In: Walsh, L. M., and Beaton, J. D.: Soil Testing and Plant Analysis. Soil Sci. Soc. Amer., Madison 1973.
[118] Karlovsky, J.: Method of Assessing the Utilization of Phosphorus on Permanent Pastures. Trans. Int. Soil Sci. Conf. New Zealand, pp. 726–730 (1972).
[119] Köhnlein, J., und Knauer, N.: Die Entzugszahl als Hilfsmittel zur richtigen Bemessung der P_2O_5- und K_2O-Gabe. Z. Acker- u. Pflanzenbau 104, 329–370 (1957).
[120] Köster, W.: Beziehung zwischen Phosphorgehalt von Kartoffelkraut und Boden. Z. Pflanzenernähr. u. Bodenkde. 137, 19–31, 1974.
[121] Linser, H.: Die Rentabilität und wirtschaftliche Bedeutung der Düngung. In: Handbuch Pflanzenernährung. Vol. III/2, pp. 1390–1444, 1965 [9].
[122] Matzel, W.: Probleme der Ausnutzung des Dünger- und Bodenphosphors. Archiv Acker- u. Pflanzenbau 18. 471–487, 1974.
[123] Mitscherlich, E.: Bodenkunde, Parey, Berlin 1954.
[124] Møller-Nielsen, J., and Friis-Nielsen, B.: Evaluation and Control of the Nutritional Status of Cereals. Plant and Soil 45, 647–658 (1976).
[125] Munk, H.: Phosphatdüngung — Phosphatverfügbarkeit. Phosphorsäure 29, 35–56, 1971.

[126] Munson, R. D., and Doll, J. P.: The Economics of Fertilizer Use in Crop Production. Advanc. Agron. *11*, 133–169 (1959).
[127] Munson, R. D., and Nelson, W. L.: Principles and Practices in Plant Analysis. In: Walsh and Beaton [141].
[128] Nethsinge, D. A.: The Use of Isotopes and Radiation in Studies of the Efficient Use of Fertilizers. FAO, Rome 1977.
[129] Rauterberg, E.: Chemische und physikalisch-chemische Verfahren (zur Bestimmung des Düngebedürfnisses der Böden). In: Handbuch Pflanzenernährung. Vol. II/1, pp. 800–844, 1966 [9].
[130] Reisenauer, H. M.: Soil and Plant-tissue Testing in California. Bull. 1879, University of California 1976.
[131] Ris, J., and van Luit, B.: The Establishment of Fertilizer Recommendations on the Basis of Soil Tests. Instit. Bodemvruchtbaarheid, Haren-Gr. 1973.
[132] Saalbach, E.: Sulfur Requirements and Sulfur Removals of the Most Important Crops. Annal. Agronom. 1–55 (1972).
[133] Scharpf, H. C., und Wehrmann, J.: Die Bedeutung des Mineralstoffvorrates des Bodens zu Vegetationsbeginn für die Bemessung der N-Düngung zu Winterweizen. Landw. Forschung *32/1*, 100–114 (1975).
[134] Schüller, H.: Entwicklung und Stand der Bodenuntersuchung. In: 100 Jahre Landw.-chem. Bundesversuchsanstalt. pp. 303–321, Vienna 1970.
[135] Stamer, H.: Landwirtschaftliche Marktlehre I. Parey, Hamburg 1966.
[136] Thun, R., Herrmann, R., und Knickmann, E.: Die Untersuchung von Böden. Neumann, Radebeul 1955.
[137] Unger, H.: Preisunterschiede zwischen Mehrnährstoff- und Einzeldüngern. In: Agrarmarktstudien Bull. 4, Parey, Hamburg 1967.
[138] Verband der Landwirtschaftskammern: Düngervoranschlag über EDV. Schriften des Verbandes, Heft 21, Bonn 1977.
[139] Vetter, H.: Wieviel düngen? DLG-Verlag, Frankfurt 1977.
[140] Wallace, T.: The Diagnosis of Mineral Deficiencies in Plants. Her Majesty's Stat. Office, London 1961.
[141] Walsh, C. M., and Beaton, J. D.: Soil Testing and Plant Analysis. Soil Sci. Soc. Amer., Madison 1973.

References to Chap. 6

[142] Adlkofer, J., und Schwarzmann, M.: Gefährden Stickstoffdünger die Ozonschicht? BASF Publikation, Ludwigshafen 1976.
[143] Aereboe, F.: Neue Düngewirtschaft ohne Auslandsphosphate. Berlin 1922.
[144] Andreae, B.: Wirtschaftslehre des Ackerbaus. Ulmer, Stuttgart 1958.
[145] Ayers, R. S., and Westcot, D. W.: Water Quality for Agriculture. Irrigation and Drainage Paper 29. FAO, Rome 1976.
[146] Baden, W.: Die Kalkung von Moor und Anmoor. In: Handbuch Pflanzenernährung. Vol. III/2, pp. 1445–1516, 1965 [9].
[147] Baltin, F.: Ausbringegeräte (für Düngemittel). In: Handbuch Pflanzenernährung. Vol. III/1, pp. 35–95, 1965 [9].
[148] Berge, H.: Immissionsschäden. In: Handb. Pflanzenkrankheiten (Ed. B. Rademacher und H. Richter), Vol. 1/4. Parey, Berlin 1970.
[149] Boguslawski, E. v.: Düngungsplan und Fruchtfolge. In: Handbuch Pflanzenernährung. Vol. III/2, pp. 1540–1568, 1965 [9].

[150] Bräunlich, K.: Düngung mit flüssigem Ammoniak bzw. konzentrierten, stickstoffhaltigen Lösungen. In: Handbuch Pflanzenernährung. Vol. III/1, pp. 96–195, 1965 [9].

[151] Burghardt, H.: Blattdüngung der Kulturpflanzen. Ang. Botanik *35*, 191–214 (1961).

[152] Deecke, U., Heller, L., und Jacobs, W.: Düngungskosten frei Wurzel 1972/73. BASF, Ludwigshafen 1972.

[153] Engelstad, O. P., and Russell, D. A.: Fertilizers for Use under Tropical Conditions. Advanc. Agronomy *27*, 175–208 (1975).

[154] FAO: Effects of Intensive Fertilizer Use on the Human Environment. Soils Bull. 16, Rome 1962.

[155] FAO/UNESCO: Irrigation, Drainage and Salinity. UNESCO Paris 1973.

[156] FAO/UNESCO: Soil Map of the World. Vol. I, Legend. UNESCO Paris 1974.

[157] Finck, A.: Tropische Böden. Parey, Hamburg 1963.

[158] Finck, A.: Fruchbarkeit tropischer Böden. In: Handbuch der Landwirtschaft und Ernährung in den Entwicklungsländern. Vol. II, pp. 99–125, Ulmer, Stuttgart 1971.

[159] Finck, A.: General Aspects of Fertilization in Tropical and Subtropical Agriculture. Plant Res. and Developm. *6*, 40–63 (1977).

[160] Frohner, W.: Die Lanzendüngung im Obst- und Weinbau. In: Handbuch Pflanzenernährung. Vol. III/1, pp. 115–128, 1965 [9].

[161] Frohner, W.: Die Technik der Blattdüngung. In: Handbuch Pflanzenernährung. Vol. III/1, pp. 128–154, 1965 [9].

[162] Fruhstorfer, A.: Die Ausbringung von Düngemitteln. Allgemeines. In: Handbuch Pflanzenernährung. Vol. III/1, pp. 11–35, 1965 [9].

[163] Kloke, A.: Orientierungsdaten für tolerierbare Gesamtgehalte einiger Elemente in Kulturböden. Pers. Communication 1977.

[164] Kolenbrander, C. J.: Eutrophication from Agriculture with Special Reference to Fertilizers and Animal Waste. Stikstof *15*, 56–67 (1972).

[165] Kopetz, L. M.: Die Beregnungsdüngung. In: Handbuch Pflanzenernährung. Vol. III/1, pp. 154–173, 1965 [9].

[166] Koppe, J. G.: Ökonomie. Leipzig 1831.

[167] Krauß, A.: Einfluß der Ernährung der Pflanzen mit Mineralstoffen auf den Befall mit parasitischen Krankheiten und Schädlingen. Z. Pflanzenern. u. Bodenkde. 124, 129–147, 1969.

[168] Löcker, H., und Rank, V.: Chemische Reaktionen beim Mischen (von Düngemitteln). In: Handbuch Pflanzenernährung. Vol. II/2, pp. 1352–1355, 1968 [9].

[169] Maier-Bode, F. W.: Die drei Stufen der Düngung. Westdeutscher Verlag, Opladen 1948.

[170] Mayer, A.: Das Düngerkapital und der Raubbau. Heidelberg 1869.

[171] Mohr, E. C. J., Baren, F. A. van, and Schuylenborgh, J. v.: Tropical Soils. Mouton, The Hague 1972.

[172] Müller, G.: Bodenbiologie. Fischer, Jena 1965.

[173] Norden, J., und Schmidt, P.: Die Flüssigdüngung. Parey, Hamburg 1974.

[174] Nye, P. H., and Greenland, D. J.: The Soil under Shifting Cultivation. Commonw. Bureau of Soils, Techn. Comm. No. 51 (1960).

[175] Quade, J.: Handhabung und Lagerung loser Düngemittel. Boden und Pflanze. Bull. 15, Ruhr-Stickstoff, Bochum 1971.

[176] Sanchez, P. A.: Properties and Management of Soils in the Tropics. Wiley. New York 1976.

[177] Thaer, A.: Grundsätze der rationellen Landwirtschaft. Realschulbuchhandlung, Berlin 1809.

[178] Thünen, J. H. v.: Der isolierte Staat. 3. ed., Wiegandt, Hempel und Parey, Berlin 1875.

References to Chap. 7

[179] Arnon, J.: Mineral Nutrition of Maize. Int. Potash Institut, Bern 1974.
[180] Atanasiu, N.: Cassava, Yam, Batate. In: Handbuch der Landwirtschaft und Ernährung in den Entwicklungsländern. (Ed.: Blankenburg, P. v., und Cremer, H.). Ulmer, Stuttgart, Vol. II, pp. 299–319, 1971.
[181] Aufhammer, G., und Fischbeck, G.: Getreide. DLG-Verlag, Frankfurt 1973.
[182] BASF: 50 Jahre Nitrophoska. BASF, Ludwigshafen 1977.
[183] Bayer, L. D.: Sugar Cane (Fertilization). In: Handbuch Pflanzenernährung. Vol. III/1, pp. 740–752, 1965 [9].
[184] Becker, M.: Der Gehalt der wichtigsten Grünfutterpflanzen an Mineralstoffen. In: Handbuch der Futtermittel (Hrsg.: Becker, M., und Nehring, K.). Parey, Hamburg, Vol. I, pp. 178–185, 1969.
[185] Berger, J.: The World's Major Fibre Crops, their Cultivation and Manuring. Centre d'Etude de l'Azote, Zürich 1969.
[186] Boguslawski, E. v., und Schuster, W.: Brassica-Rüben (Düngung). In: Handbuch Pflanzenernährung. Vol. III/1, pp. 441–456, 1965 [9].
[187] Bolhuis, G.: Zuckerrohr. In: Handbuch Landwirtschaft in Entwicklungsländern. Vol. II, pp. 481–495, 1971 [180].
[188] Büring, W.: Maisanbau lohnend gemacht. Guano-Werke, Hamburg 1970.
[189] Ferwerda, J.-D.: Hevea. In: Handbuch Landwirtschaft in Entwicklungsländern. Vol. II, pp. 591–606, 1971 [180].
[190] Geus, J. G. de: Fertilizer Guide for the Tropics and Subtropics. Centre d'Etude de l'Azote, Zürich 1973.
[191] Gisiger, L.: Die Düngung im Futterbau. In: Handbuch Pflanzenernährung. Vol. III/1, pp. 457–503, 1965 [9].
[192] Gökgöl, M.: Erdnuß. In: Handbuch Pflanzenernährung. Vol. III/1, pp. 720–740, 1965 [9].
[193] Hanus, H.: Anforderungen des Pflanzenbaus an die N-Bedarfsdiagnose. Landw. Forschung, Sonderh. 33/II, 52–57, 1976.
[194] Hasler, A., und Hofer, H.: Düngungslehre. Wirz, Aarau 1975.
[195] Heinemann, C.: Die Ölpalme, die Kokospalme, der Ölbaum (Düngung). In: Handbuch Pflanzenernährung. Vol. III/1, pp. 625–678, 1965 [9].
[196] Heinemann, C.: Die Baumwolle, Sisal; Kautschukpflanzen (Düngung). In: Handbuch Pflanzenernährung. Vol. III/1, pp. 519–562 and 599–624, 1965 [9].
[197] Hiepko, G.: Erdnuß. In: Handbuch Landwirtschaft in Entwicklungsländern. Vol. II, pp. 362–371, 1971 [180].
[198] Husz, G. S.: Sugar Cane. Monographs Ruhr-Stickstoff, Bochum 1972.
[199] Ishizuka, Y.: Potential for Increasing Rice Yields. Int. Rice Res. Conf. IRRI, Los Baños 1977.
[200] Jacob, A., and Uexküll, H. v.: Fertilizer Use. Verlagsges. Ackerbau, Hannover 1958.
[201] Jahn-Deesbach, W.; Lein und Hanf (Düngung). In: Handbuch Pflanzenernährung. Vol. III/1, pp. 562–599, 1965 [9].
[202] Kee, N. S.: The Oil Palm, its Culture, Manuring and Utilisation. Int. Potash Inst., Berne 1972.
[203] Kellner. O., und Becker, M.: Grundzüge der Fütterungslehre. Parey, Hamburg 1971.
[204] Kemmler, G.: Modern Aspects of Wheat Manuring. IPI-Bulletin Nr. 1, Int. Potash Institut, Berne 1974.
[205] Kirchgeßner, M.: Tierernährung. DLG-Verlag, Frankfurt 1975.

[206] Klapp, E.: Wiesen und Weiden. Parey, Berlin 1971.

[207] Knapp, O. E.: Rizinus. In: Handbuch Landwirtschaft in Entwicklungsländern. Vol. II, pp. 355-362, 1971 [180].

[208] Kürten, P. W.: Reis, Mais (Düngung). In: Handbuch Pflanzenernährung III/1, 316-352, 1965 [9].

[209] Lüdecke, H., und Müller, A. v.: Zuckerrübe, Futterübe, Maniok und Batate, Topinambur (Düngung). In: Handbuch Pflanzenernährung. Vol. III/1, pp. 382-415 and 752-763, 1965 [9].

[210] Malavolta, E., Haag, H. P., Mello, F. A. F., and Brasil Sobr., M. O. C.; On the Mineral Nutrition of some Tropical Crops. Internat. Potash Institut, Berne 1962.

[211] Müller, G.: Cotton. Monographs, Ruhr-Stickstoff, Bochum 1968.

[212] Nehring, K., und Becker. M.: Wurzeln und Knollen. In: Handbuch der Futtermittel. Vol. I, pp. 361-454, 1969 [184].

[213] Nieschlag, F.: Die Düngung in der Praxis. Parey, Hamburg 1963.

[214] Ochse, J. J., Soule, M. J., Dijkman, M. J., and Wehlburg, C.; Tropical and Subtropical Agriculture. Macmillan, New York 1961.

[215] Plarre, W.: Sesam. In: Handbuch Landwirtschaft in Entwicklungsländern, Vol. II, pp. 349-355, 1971 [180].

[216] Primost, E.: Weizen, Roggen, Hafer (Düngung). In: Handbuch Pflanzenernährung. Vol. III/1, pp. 174-269 and 283-315, 1965 [9].

[217] Raymond, W. F.: The Nutritive Value of Forage Crops. Advances in Agronomy *21*, 1-108 (1969).

[218] Rehm, S., und Espig, G.: Die Kulturpflanzen der Tropen und Subtropen. Ulmer, Stuttgart 1976.

[219] Rüther, H.: Die Düngung der Hülsenfrüchte. In: Handbuch Pflanzenernährung. Vol. III/1, pp. 504-518, 1965 [9].

[220] Ruhr-Stickstoff AG: Faustzahlen für Landwirtschaft und Gartenbau. Ruhr-Stickstoff, Bochum 1977.

[221] Schulze, E.: Anwendung und Wirkung der Stickstoffdünger bei Feldfrüchten und Dauergrünland. In: Der Stickstoff. Fachverband Stickstoffindustrie. Düsseldorf 1961.

[222] Schuster, W.: Raps und Rübsen (Düngung). In: Handbuch Pflanzenernährung. Vol. III/1, pp. 678-708, 1965 [9].

[223] Stählin, A.: Grünfutter und Heu. In: Handbuch der Futtermittel. Vol. 1, 1-117, 1969 [184].

[224] Steineck, O.: Kartoffel (Düngung). In: Handbuch Pflanzenernährung. Vol. III/1, pp. 415-441, 1965 [9].

[225] Sturm, H.: Trends in der Pflanzenernährung, bezogen auf Stickstoff. BASF Mitteil. f. Landbau *2*, 159-187 (1976).

[226] Tanaka, A.: Present Problems of Fertilizer Use. In: Better Exploitation of Plant Nutrients. FAO, Rome 1977.

[227] Tanaka, A., and Yoshida, S.: Nutritional Disorders of the Rice Plant in Asia. Int. Rice Res. Inst. (IRRI), Los Baños 1970.

[228] Taysi, V.: Hirse (Düngung). In: Handbuch Pflanzenernährung. Vol. III/1, pp. 353-381, 1965 [9].

[229] Uexküll, H. R. v.: Aspects of Fertilizer Use in Modern, High yielding Rice Culture. Int. Potash Institut, Berne 1976.

[230] Ulrich, A., and Hills, F. J.: Plant Analysis as an Aid of Fertilizing Sugar Crops I. Sugar beets. In: Walsh, C. M., and Beaton, J. D.: Soil Testing and Plant Analysis. Soil Sci. Soc. Amer., Madison 1973.

[231] Yamasaki, T.: Rice Cultivation in Japan — an Example of Intensive Agriculture. Intern. Potash Institut, Berne 1971.

[232] Zürn, F.: Neuzeitliche Düngung des Grünlandes. DLG-Verlag, Frankfurt 1968.

References to Chap. 8

[233] Becker-Dillingen, J.: Handbuch der Ernährung der gärtnerischen Kulturpflanzen. Parey, Berlin 1943.

[234] Brüning, D., und Müller, W.: Die Düngung der Teiche. In: Handbuch Pflanzenernährung. Vol. III/2, pp. 1517–1539, 1965 [9].

[235] Childers, N. F. (Ed.): Nutrition of Fruit Crops. Somerset, New Jersey 1966.

[236] Delfs-Fritz, W.: Citrus. Monographs Ruhr-Stickstoff, Bochum 1970.

[237] Fiedler, H. J., Nebe, W., und Hoffmann, F.: Forstliche Pflanzenernährung und Düngung. Fischer, Stuttgart 1973.

[238] Fritz, D., und Venter, F.: Gemüsebau. In: Storck, H.: Gartenbau. Ulmer, Stuttgart 1969.

[239] Gärtel, W.: Die Mikronährstoffe — ihre Bedeutung für die Rebenernährung. Weinberg and Keller *21*, 435–508 (1974).

[240] Gruppe, W.: Die Düngung im Obstbau. In: Handbuch Pflanzenernährung. Vol. III/2, pp. 843–893, 1965 [9].

[241] Gussone, H.-A., Rehfuß, K. E., und Ulrich, B.: Entwicklungstendenzen der Forstdüngung. Allg. Forst- und Jagdzeitung *143*, 41–48 (1972).

[242] Haas, P. de: Marktobstbau. BLV, München 1957.

[243] Heinemann, C.: Kaffee, Kakao, Tee (Düngung). In: Handbuch Pflanzenernährung. Vol. III/2, pp. 1130–1217, 1965 [9].

[244] Hoffmann, G., und Vogelmann, A.: Substratverwendung und Düngung (im Zierpflanzenbau). In: Storck, H.: Gartenbau. Ulmer, Stuttgart 1969.

[245] Hunte, W.: Champignonanbau. Parey, Hamburg 1973.

[246] Jung, J.: Die Düngung der Forstpflanzen. In: Handbuch Pflanzenernährung. Vol. III/2, pp. 986–1021, 1965 [9].

[247] Jung, J., und Riehle, G.: Beurteilung und Behebung von Ernährungsstörungen bei Forstpflanzen. BLV Verlags-Ges., München 1969.

[248] Knickmann, E., und Tepe, W.: Pflanzenernährung im Gartenbau. Ulmer, Stuttgart 1966.

[249] Linser, H., und Schmid, K.: Tabak (Düngung). In: Handbuch Pflanzenernährung. Vol. III/2, pp. 1065–1096, 1965 [9].

[250] Mappes, F. M., und Will, H.: Die Düngung im Gemüsebau. In: Handbuch Pflanzenernährung. Vol. III/1, pp. 796–842, 1965 [9].

[251] Neumann, K.-H.: Citrus, Banane (Düngung). In: Handbuch Pflanzenernährung. Vol. III/2, pp. 1217–1230 and 1250–1259, 1965 [9].

[252] O'Kelley: Mineral Nutrition of Algae. Ann. Rev. Plant Physiol. *19*, 89–112 (1968).

[253] Penningsfeld, F.: Die Ernährung im Blumen- und Zierpflanzenbau. Parey, Hamburg 1960.

[254] Penningsfeld, F., und Forchthammer, L.: Die Düngung im Blumen- und Zierpflanzenbau. In: Handbuch Pflanzenernährung. Vol. III/2, pp. 917–985, 1965 [9].

[255] Pissarek, H. P.: Untersuchungen über die Innenblattnekrose von Chinakohl. Manuscript, Kiel 1978.

[256] Schröder, H.: Die Düngung von Arznei- und Gewürzpflanzen. In: Handbuch Pflanzenernährung. Vol. III/2, pp. 1022–1064, 1965 [9].

[257] Siegel, O.: Die Düngung im Weinbau. In: Handbuch Pflanzenernährung. Vol. III/2, pp. 894-916, 1965 [9].

[258] Storck, H.: Gartenbau. Ulmer, Stuttgart 1969.

[259] Ulrich, B.: Stoffhaushalt von Wald-Ökosystemen. Inst. f. Bodenkunde u. Waldernährg., Göttingen 1976.

[260] Wehrmann, J.: Methodische Untersuchungen zur Durchführung von Nadelanalysen in Kiefernbeständen. Forstwiss. Centralblatt *78*, 65-128 (1959).

[261] Wilde, S. A.: Forstliche Bodenkunde. Parey, Hamburg 1962.

[262] Zattler, F.: Hopfen (Düngung). In: Handbuch Pflanzenernährung. Vol. III/2, pp. 1097-1133 (1965) [9].

[263] Zöttl, H.: Waldstandort und Düngung. Centralblatt f. Forstwesen *81*, 1-24 (1964).

References to Chap. 9

[264] Abele, U.: Vergleichende Untersuchungen zum konventionellen und biologisch-dynamischen Pflanzenbau unter besonderer Berücksichtigung von Saatzeit und Entitäten. PhD-Thesis, Gießen 1973.

[265] Aehnelt, E., und Hahn, J.: Fruchtbarkeit der Tiere – eine Möglichkeit zur biologischen Qualitätsprüfung von Futter- und Nahrungsmitteln? Tierärztliche Umschau *4*, 155f. (1973).

[266] Allaway, W. H.: The Effect of Soils and Fertilizers on Human and Animal Nutrition. Agric. Inform Bull. *378*, USDA, Washington 1975.

[267] Amberger, A.: Düngung und Nahrungswert pflanzlicher Produkte. Landw. Forschung *30/I*, Sonderh. 10-20 (1973).

[268] Bersin, Th.: Biochemie der Mineral- und Spurenelemente. Akad. Verlagsges., Frankfurt 1963.

[269] Brugger, G.: Gedanken zum "Biologischen" Landbau. Informat. Landw.-Beratung Baden-Württ. 4 (1974).

[270] CEA (Centre d'Etude de l'Azote) u. a.: Handbuch: Umweltaspekte der Düngemittelanwendung. CEA Zürich 1976.

[271] Catel, W.: Über den Einfluß der Verfütterung verschieden gedüngter Nahrungspflanzen auf das Gedeihen von Säuglingen. Landw. Forschung *1*, 220-223 (1949).

[272] Cremer, H. D.: Die Ernährung der Bevölkerung. Ernährungsverhältnisse und Ernährungszustand. In: Handbuch der Landwirtschaft und Ernährung in den Entwicklungsländern. Vol. 1, pp. 498-524, Ulmer, Stuttgart 1967.

[273] Dambroth, M., et al.: Alternativen im Landbau (Statusbericht). Landwirtschaftsverlag, Münster 1978.

[274] Deutsche Gesellschaft für Ernährung: Ernährungsbericht Deutsch. Ges. f. Ernährung, Frankfurt 1976.

[275] Food and Nutrition Board (USA): Richtlinien für die Deckung des Nährstoffbedarfs. Umschau Verlag, Frankfurt 1967.

[276] Finck, A.: Düngung und Nahrungsqualität in unterschiedlichen Anbausystemen. Landw. Forschung Sonderh. 35, 122-132, 1978.

[277] Glatzel, H.: Sinn und Unsinn in der Diätetik. XII. „Biologisch" oder mineralgedüngt? Medizin. Welt *28*, 253-260, 307-311 (1977).

[278] Harper, H. A., Löffler, G., Petrides, P. E., und Weiß, L.: Physiologische Chemie. Springer, Berlin 1975.

[279] Heinze, H.: Qualitätsforschung und Qualitätsmaßstäbe zur Prüfung pflanzlicher Nahrungsmittel „Lebendige Erde" *3*, 87-92, 1970.

424   *References*

[280] Karlson, P.: Kurzes Lehrbuch der Biochemie. Thieme, Stuttgart 1974.
[281] Koepf, H., Petterson, B., und Schaumann, W.: Biologische Landwirtschaft. Eine Einführung in die biologisch-dynamische Wirtschaftsweise. Ulmer, Stuttgart 1974.
[282] Kolisko, E., und Kolisko, L.: Die Landwirtschaft der Zukunft. Kolisko, Schaffhausen 1953.
[283] Kübler, W.: Umwelt und Ernährung. Schriftenreihe Agrarwiss. Fachbereich Kiel *53*, 141–156 (1975).
[284] Kraut, H., und Wirths, W.: Die Bedeutung der Düngung für die menschliche Ernährung. In: Handbuch Pflanzenernährung Vol. III/2, pp. 1355–1379, Springer, Vienna 1965 [9].
[285] Kühnau, J.: Unterschiede in der ernährungsphysiologischen Bedeutung pflanzlicher und tierischer Lebensmittel für den Menschen. Landw. Forschung *32/II*. Sonderh. 168–179 (1975).
[286] Lang, K.: Biochemie der Ernährung. Steinkopff, Darmstadt 1974.
[287] Lehninger, A. L.: Biochemie, Verlag Chemie, Weinheim 1975.
[288] Lindner, E.: Toxikologie der Nahrungsmittel. Thieme, Stuttgart 1974.
[289] Metzner, H.: Biochemie der Pflanzen. Enke, Stuttgart 1973.
[290] Mohler, H.: Sinn und Unsinn unserer Ernährung. Sauerländer, Frankfurt 1972.
[291] Müller, K.: Zum Einfluß der Mineraldüngung auf die Qualität pflanzlicher Erzeugnisse. Kali-Briefe, Fach 11, 1975.
[292] Nehring, K.: Düngung, Qualität und Futterwert. In: Handbuch Pflanzenernährung. Vol. III/2, pp. 1260–1354, 1965 [9].
[293] Pfeiffer, E.: Die Fruchtbarkeit der Erde. Geering. Dornach (Switzerland) 1956.
[294] Preuschen, G., Brauner, H., Storhas, R., Willi, J.: Gesunder Boden = Leistungsstarker Betrieb. Stocker-Verlag, Graz 1977.
[295] Richter, G.: Stoffwechselphysiologie der Pflanzen. Thieme, Stuttgart 1969.
[296] Rusch, H.: Bodenfruchtbarkeit. Eine Studie biologischen Denkens. Haug, Heidelberg 1974.
[297] Scheidegger, W., und Müller, B.: Biologischer Landbau. Biofarm Verlag, Bern 1976.
[298] Seifert, A.: Gärtnern, Ackern — ohne Gift. Biederstein, Munich 1975.
[299] Scheunert, A., Sachse, M., und Specht, M.: Über die Wirkung fortgesetzter Verfütterung von Nahrungsmitteln, die mit oder ohne künstlichen Dünger gezogen sind. Biochem. Zeitschr. *274*, 372–296 (1934).
[300] Schuphan, W.: Mensch und Nahrungspflanze. Der biologische Wert der Nahrungspflanze in Abhängigkeit von Pestizideinsatz, Bodenqualität und Düngung. Junk, the Hague 1976.
[301] Steiner, R.: Geisteswissenschaftliche Grundlagen zum Gedeihen der Landwirtschaft (1924), Steiner Verlag, Dornach 1975.
[302] Voisin, A.: Boden und Pflanze. Schicksal für Tier und Mensch. Bayer. Landw.-Verlag, Munich 1959.
[303] Voisin, A.: Grundgesetze der Düngung. Bayer. Landw.-Verlag, Munich 1966.
[304] Vogtmann, H.: Biologischer Landbau — eine Alternative? Schweiz. Stiftung z. Förd. biol. Landbaus, Winterthur 1973.
[305] WHO (World Health Org.): Trace Elements in Human Nutrition. WHO Techn. Report 532, Geneva 1973.
[306] Wortmann, M.: Konventionelle und biologische Landwirtschaft im Spannungsfeld von Ökonomie und Ökologie. Verlag „Lebendige Erde", Darmstadt 1977.
[307] Zimmermann, W.: Steine geben Brot. Cohrs, Rotenburg 1975.

Index

Auxiliary substance 172
Availability of nutrient 10
A-value 198
Avitaminose 400
Avocado 355
Azalea 346
Azaleas 345
Azotobacter 145, 174

B (boron) 102
Bacterial cultures, yeast 378
Baking quality 393
Ballast salts 84
– substances 400
Balm mint 373
Banana 354 f
Barley 295 f
– K-fertilization 215
– N-fertilization 209
– P-fertilization 213
Base 409
Basic active substance 143
– nutrient 9
Batata 312
Bayfolan 130
BBA-scale 291
Beans 321
– K-fertilization 215
– P-fertilization 213
Beech 360, 364, 366
Beet, Mg-fertilization 217
Beetroot 341
Beneficial element 9
– nutrients 134
Berger and Truog method 220
Berries 352
Betaine 392
B-fertilizer 102
Bilberry 352
Biological agriculture 384
Biotic value 391
Birch 364, 366
Biuret 42, 43
Black currant 352
– earth 257
– henbane 375
– spottiness 396
Blackberry 352
Blast furnace lime 141 f, 144, 146

Blood powder 164
B-Nine 174
Bog fertilizer 70
– soils 266 f
Bone meal 164
Borax 102 f
Boric acid 102 f
Boron (B) 102
– fertilization 103 f
– fertilizer 102
– food quality 398
– requirement 220
– silicate 102
Boussingault 24, 25
Bromelia 346
Bromine 336
Brown earth 257
Brown-spot disease 90, 396
Brussels sprouts 341
Bulk fertilizer 241
B-value 202

Ca (calcium) 89
Cabbage 331, 341
– boron fertilization 220
– K-fertilization 215
Cadmium 283
Ca-fertilizer 89 ff
– application 90
Caffeine 367
Calcium-ammonium nitrate 41
Calcium carbonate 143, 144
– chloride 90
– cyanamide 44 f
– fertilizer 89 ff
– food quality 396
– molybdate 104
– nitrate 38 f, 90
– phosphate 62
– sulfate 90
CAL-method 198
Cal-Nitro 41
Camellia 345
Camomile 373
CaO, reference base 143
Caraway 373
Carbamate 43
Carbamide 42 f
Carbon dioxide 133, 137